PROCEEDINGS OF THE FIRST CLEVELAND SYMPOSIUM ON MACROMOLECULES

Structure and Properties of Biopolymers

PROCEEDINGS OF THE FIRST CLEVELAND SYMPOSIUM ON MACROMOLECULES

Structure and Properties of Biopolymers

edited by

A.G. WALTON

ELSEVIER SCIENTIFIC PUBLISHING COMPANY

AMSTERDAM – OXFORD – NEW YORK 1977

ELSEVIER SCIENTIFIC PUBLISHING COMPANY
335 Jan van Galenstraat
P.O. Box 211, Amsterdam, The Netherlands

Distributors for the United States and Canada:

ELSEVIER NORTH-HOLLAND INC.
52, Vanderbilt Avenue
New York, N.Y. 10017

ISBN: 0-444-41561-0

Printed in The Netherlands

CONTENTS

Preface vii

The Structure of Fibrous Proteins 1
R.D.B. Fraser

Polypeptides as Structural Models for Proteins 23
W. Traub

Structure of Poly(D-L-Peptides) 39
B. Lotz, F. Heitz and G. Spach

Molecular Architecture of the Animal and Some Microbial Extracellular Polysaccharides 61
E.D.T. Atkins

Secondary Structure of Polynucleotides 87
S. Arnott

Structure, Morphology and Mechanical Properties of Bioplastics 105
P.H. Geil, S. Barenberg and W.M. Wong

Structural Aspects of Edible Materials 135
W.E. Marshall

Properties of Modified Collagenous Materials 153
E.J. Kramer

Mechanical Properties of Native Fibrous Materials 177
L. Rebenfeld

Materials from Polysaccharides 203
S.H. Carr

Polynucleotides as Inducers of Interferon 217
E. DeClerq

Immunology of Synthetic Polypeptides 245
P.H. Maurer

Polypeptide Complexes with Complement 269
B.J. Johnson

Interaction of Synthetic Polypeptides with Blood 286
A.G. Walton

PREFACE

The First Cleveland Symposium on Macromolecules, held on the campus of Case Western Reserve University, Oct. 11th-15th, 1976, had as its theme "Structure and Properties of Biopolymers". The objective of the organizers was to put together a meeting which was somewhat unconventional in its scope, bringing together researchers in three areas of biopolymers: proteins/polypeptides, nucleic acids/polynucleotides and polysaccharides. Topics discussed covered a wide range of fundamental and applied aspects. Three particular foci were identified -- basic structure, materials properties and medical applications. This book is divided into three sections which cover the invited lectures in these areas. The invited lecturerswere asked to review the field in the area which they were covering as well as including certain new features of their own work. These lectures then acted as nuclei around which contributed original papers were clustered. (The contributed papers are appearing elsewhere in international journals.) Whereas, in several cases, the space available for publication has forced an abbreviation of the review articles, nevertheless this volume does reflect an up-to-date record of a broad swath of biopolymer and biological materials research.

The organizers are most pleased to acknowledge the financial support of several organizations, particularly the U. S. National Science Foundation, General Foods, Inc. and Beckman Instrument Corp.

Alan Walton

October 15, 1976

THE STRUCTURE OF FIBROUS PROTEINS

R.D.B. FRASER
CSIRO Division of Protein Chemistry, Melbourne, Australia

SUMMARY

Evidence on chain conformation in fibrous proteins and on the structures of naturally occurring assemblies of fibrous proteins is reviewed and some general principles of design are identified.

INTRODUCTION

In early studies of protein structure it was customary to distinguish between two classes of protein. The fibrous proteins, typified by insect silks, collagen and keratin, were characteristically solid in their functional form, whilst the globular proteins, typified by hemoglobin and insulin, were characteristically in solution in their functional form. Globular protein molecules were visualized as having a compact structure whilst fibrous protein molecules were thought to be elongated.

The classification into globular and fibrous proteins is still widely used and is largely valid, but the belief that the classification might be related to some fundamental difference in molecular design is untenable in the light of present day knowledge. For example, the framework of the thin filaments of muscle (F-actin) is formed by the polymerization of a molecule (G-actin) with typical globular properties and sickle cell hemoglobin, which differs by only one amino acid residue in the beta chain from normal hemoglobin, forms a variety of fibrous structures [1]. Some authors have preferred therefore to define a class of proteins, termed structure proteins, which includes all proteins that serve a structural function. In the present context the term fibrous protein will be used in its broadest context and taken to be synonymous with structure protein.

Fibrous proteins frequently occur in ordered assemblies. These are amenable to study by electron microscopy, infrared spectrometry and diffraction methods but there are many obstacles to the direct interpretation of the data in terms of chain conformation and mode of assembly. Much of our knowledge concerning these topics has been derived indirectly through studies of synthetic polypeptides and artificially produced assemblies of fibrous proteins. In recent years some progress has been made in the interpretation of data obtained from the naturally occurring assemblies and certain general principles in the design of fibrous proteins have emerged. These will be reviewed in the present contribution and where appropriate, illustrated with examples from recent studies of naturally occurring assemblies. Comprehensive reviews of structural studies of fibrous proteins are available [2,3,4,5].

CHAIN CONFORMATION

In synthetic polypeptides composed of only one type of residue the polypeptide chain commonly adopts a regular conformation in which all the amino acid residues have equivalent environments. This is a natural consequence of the fact that the residues are identical and in the general case leads to a helical conformation [6]. In an infinite helical array the repeating units are related by symmetry and any particular unit may be generated from the preceding unit by a translation h parallel to the helix axis and a rotation t about the helix axis. These parameters, which are termed the unit height and unit twist respectively, are related to the pitch of the helix P and to the number of units per turn n by the expressions

$$P = 2\pi h/t\ , \qquad n = 2\pi/t \tag{1,2}$$

It is convenient to restrict h to positive values and to assign a positive value to t for right-handed helices and negative values for left-handed helices. Since there will always be a family of helices which pass through all the repeating units, the unit twist is not uniquely defined. The particular helix which minimizes the magnitude of t is termed the basic helix [7].

Synthetic polypeptides containing only one type of residue adopt a variety of helical conformations [5]. The form of the helix is determined principally by the type of sidechain but is also influenced by chain length and by environment. The stability of the helical conformation relative to a disordered form is also a function of the same parameters.

In all except the simplest types of fibrous protein there is a diversity of residues. Nevertheless, segments of the polypeptide chain adopt ordered conformations which closely resemble those observed in synthetic polypeptides containing only a single type of residue. In these ordered conformations the spatial relationship between the mainchain portions of the residues (-NH.CH.CO-) approximates to helical symmetry. A common feature to emerge from studies of the primary structure of fibrous proteins is a sharp division of the sequence into segments in which the character and arrangement of the residues are markedly different. As might be expected from studies of synthetic polypeptides, this results in abrupt transitions from one type of conformation to another along the mainchain. The usual type of change appears to be from a form in which the mainchain has a helical conformation to a form in which it has a non-helical conformation or vice-versa, rather than from one helical form to another. The helical segments serve a primarily structural function and are generally associated with some form of repetitive pattern in the sequence of residues. In some cases the non-helical segments also have a structural function, as for example a matrix in filament-matrix composites, but may also be associated with biochemical control of assembly or functional properties.

The proportion of the polypeptide mainchain that is in a helical conformation varies widely between different structures. In collagen and in tropomyosin almost the entire mainchain has a helical form whilst elastin and the matrix proteins of mammalian keratins appear to be devoid of helical conformations. Quantitative studies of the proportion of mainchain with a helical conformation have been made for a number of fibrous proteins in their native state using infrared spectrometry and examples of the estimates obtained are 30% for *Bombyx mori* silk fibroin and 25% for feather rachis keratin [5]. There are formidable difficulties in carrying out such analyses and there is a great need for the development of reliable methods for the quantitative estimation of the proportions of different helical conformations in naturally occurring assemblies of fibrous proteins.

On the basis of high-angle X-ray diffraction studies, Astbury and co-workers classified the ordered mainchain conformations that occur in fibrous protein assemblies into three main types. The α-conformation was characterized by a strong axial periodicity of 5.15Å, the β-conformation by an axial periodicity of 3.1 - 3.5Å and the collagen conformation by an axial periodicity of 2.86Å. A fourth type of ordered mainchain conformation occurs in *B. mori* silk prior to spinning, but this has not been studied in such detail as the other three [5].

The α-Conformation

Specimens of poly(L-alanine) can be prepared in which the majority of the polymer has a conformation that has been shown [8] to be very close to that of the α-helix described by Pauling and Corey. In poly(L-alanine) the amino acid residues are arranged on a helix with unit height h = 1.496Å and unit twist t = 99.6° and the helix is stabilized by the formation of intrachain hydrogen bonds.

The high-angle X-ray diffraction patterns yielded by several ordered assemblies of fibrous proteins, including hair, epidermis, honey-bee silk and muscle fibres, contain some of the features predicted for an α-helix. In particular, they all exhibit a strong meridional reflection of spacing around 1.5A°, but in addition have a prominent meridional reflection of spacing around 5.15Å which is not predicted for the α-helix. It was shown independently by Crick [9] and by Pauling and Corey [10] that this additional reflection could be accounted for by supposing that the α-helices in these materials were regularly distorted to give what has been termed a coiled-coil conformation. In the tropomyosin component of the thin filaments of vertebrate muscle two such distorted α-helices are twisted around each other to form a two-strand rope. In this particular material the axis of the rope itself describes a slow twist as it follows the line of actin molecules in the framework of the thin filament and so the chain conformation is in fact that of a coiled-coiled-coil [11].

Pauling and Corey [10] envisaged the distortion of the α-helix into a coiled-coil conformation as arising from a repeating sequence of amino acid residues in which the individual residues formed hydrogen bonds of different lengths. The origin of the distortion was thus thought to be intrahelical. On the other hand Crick [9] suggested that two α-helices could pack together more closely if they were distorted by being twisted together about a common axis to give a rope-like structure. The amount of twist required to achieve optimum packing depends upon the helical parameters of the undistorted α-helices [5]. Crick [9] calculated that for right-handed α-helices with a unit height of h = 1.5Å and a unit twist of t = 100° the pitch of the distorted α-helix axis for an interhelical distance of 10Å would be -186Å for optimum packing, where the negative sign indicates a left-handed twist. The actual pitch in fibrous protein structures is difficult to determine but would appear to be commonly around -140Å [12]. The values of h and t requiring a pitch of this magnitude for optimum packing lie well within the observed range for α-helices in synthetic polypeptides [5].

Crick [9] pointed out that the sidechains constituting the contacting surfaces between the two distorted α-helices in a coiled-coil rope structure would be expected to have a hydrophobic character since they are shielded from contact with the solvent in an aqueous environment. The geometry of the coiled-coil conformation (Fig. 1) is such that this implies a repeating heptapeptide sequence of the type $(a\text{-}b\text{-}c\text{-}d\text{-}e\text{-}f\text{-}g\text{-})_n$ where a and d are non-polar residues, such as alanine, leucine or valine.

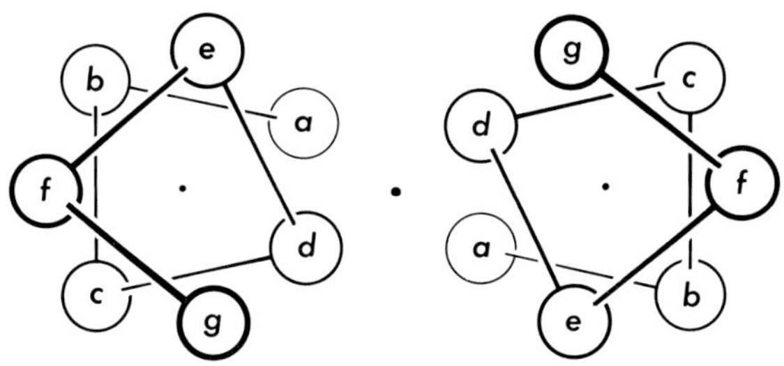

Fig. 1. Helical projection of the seven residue repeating units in a two-strand coiled-coil rope.

No specific limitation is imposed on the remaining residues in the heptad. Almost the entire sequence of α-tropomyosin has a pattern of this type [13] and an extract is shown in Fig. 2a.

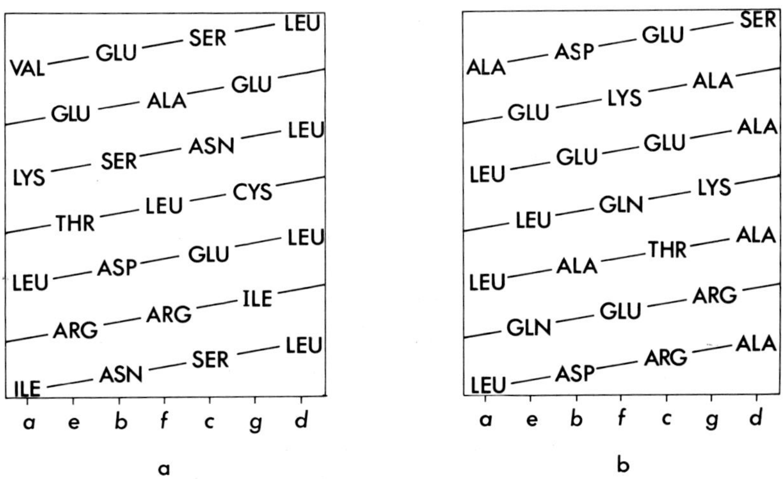

Fig. 2. Extracts from the amino acid sequences of (a) α-tropomyosin [13] and (b) a helix-rich segment from a low-sulphur wool protein [14] showing the 7-residue periodicity in hydrophobic residues at positions a and d, which are located in the inner region of the rope.

The so-called low-sulphur proteins of hard mammalian keratins contain segments of chain that have a coiled-coil conformation [5] and again a heptapeptide repeating pattern of the type envisaged by Crick is evident in the sequences of these segments [14]. An extract from the sequence of one of the segments is given in Fig. 2b.

The tropomyosin molecule is made up of two very similar or identical chains which are similarly directed as regards the progression of the mainchain atoms -NH.CH.CO-. A detailed study of possible interactions between the two chains has been carried out by McLachlan and Stewart [15]. In addition to the interactions between the hydrophobic residues occupying positions *a* and *d* in one chain (Fig. 1) and corresponding positions in the other chain it was found that interactions involving residues in positions *e* and *g* also influenced the stability of a two-strand rope. In the case of the tropomyosin molecule maximum stabilization was predicted to occur when the two chains were in longitudinal register.

If the axes of the two α-helices in tropomyosin follow a helical path of constant radius the interactions between the hydrophobic residues in positions *a* and *d* in one chain and the corresponding positions in the other chain are not particularly favourable when the chains are in register [16]. However, McLachlan and Stewart [15] suggested that by introducing small local variations in the distance between the two α-helices and in the α-helix parameters it would be possible to achieve a close-packed core with the chains in register. Such variations would reduce the coherent length as regards diffraction phenomena in line with the findings for paramyosin-containing muscle and porcupine quill [5].

In the case of the tropomyosin molecule interactions between charged residues in the clefts of the two-strand rope (positions *e* and *g* in Fig. 1) appear to play an important part in favouring a rope structure in which the chains are in longitudinal register [15]. Similar studies of interactions between the α-helical segments of wool keratin proteins suggest that ionic interactions between residues occupying positions *e* and *g* must play an important part in determining not only chain stagger, but also mutual orientation (i.e. parallel or antiparallel) and the segment composition of ropes containing more than one type of α-helical segment.

There is a good deal of circumstantial evidence that the molecules of tropomyosin, myosin and paramyosin all contain two-strand rope structures. In the case of paramyosin-containing muscle this is supported by quantitative X-ray diffraction studies [5]. Crick [9] also described a three-strand rope and suggested this as a model for the ordered mainchain conformation in α-keratin. Evidence for the existence of three-strand ropes is less definite than that for two-strand ropes. The repeating pattern of hydrophobic residues predicted for a three-strand rope is exactly the same as that for a two-strand rope and so it is not possible to determine the number of strands in the rope from this feature. It seems likely that the pattern of charged residues in positions *e* and *g* (Fig. 1) will have a strong influence on the relative stabilities of the two-strand and three-strand rope structures.

The fibrinogen molecule is believed to contain rod-like segments composed of three chains [17]. Films of fibrinogen give the characteristic α-type X-ray diffraction pattern and it has been suggested that the rod-like segments have a coiled-coil conformation. So far it has not been established directly however that a three-strand rope is present. A similar situation exists in the case of epidermal keratin, which has a three-chain unit of structure [18] and gives an α-type X-ray diffraction pattern. Again, it has not been established directly that a three-strand rope structure is present.

The chain structure of the so-called hard α-keratin of hair, wool, quill, etc. for which Crick proposed the three-strand rope as a model is unknown at present. The X-ray diffraction pattern has been recorded and compared with that predicted for two-strand and three-strand ropes with various combinations of chain directions [5]. In no case was the agreement completely satisfactory although the three-strand parallel chain model gave marginally better agreement. Subsequent calculations in which allowance was made for the effects of the non-helical matrix in this material showed that the differences between the patterns of two-strand and three-strand ropes are less marked in this case. When combined with other complicating features such as the modulation of intensity by an inter-rope interference function the assignment to a two-strand or three-strand rope structure became impracticable.

The X-ray diffraction pattern yielded by the silk of the honey bee *Apis mellifera* is of the α-type and has been studied in some detail [19]. On the basis of the distribution of intensity in the pattern it has been suggested that this material contains four-strand ropes.

In principle it should be possible to decide the number of strands in the rope structure in a fibrous protein assembly by comparing the observed intensity data with that predicted for the various models. In no case, however, has it proved possible to obtain satisfactory agreement between the observed and calculated intensities for the entire pattern. The reason for this is believed to be that the naturally occurring rope structures are generally distorted in some way and that the rope pattern is modulated by a (generally unknown) inter-rope interference function which may be quite complex in the case of a filament with helical symmetry. There is considerable scope for the study of model polypeptides with repeating sequences designed to stabilize the coiled-coil conformation. Quantitative X-ray diffraction studies of such model compounds should enable the precise arrangement of the amino acid residues in the coiled-coil conformation to be determined and also permit features diagnostic of rope parameters such as number of chains and chain directions to be identified.

β-Conformation

Specimens of poly(L-alanine) can be prepared in which the polypeptide chains have a near-extended conformation and it has been shown [20] that the crystallites contain antiparallel-chain pleated sheets with a structure very close to the model proposed by Pauling and Corey [21]. The repeating unit in the sheet contains portions of two oppositely directed chains. The individual chains have a helical conformation with a unit height h = 3.45Å and a unit twist $t = 180^{\circ}$ and the sheet is stabilized by a network of interchain hydrogen bonds. The β-conformation occurs naturally in many insect silks and also in the hard keratins of birds and reptiles.

Tussah, or wild silk, is the product of moths belonging to the genus *Antheraea* and the fibroins are uniformly rich in alanyl residues. The fibroin of *Antheraea mylitta* for example contains 41 alanyl residues per 100 residues [22]. The X-ray diffraction patterns obtained from Tussah fibroins are almost indistinguishable from the pattern obtained from poly(L-alanine) and it seems likely that the crystallites in these silks have a structure very similar to that found in the synthetic polypeptide. Since the reflections are sharp in the fibre axis direction the crystallites must have an appreciable length parallel to the chain direction and it follows that there must be lengthy runs of the sequence that contain only alanyl residues, that is, a repeating pattern of 1 residue.

The cocoon silk produced by the larvae of the moth *Bombyx mori*, which is widely used as a textile fiber, also contains crystallites in which the chains have a β-conformation but the sequence in these regions has a repeating pattern of the form $(X\text{-}Gly)_n$ where X is either an alanyl or a seryl residue [23]. The alanyl and seryl residues are also ordered and over much of the sequence in the crystalline regions there is a hexapeptide repeat of the form (Ser-Gly-Ala-Gly-Ala-Gly)$_n$. Since the polypeptide chain has a unit twist of 180° all the glycyl residues in any one chain project on one side of the sheet and all the alanyl and seryl residues on the other. The spacings and relative intensities of certain prominent equatorial reflections in the X-ray diffraction pattern obtained from *B. mori* fibroin can be accounted for by supposing that all the polypeptide chains are oriented in a like manner about their axes so that all the glycyl residues project on one side of the sheet [24]. The sheets are then packed so that glycyl surfaces contact glycyl surfaces and this results in an alternation in intersheet spacing in the crystal. Synthetic polypeptides with repeating sequences have been synthesized as models for the structure of *B. mori* fibroin. The polymers (Ala-Gly)$_n$ and (Ala-Gly-Ala-Gly-Ser-Gly)$_n$ both form crystallites in which the β-sheets have all the glycyl residues projecting from one surface and pack together so that the glycyl surfaces are in contact [5].

A specialized modification of the β-conformation, termed the cross-β conformation, occurs in the egg stalks produced by the green lace-wing fly, *Chrysopa flava* [25]. In this material the sheets are formed by a single chain looping backwards and forwards at regular intervals and the sheets are oriented so that the chains are perpendicular to the axis of the stalk. There is evidence that the amino acid sequence contains a 16 residue repeating unit, and in the model proposed by Geddes *et al* [25] this is supposed to correspond to one complete loop consisting of a pair of oppositely directed chain segments in the β-sheet.

The hard keratins of birds and reptiles have a completely different structure from the mammalian α-keratins discussed earlier. Feather rachis has been shown to contain ordered assemblies of keratin molecules in which about one-third of the polypeptide chain has a distorted β-conformation and the remainder is non-helical [5]. The β-sheets in feather keratin are much smaller than is general in insect silk and are twisted in a similar fashion to the small β-sheets that occur in globular proteins [26].

Globular protein molecules in which the only regular secondary structure is the β-sheet generally have an alternating (antiparallel) rather than a parallel arrangement of chains in the sheet [27]. Infrared studies of feather rachis suggest that a similar situation exists in feather keratin [5]. The complete amino acid sequences of feather keratin molecules from two species of birds have been determined and both are sharply divided into three segments as regards residue type. The central segment is almost devoid of charged residues and is rich in residues that favour the β-conformation. The terminal segments contain the bulk of the charged residues and the cysteinyl residues. It seems likely that the portion in the β-conformation resides in the central segment of the molecule. There are prominent 8-residue periodicities in both β-favouring and bend-favouring residues, which are 180° out of phase [28]. This suggests that the β-sheets are formed by the chain looping backwards and forwards every 8 residues to give an anti-parallel chain sheet which is topologically similar to that believed to exist in the egg stalk of *Chrysopa flava* discussed earlier.

So far there is no evidence for the occurrence of the parallel-chain pleated sheet in fibrous protein structures. In globular proteins parallel-chain pleated sheets occur quite commonly in a class of structures in which a length of β-conformation alternates with a length of α-helix but is rare in other classes [27]. In native fibrous proteins examined so far no evidence has been obtained for the co-existence of the α-conformation and the β-conformation in the same molecule and so the correlations established for globular proteins are valid at least in part for fibrous proteins.

Collagen Conformation

The polypeptide mainchain in collagens has a distinctive conformation that is associated with a repeating sequence of the form $(\text{gly-X-Y})_n$ and the frequent occurrence of prolyl and hydroxyprolyl residues in the second and third positions respectively in the triplet. The presence of imino acid residues greatly restricts the range of regular helical conformations that are energetically possible. They cannot, for example, be introduced into either an α-helix or a β-sheet without disruption of the helix.

Poly(L-proline), poly(L-hydroxyproline) and poly(glycine) can all be crystallized in forms where the polypeptide chain has a helical conformation with a unit height $h \simeq 3.1$Å and a unit twist of $t = -120^{\circ}$ [5]. In the case of poly(glycine) the crystal is stabilized by the formation of a 3-dimensional network of interchain hydrogen bonds.

The collagen molecule contains three chains in which the tripeptide repeat persists, apart from a short section at each end, over the entire length of the molecule. The three chains are similarly directed and are wound around each other to form a three-strand rope with a pitch of about 86Å. The conformation of the individual chains is similar to that found in the synthetic polypeptides except that the rope twist converts them from simple helices to coiled-coils. The mean unit height is reduced to 2.86Å by the rope twist. The precise form of the chain conformation in collagen has not yet been determined. The structure is stabilized by the formation of interstrand hydrogen bonds and three different topological arrangements have been proposed [5]. In the first two, proposed by Rich and Crick, there is one hydrogen bond per three residues and in the third, proposed by Ramachandran and Kartha, there are two hydrogen bonds per three residues. In the latter structure it is not stereochemically feasible to accommodate an imino acid in the second position in the triplet and it has been suggested that the chains adopt one of two alternative conformations depending on whether there is an imino acid residue in this position. In the models proposed by Rich and Crick imino acids can be accommodated in either the second or third positions in the triplet without difficulty.

A wide range of synthetic polypeptides with repeating tripeptide and hexapeptide sequences has been synthesized as models for collagen [5]. In many cases it has been possible to demonstrate that three-strand rope structures are present but in no case has it proved possible to collect sufficient X-ray data to provide an objective assessment of the various proposed structures. In the α1 chain of calf skin collagen the C-terminal portion has the sequence (Gly-Pro-Hyp)$_4$-Gly-Pro-Pro [29]. Due to the restrictions placed on chain conformation by imino acid residues this section of chain is likely to have a conformation very similar to that of the synthetic polypeptide (Gly-Pro-Pro)$_n$ studied by Yonath and Traub [30]. The hydrogen bond topology in this polymer was shown to be that of the type II model proposed by Rich and Crick and a detailed model for (Gly-Pro-Pro)$_n$ was proposed on the basis of a combined X-ray and stereochemical study. This model probably represents the closest approach so far to the determination of the nature of the mainchain conformation in collagen.

Non-Helical Conformations

There are several fibrous proteins in which no regular secondary structure has so far been identified. These include elastin and resilin, which occur in highly elastic structures in vertebrates and in insects respectively, and the cystine-rich proteins which form the matrix component of the filament-matrix texture in hard mammalian keratins such as hair. In each case the naturally occurring structures are believed to be stabilized by a three-dimensional network of specialized crosslinkages between sidechains.

Sequence studies of elastin [31] reveal frequent repetitions of a tetrapeptide (Gly-Gly-Val-Pro), a pentapeptide (Pro-Gly-Val-Gly-Val), a hexapeptide (Pro-Gly-Val-Gly-Val-Ala) and other evidence of regularities but their influence on the conformation of the polypeptide chain has not been established so far. The cystine-rich proteins of the matrix in α-keratin have prominent pentapeptide [32] and decapeptide [33] periodicities in their sequence but again their significance is not understood.

The elucidation of the chain conformation in materials such as elastin presents great technical difficulty. In the case of a well ordered structure with a regular conformation the X-ray diffraction pattern barely contains enough information to solve the structure. In poorly ordered molecular assemblies the information content of the X-ray diffraction pattern is very low, even though there may be considerable regularity at the molecular level. The elucidation of the chain conformation in these cases appears, at present, to be an intractable problem. The best hope of obtaining information is through the isolation of purified protein species and the artificial induction of ordered assemblies suitable for analysis.

ORDERED ASSEMBLIES

In their functional form many fibrous proteins occur in ordered assemblies. These are built from an asymmetric structural unit or protomer, consisting of one or more molecules, which is repeated throughout the assembly in such a way that each protomer is bonded to its neighbors in an identical manner. The units are therefore equivalent, in a crystallographic sense, and are related by spatial symmetry operations. For an enantiomorphic structure such as a protein the only types of symmetry operation that are possible are translations, rotations and combinations of these two operations [6, 34].

Helical Line Groups

The most general operation is that of screw symmetry, which involves both a rotation about an axis and a translation parallel to the same axis, and the continued repetition of this operation generates a helical structure. Possible combinations of a screw operation and additional rotations have been listed by Klug, Crick and Wyckoff [7] and these constitute the enantiomorphic helical line groups (Table 1). In the simplest form of helix (line group *s*) the repeating unit consists of a single asymmetric unit or protomer but in line groups of higher symmetry the repeating unit in the helix contains a number of symmetry related protomers.

Many ordered assemblies of fibrous proteins have a filamentous form. Crane [6] predicted that any biological structure that is rod-like is likely to have a screw axis of symmetry and this has turned out to be the case in all the filaments which have been studied so far. A well substantiated example is found in the framework of the thin filaments of vertebrate skeletal muscle. The framework consists of actin molecules related by a screw axis which gives a left-handed basic helix with a unit height of h = 27Å and a unit twist t = -166° (Fig. 3a).

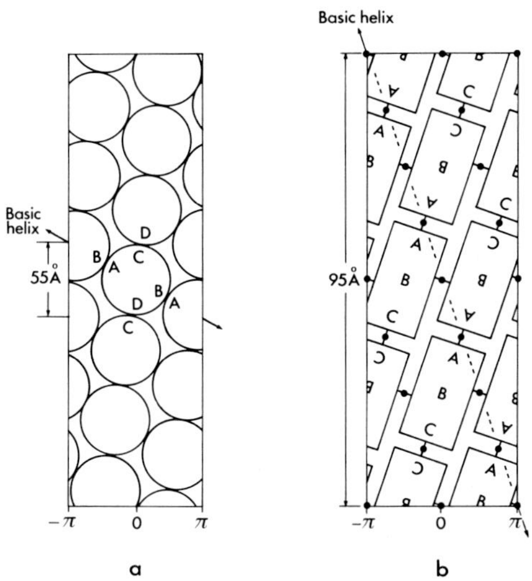

Fig. 3. Diagrammatic radial projection illustrating the disposition of the molecules and the domains of bonding between them in (a) the actin framework of the thin filament in muscle and (b) the microfibril in feather keratin. The dots in (b) represent two-fold rotation axes.

The line group is $\underline{s}$ and the asymmetric unit or protomer is a single actin molecule. As befits its function the filament is polar in the sense that a rotation of 180° about an axis normal to axis of the filament produces a physically distinguishable object. The thick filaments of skeletal muscle are bipolar and do not possess true helical symmetry but the myosin molecules in each half of the filament are related by screw symmetry. The line group in the half filament is $\underline{sr}$.

Feather keratin contains filaments, termed microfibrils, in which the protomer again consists of a single molecule but the repeating unit of the helix consists of a pair of protomers related by a two-fold rotation axis perpendicular to the screw axis (Fig. 3b). The structure therefore belongs to the line group <u>s2</u> (Table 1) and the filament is consequently non-polar in the sense that a rotation of 180° about one of the two-fold axes does not produce a physically distinguishable object. If the pleated sheets in feather have the same right-handed twist as found in globular protein molecules [26] the basic helix of the microfibril will be left-handed [28] with a unit height h = 24Å and a unit twist t = -90°. The keratin of mammalian hair contains larger microfibrils that have a helical structure in which h = 470Å, $|t|$ = 49.1° and the line group is <u>s</u> or <u>s2</u>. The keratin obtained from mammalian epidermis contains microfibrils of similar diameter to those in hair but the nature of the helical symmetry is unknown.

There is indirect evidence that collagen forms filaments about 40Å in diameter but these have not been detected so far in electron micrographs. The precise nature of the molecular packing in the filament is uncertain but the model in best accord with the available data appear to be that suggested by Smith, as modified by Miller and Parry [35]. In the Smith model the çollagen molecules are related by screw symmetry on a basic helix with unit height h = 670Å and unit twist $|t|$ = 360/5°. The line group is <u>s</u> (Table 1) and the protomer or asymmetric unit consists of a single molecule. The molecules are arranged in five strands and the distance between equivalent points in the same strand if 5 x 670 = 3350Å. In the modification suggested by Miller and Parry the microfibril is twisted so that it has a four-fold rather than a five-fold screw axis and the value of $|t|$ is therefore increased to 90°. The conversion from five-fold to four-fold screw symmetry is not a unique operation. If the final basic helix is right-handed the pitch of the molecular strands will be

$$P = h/(n + 1/20) \qquad (3)$$

for an initial helix of the same hand and

$$P = h/(9/20 - n) \qquad (4)$$

for an initial helix of opposite hand, where n is an integer. On the basis of their X-ray diffraction studies Miller and Parry favoured the solutions obtained from expression (3) by setting n = ±1. With h = 670Å these are P = 638 and -705Å. More recent studies (Fraser, MacRae & Miller, unpublished data) suggest that a value of P = -1218Å, obtained from expression (4) by setting n = 1, gives better agreement with the intensity distribution in the near-equatorial region of the X-ray diffraction pattern. In these models with $|t|$ = 90° the axis of the collagen molecule follows a helical path and so, as in tropomyosin, the chain conformation is that of a coiled-coiled-coil. Since the line group of the microfibril is <u>s</u> the structure is polar.

Physical studies of elastin have so far failed to produce any convincing evidence of the presence of ordered assemblies of protein molecules in this material. The fibrillar aggregates observed by electron microscopy after treatment with ultrasonics or mechanical disintegration are possibly due to cellulosic contamination [36].

Classification of Protomer Interactions

Monod, Wyman and Changeux [37] define the spatially organized collection of the groups or residues of one protomer which are involved in its binding to another protomer as a "binding set". In the present context the term protomer will be taken to be synonymous with asymmetric unit, that is, the smallest unit from which the structure may be generated by symmetry operations. The two linked binding sets are called the "domain of bonding" of the pair of protomers.

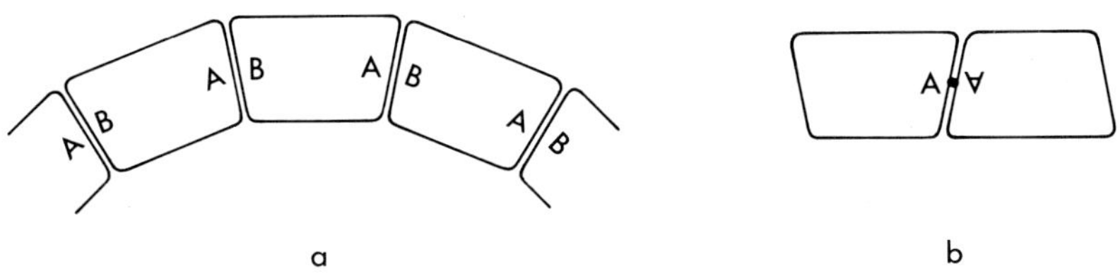

Fig. 4. Modes of association of protomers; (a) heterologous, (b) homologous. In (b) the dot represents a two-fold rotation axis.

Two modes of association are distinguished; in "heterologous associations" the domain of bonding comprises two different binding sets (Fig. 4a) whilst in "homologous associations" the domain of bonding involves two identical binding sets related by a two-fold rotation axis (Fig. 4b). The number of binding sets in the protomer is therefore equal to the number of homologous associations plus twice the number of heterologous associations in which the protomer is involved.

There is an implied relationship between the helical line group of a filament and the types of association between the protomers which constitute the filament. In the case of line group <u>s</u> there will be at least one heterologous association between the protomers but there can be no strictly homologous associations. For example in the actin framework of the thin filament (Fig. 3a) an actin molecule in positicn i appears to be in contact with other actin molecules in positions $i + 1$ (A-B), $i - 1$ (B-A), $i + 2$ (C-D) and $i - 2$ (D-C) in the basic helix. This implies the existence of four binding sets and two types of heterologous association.

When contacting protomers are related by a two-fold rotation axis, as is the case in the feather keratin microfibril (Fig. 3b), a homologous association between them is implied (Fig. 4b). Heterologous associations may or may not be present. In the feather keratin microfibril each protomer makes three prominent contacts with other protomers that in each case are related to it by a two-fold rotation axis. Thus there will be at least three binding sets on each protomer and each protomer will form at least three homologous associations. Heterologous associations probably play only a minor part in this particular filament.

The minimum number of heterologous and homologous associations implied by the four helical line groups are listed in Table 1.

Table 1

Enantiomorphic helical line groups, number of protomers in the repeating unit and implied associations of the protomer

Line group symbol*	Protomers in repeating unit	Implied associations			Other possibilities	
		Homologous	Heterologous	Unspecified	Heterologous**	Unspecified**
s	1		1		m	
s2	2	1		1		m
sr (N = 2)	2	1	1			m
(N > 2)	N		2		m	
sr2 (N = 2)	4	2		1		m
(N > 2)	$2N$	1	1	1		m

* s indicates a screw axis, 2 indicates a two-fold rotation axis normal to the screw axis, r indicates an N-fold rotation axis parallel to the screw axis.

** m signifies a positive integer or zero.

Nature of Protomer Interactions

Ordered assemblies of fibrous proteins are very precise structures and it follows that the bonding between protomers must be highly specific. Crane [6] reasoned that this property was likely to be associated with domains of bonding that consisted of a large number of relatively weak bonds. For the interaction to be strong all the interacting groups in the two binding sets constituting the domain of bonding must simultaneously approach closely and so there is a requirement for the interacting surfaces to be spatially complementary. This complementarity was considered to be the basis of the high specificity and it was suggested that van der Waals interactions would be the main type of bond, with the possibility that hydrogen bonds might also be involved.

Recent studies of the forces that stabilize oligomeric assemblies of globular proteins have shown that the main interactions are between non-polar residues such as leucine, isoleucine and valine. Some additional interactions occur between residues such as serine, threonine, glutamine and asparagine through the formation of hydrogen bonds. It seems likely that the same types of interactions will be important in the stabilization of ordered assemblies of fibrous proteins. In mammalian keratin the presence of a pattern of uncharged residues on the surface of the coiled-coil ropes can be correlated with inter-rope contacts. The types of residue in the contacting areas are similar to those present in the areas of contact between protomers in oligomeric assemblies of globular proteins.

There is evidence that charged residues play an important part in the assembly of ordered aggregates. Studies of synthetic aggregates prepared *in vitro* from solubilized fibrous proteins have shown that the structures of the aggregates are very sensitive to such factors as the pH and ionic strength of the solutions from which they are precipitated. In some instances a single protein species may form many polymorphic aggregates [5, 38, 39].

The part played by sequence regularities in the determination of chain conformation and the formation of multichain structures has already been remarked upon. Longer range regularity, particularly in the distribution of charged and uncharged residues, has been noted in a number of fibrous proteins. In the case of tropomyosin this distribution can be correlated with areas of contact between the tropomyosin molecule and the actin framework of the thin filament [40]. In collagen, long-range periodicities in sequence can be correlated with the stagger between adjacent molecules in the microfibril [41] and also with the intermolecular contacts [42].

Quasi-equivalence

The unifying feature in the development of a fibrous texture in ordered aggregates of fibrous proteins is the tendency for like residues, like chains and like molecules to occupy equivalent positions in the structure. A structure built from protomers that are related by symmetry operations is not necessarily one of minimum free energy and it may be possible to arrive at a structure of lower free energy by some systematic deformation of the idealized form. The individual protomers will then no longer be equivalently related, but if the domains of bonding between the protomers are preserved the protomers may be said to be quasi-equivalently related [43]. An important feature of structures deformed in this way is that they are still completely regular but the new asymmetric unit will contain a number of the original protomers. Systematic deformation leading to a quasi-equivalent relationship between protomers may be due to intrinsic factors within a filament or may arise from extrinsic factors such as the packing together of fibrils on a lattice which does not provide equivalent environments for all the protomers. This latter situation is common in crystallites of synthetic polypeptides [5] and has also been shown to occur in feather keratin.

Crystalline Assemblies

Several fibrous proteins or fragments of fibrous proteins have been crystallized, including tropomyosin, proteolytically modified fibrinogen and fragments of the collagen molecule. The filaments in striated muscle and many other naturally occurring structures form regular arrays and in the case of collagen the packing of the microfibrils in tendon appears to be truly crystalline [44, 45]. The unit cell in collagen is very large and recent studies (Fraser, MacRae & Miller, unpublished data) suggest that in decrimped tendon it is triclinic with $a = b = 79$Å, $c = 2680$Å, $\alpha = \beta = 93.2^{\circ}$ and $\gamma = 89.8^{\circ}$. Four microfibrils pass through the cell, which contains 16 collagen molecules. After suitable treatment the triclinic form can be transformed to the so-called orthomorphic form [44] in which the cell is tetragonal. In the triclinic form the asymmetric unit is a single molecule so that there are sixteen different environments and the molecules in the cell are only quasi-equivalently related. In the orthomorphic form the space group is $P4_1$ or $P4_3$ and there are only four different environments.

Like most other protein crystals the crystalline assemblies of fibrous proteins so far discussed contain appreciable proportions of water and the crystallinity is either lost or seriously degraded if the water is removed by drying. Tulloch (unpublished data) has recently shown however that the highly ordered ribbons of fibrous protein isolated from the ootheca of the praying mantis [46] retain their crystallinity even after the removal of water. Moreover the order within the dried ribbons (Fig. 5a) approaches that which exists in fully hydrated crystals of globular proteins.

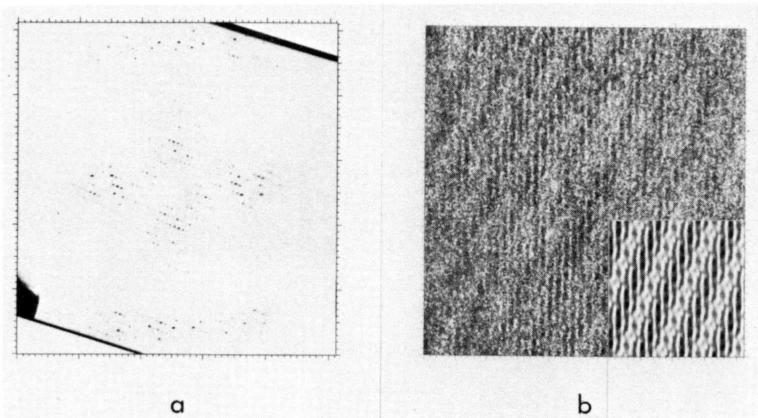

Fig. 5. (a) Electron diffraction pattern obtained from a ribbon of fibrous protein isolated from the ootheca of the praying mantis *Tenodera australasiae* and stained with uranyl acetate. The high-angle α-pattern is sampled by the crystal lattice. (b) Electron micrograph from a similar ribbon. The inset shows an averaged image obtained by digital processing. The periodicity of the broad inclined banding pattern is 136Å.

An electron micrograph of a ribbon stained with uranyl acetate, together with a reiterated averaged image of the projected unit cell, is shown in Fig. 5b. The unit cell dimensions are known to be vary large [47]; the lattice parameters in the plane of the ribbon, measured from Fig. 5b, are a = 136Å, c = 84 Å and $\beta = 92^{\circ}$.

REFERENCES

1 R. Josephs, H.S. Jarosch and S.J. Edelstein, J. Mol. Biol., 102(1976)409.

2 S. Seifter and P. Gallop, in H. Neurath (Editor), The Proteins, Vol. 4, Academic, New York, 2nd ed., 1966, p. 153.

3 M. Florkin and E.H. Stotz (Editors), Comprehensive Biochemistry, Vol. 26B, Elsevier, Amsterdam, 1968.

4 M. Florkin and E.H. Stotz (Editors), Comprehensive Biochemistry, Vol. 26C, Elsevier, Amsterdam, 1971.

5 R.D.B. Fraser and T.P. MacRae, Conformation in Fibrous Proteins, Academic, New York, 1973.

6 H.R. Crane, Sci. Mon., 70(1950)376
7 A. Klug, F.H.C. Crick and H.W. Wyckoff, Acta Crystallogr., 11(1958)199.
8 S. Arnott and S.D. Dover, J. Mol. Biol., 30(1967)209.
9 F.H.C. Crick, Acta Crystallogr., 6(1953)685,689.
10 L. Pauling and R.B. Corey, Nature (London), 171(1953)59
11 C. Cohen, D.L.D. Caspar, D.A.D. Parry and R.M. Lucas, Sympos. Quant. Biol., 36(1971)205.
12 D.A.D. Parry, Nature (London), 256(1975)346.
13 D. Stone, J. Sodek, P. Johnson and L.B. Smillie, Proc. IX FEBS Meeting, Budapest, 31 (1975) 125.
14 W.G. Crewther, L.M. Dowling, K.H. Gough, A.S. Inglis, N.M. McKern, L.G. Sparrow and E.F. Woods, Proc. Int. Wool Text. Res. Conf., 5th, Aachen, 1976, (in press).
15 A.D. McLachlan and M. Stewart, J. Mol. Biol., 98(1975)293
16 J. Sodek, R.S. Hodges, L.B. Smillie and L. Jurasek, Proc. Nat. Acad. Sci. U.S.A., 69(1972)3800.
17 R.F. Doolittle, Advan. Protein Chem., 27(1973)1.
18 D. Skerrow, Biochem. Biophys. Res. Commun., 59(1974)1311.
19 E.D.T. Atkins, J. Mol. Biol., 24(1967)139.
20 S. Arnott, S.D. Dover and A. Elliott, J. Mol. Biol. 30(1967)201.
21 L. Pauling and R.B. Corey, Proc. Nat. Acad. Sci. U.S.A., 39(1953)253.
22 F. Lucas and K.M. Rudall, in M. Florkin and E.H. Stotz (Editors), Comprehensive Biochemistry, Vol. 26B, Elsevier, Amsterdam, 1968, p.475.
23 F. Lucas, J.T.B. Shaw and S.G. Smith, Advan. Protein Chem., 13(1958) 107.
24 R.E. Marsh, R.B. Corey and L. Pauling, Acta Crystallogr., 8(1955)710.
25 A.J. Geddes, K.D. Parker, E.D.T. Atkins and E. Beighton, J. Mol. Biol., 32(1968)343.
26 C. Chothia, J. Mol. Biol., 75(1973)295.
27 M. Levitt and C. Chothia, Nature (London), 261(1976)552.
28 R.D.B. Fraser and T.P. MacRae, Proc. 16th Int. Ornith. Congr., 1976 (in press).
29 P.P. Fietzek and K. Kühn, Mol. Cell. Biochem., 8(1975)141.
30 A. Yonath and W. Traub, J. Mol. Biol., 43(1969)461.
31 L. Sandberg, Int. Rev. Connect. Tissue Res., 2(1976)159.
32 L.S. Swart and D. Parris, Nature (London), 249(1974)580.
33 T.C. Elleman, H. Lindley and R.J. Rowlands, Nature (London), 246(1973) 530.
34 A. Klug, Symp. Int. Soc. Cell Biol., 6(1967)1.

35 A. Miller, in E.D.T. Atkins and A. Keller (Editors), Structure of Fibrous Biopolymers, Butterworths, London, 1975, p. 127.

36 G.R. Millward, J. Cell. Biol., 42(1969)317.

37 J. Monod, J. Wyman and J. Changeux, J. Mol. Biol., 12(1965)88.

38 Caspar, D.L.D. and C. Cohen, in A. Engström and B. Strandberg (Editors), Symmetry and Function of Biological Systems at the Macromolecular Level, Wiley, New York, 1969, p. 393.

39 B.B. Doyle, D.W.L. Hukins, D.J.S. Humes, A. Miller and J. Woodhead-Galloway, J. Mol. Biol., 91(1975)79.

40 A.D. McLachlan and M. Stewart, J. Mol. Biol., 103(1976)271.

41 D.J.S. Humes, A. Miller, D.A.D. Parry, K.A. Piez and J. Woodhead-Galloway, J. Mol. Biol., 79(1973)137.

42 L.W. Cunningham, H.A. Davies and R.G. Hammonds, Biopolymers, 15(1976) 483.

43 D.L.D. Caspar and A. Klug, Cold Spring Harbor Symp. Quant. Biol., 27 (1962)1.

44 A. Miller and J.S. Wray, Nature (London), 230(1971)437.

45 A. Miller and D.A.D. Parry, J. Mol. Biol., 75(1973)441.

46 W. Kenchington and N.E. Flower, J. Microsc., 89(1969)263.

47 C.J. Rattew, D. Phil. Thesis, Oxford, 1974.

POLYPEPTIDES AS STRUCTURAL MODELS FOR PROTEINS

W. TRAUB
Structural Chemistry Department, Weizmann Institute, Rehovot, Israel

SUMMARY

The role of X-ray studies of polypeptides in elucidating the structures of native proteins is reviewed in this article.

Homopolypeptides have been found to model the main regular polypeptide chain conformations which occur in fibrous and globular proteins, but they are not good models for specific assemblies of chains. These aspects of structure can, however, be studied with the aid of polypeptides of ordered amino acid sequence. Such compounds have provided useful models for silk and for collagen and have recently been applied to structural investigations of multistranded α-helices and the "oiled coils" of elastin. They have also provided better understanding of relationships between primary sequence and conformation.

The crystallization of ordered sequence oligopeptides has opened the prospect of more precise and detailed X-ray structure determinations of features both of fibrous and of globular proteins. However, increased understanding of protein structures, and improved possibilities of determining them directly, may cause a shift of interest from polypeptides which model proteins to those which in themselves have interesting and useful properties.

INTRODUCTION

Protein molecules owe their enormous functional versatility to the fact that under appropriate conditions they spontaneously fold into complicated and distinctive conformations determined by their particular amino acid sequences. Understanding how sequence determines conformation, perhaps eventually to the point of being able to predict detailed structure from sequence, is one of the most fascinating and important problems in protein chemistry. As an aid to unravelling these sequence-structure relationships, much interest has focused on synthetic polypeptides as simple models for proteins, and, over the last quarter century, their study has undoubtedly contributed very greatly to elucidating various aspects of protein structure and relating these to their physicochemical and biological properties. Yet the role of polypeptides as models logically implies an eventual shifting of interest towards the greater complexities of naturally occurring proteins, and in recent years many, including myself, have transferred our main efforts from studies of polypeptides to research on "real proteins". This trend reflects a belief that the main structural features of such simple models have already been determined and also the greatly improved possibilities of studying protein structures directly.

I feel this is an excellent forum for reassessing the usefulness of structural studies of polypeptide models. Are there still important areas of incompleteness in our understanding of their structures, or will further work be governed by the law of diminishing returns? For my part, I feel that further major advances in this field depend on the isolation and characterization of new proteins to be modeled or new technical developments which can provide answers to new kinds of questions. I see some evidence of both these trends. However, before speculating about future prospects, it is appropriate to consider the past contributions of polypeptide studies in elucidating protein structures, as well as their limitations.

CONFORMATIONS OF HOMOPOLYPEPTIDES

The first major achievement of the indirect approach to protein structure determination was to establish the dimensions of the amide group from a series of single crystal structure analyses of small peptides and amino acids [1] . These dimensions, which have been slightly revised in the light of more extensive and precise peptide structure analyses [2,3] , were used by Pauling and Corey in their investigations of possible regular extended polypeptide conformations which might be stabilised by systematic hydrogen bonding. These studies led them to propose the α helix [4,5] and the β pleated sheet [6,7] , as well as several other conformations [8] now generally forgotten. These were essentially theoretical models, incorporating dimensions of planar amide groups, hydrogen bonds and van der Waals contacts found in small molecules, and evaluated by rather crude energy estimations. Two forms of pleated sheets were considered, depending on whether neighbouring polypeptide chains are parallel or antiparallel. Both have twofold screw axes, but the repeat distances were estimated as 6.5 Å for the parallel case and 7.0 Å for the antiparallel. The α helix incorporates an axial translation of 1.50 Å and a rotation of 100° per residue and can be built in either a right-handed or a left-handed form, though the latter was considered less stable because of a rather short contact (2.64 Å) between the β-carbon atom and the carbonyl oxygen of the same residue.

Experimental support for the existence of the α helix came from the observation of a meridional 1.5 Å reflection, corresponding to the residue translation, in the X-ray diffraction pattern of poly-γ-benzyl-L-glutamate fibres [9] and from the appearance of strong layer lines in the poly-γ-methyl-L-glutamate pattern [10] , which were shown to occur at the spacings to be expected for an α helix [11] . Later, a more detailed structure analysis was made of the α form of poly-L-alanine, which shows a particularly rich X-ray pattern and has only a single methyl group as a side chain. It was shown that the X-ray intensities could be well accounted for by right-handed α helices with a random chain sense [12] and this structure was later subjected to quantitative refinement [13,14] .

Polypeptides were also found to show X-ray fibre patterns indicative of β pleated-sheet structures, and some, including poly-L-alanine and poly-γ-methyl-L-glutamate, were obtained in both α-helical and β forms [8,15] . Most of the polypeptide β X-ray patterns indicate repeat distances intermediate between those expected for parallel and antiparallel chains, though they are generally closer to the latter and that of polyglycine I is, in fact, 7.0 Å [16] . The only β structure that has been analysed in detail is that of β poly-L-alanine for which an antiparallel pleated-sheet structure, with neighbouring sheets randomly displaced, has been proposed [17] .

Polyamino acids have also been found to exist in several other conformations. Poly-β-benzyl-L-aspartate forms left-handed helices [18,19] , which in the solid state can conform either to α-helical dimensions or to those of the rather similar ω helix, with a translation of 1.3 Å and rotation of 90° per residue [20] . This latter form, which may well be related to the occurrence of bulky groups on the β-carbon atom, has also been reported for poly-S-benzylthio-L-cysteine [21] . Another kind of left-handed helix, with 3.1 Å translation and 120° rotation per residue, occurs in poly-L-proline II [22,23] , poly-L-hydroxyproline A [24] and polyglycine II [25,26] . Poly-L-proline I also exists in a right-handed helical form, which has the unusual cis configuration about all the peptide bonds [27,28] .

X-ray studies of solvated polyamino acids in liquid crystals have indicated that their conformations do not vary appreciably over wide ranges of solvation [29,30,31] . Sharp solvent-induced transitions between different conformations have been found to occur in a number of cases [31-37] , but no previously unobserved conformations were discovered.

MODELS FOR FIBROUS PROTEINS

It has proved much easier to determine the detailed structures of polypeptides than those of natural proteins. Structure analyses from X-ray fibre patterns, such as are obtained from polypeptides or fibrous proteins generally proceed through the following stages:- measurement of helical parameters and unit cell dimensions from the X-ray pattern; consideration of possible molecular models consistent with the helical parameters, any other relevant physicochemical data, and appropriate bond lengths, bond angles and van der Waals distances; consideration of how various possible molecular conformations could fit into the unit cell; comparison of observed X-ray intensities with those calculated for the various possible structures. After choosing an approximate conformation and mode of packing, it is possible to refine several parameters in terms of minimum discrepancies between observed and calculated X-ray intensities [13,14, 17] . For these procedures, polypeptides have several advantages over fibrous proteins. They have simple known amino acid sequences, they more often pack in small well-defined unit cells, and their X-ray patterns are unlikely to be complicated by the effects of long-range distortions and extraneous components. Hence, one can calculate interatomic distances and X-ray intensities corresponding to various possible conformations and, by these criteria, establish the structure with far less ambiguity than is possible in direct studies of fibrous proteins. Corresponding features of the X-ray patterns or other physico-chemical observations of the proteins are then interpretable by analogy with the polypeptides.

How well do the various polypeptide conformations described above represent the structures of fibrous proteins? The α helix, β pleated sheet and polyproline II helix have been considered, respectively, as models for the α keratin, β keratin and collagen conformations, the three main types of fibrous protein structures classified by Astbury according to their high-angle X-ray patterns [38] . It turns out that these different types of polypeptides and fibrous proteins do indeed have similar conformations, but the greater simplicity of the homopolypeptides limits their relevance as models to only some aspects of protein structure.

Thus, various observations on α keratin and myosin, including the occurence of a meridional 1.5 Å reflection, indicate that they contain α helices, presumably similar in conformation to those of α polyalanine. However, these fibrous proteins also show a strong 5.1 Å meridional reflection never yet observed in X-ray patterns of synthetic polypeptides, which apparently arises from the specific assembly of the protein α helices to form coiled coils. [39-45]

It also appears that the structures of some α-helical fibrous proteins, notably paramyosin [41,42] , may be perturbed by side-chain interactions. Such interactions also appear to occur in the structures of polypeptides with large side chains, the X-ray patterns of which show periodicities additional to those due to the α-helical polypeptide backbone [35,46,47] .

β-Polyalanine appears to be a very good model for Tussah silk fibroin, which shows an almost identical X-ray pattern [48,17] and, in fact, contains some 39% alanine, as well as 24% glycine and 11% serine [49] . The structure reportedly consists of antiparallel pleated sheets, equally spaced, but randomly displaced in the interchain direction [17] . However, the X-ray pattern of Bombyx mori silk fibroin, which contains 44% glycine, 29% alanine and 12% serine [50] , is significantly different [51] . Sequence analyses have indicated that, in most of this protein, glycine residues alternate with those of other amino acid residues, and it was found that the synthetic polydipeptide poly(Ala-Gly) shows an X-ray pattern similar to that of this silk. The structure of poly(Ala-Gly), and apparently also of Bombyx mori, consists of antiparallel pleated sheets with alternate narrow inter-sheet spacings, accommodating only glycine residues, and wider spacings, accommodating side chains of other residues [52] . The conformational specificity of the poly(Ala-Gly) sequence has been emphasized by studies of a series of ordered copolymers of alanine and glycine, which have been found to form a variety of β and non-β structures [53,54,55] . However, an even closer resemblance to the X-ray pattern of Bombyx mori is shown by the polyhexapeptide poly(Ala-Gly-Ala-Gly-Ser-Gly), which corresponds to a commonly occurring sequence in this silk [50,56] .

Polypeptides of ordered sequence have played an invaluable role in elucidating the structure of collagen [57] . Although it was established as long ago as 1955 [58] that the collagen molecule consists of three helical polypeptide chains coiled about a common axis and held together by NH···OC hydrogen bonds, controversy continued for many years as to the detailed conformation and the number and mode of interchain hydrogen bonds. Because of the rather poor X-ray pattern, indicating no unit cell, and the limited knowledge of the sequence, decisive criteria, enabling a clear choice of structure from direct studies of collagen, proved very elusive, and attention was directed to model compounds.

The threefold left-handed helices formed by homopolypeptides of proline [22,23] , hydroxyproline [2,4] , or glycine [25,26] do indeed resemble individual strands of the collagen molecule, but they do not form coiled coils. However, various polytripeptides, having, like collagen, glycine as every third residue and one or both of the imino acids proline and hydroxyproline, show X-ray patterns with all the main features of that of collagen, including the meridional 2.9 Å spacing indicative of a rope-like triple helix [59-64] . A detailed and systematic structure analysis of $(Gly\text{-}Pro\text{-}Pro)_n$ has led to an unambiguous conformation which has been refined within narrow limits [63] . Similar analyses of the polyhexapeptides $(Gly\text{-}Ala\text{-}Pro\text{-}Gly\text{-}Pro\text{-}Pro)_n$, $(Gly\text{-}Pro\text{-}Ala\text{-}Gly\text{-}Pro\text{-}Pro)_n$, $(Gly\text{-}Ala\text{-}Pro\text{-}Gly\text{-}Pro\text{-}Ala)_n$, and $(Gly\text{-}Ala\text{-}Ala\text{-}Gly\text{-}Pro\text{-}Pro)_n$, chosen for their variety of potentially H-bonding NH positions, all revealed conformations close to that found for $(Gly\text{-}Pro\text{-}Pro)_n$ [64,65] . Furthermore, the most detailed collagen X-ray pattern, that of stretched rat tail tendon, shows good agreement with that calculated from the $(Gly\text{-}Pro\text{-}Pro)_n$ conformation [66, 67] .

The polypeptide studies have also served to illuminate the structural significance of sequence regularities in collagen. The failure of poly-(Ala-Pro-Pro), poly(Ala-Hyp-Hyp), and random copolymers of glycine and imino acids [61,68] to form collagen-like structures accords with a position for every third residue near the axis of the triple helix where there is room only for glycine, the smallest amino acid. Furthermore, the poly(Gly-Pro-Pro) conformation indicates severe steric restrictions for phenylalanine, tyrosine, and leucine residues in position Y of the poly(Gly-X-Y) sequence, but stabilizing hydrogen-bonding interactions for hydroxyproline, threonine, glutamine, arginine, and lysine residues in this position [69,70,71]. Apparently, appreciable amounts of conformationally restrictive imino acid residues are required to prevent collagen assuming energetically more favoured α-helical or β-pleated-sheet conformations, as have been found for poly(Gly-Ala-Ala) [55] and poly(Gly-Ala-(OEt)) [72]. Polytripeptides of the form $(Gly\text{-}Amino\text{-}Imino)_n$ have been found to form triple helices less readily than those of the form

(Gly-Imino-Amino)$_n$, possibly because of different interactions with water [57,73]. Structural investigations of homopolypeptides [74] and, more recently, of several series of related sequential polypeptides [75] have also served to indicate tendencies of various amino acid residues to favor α-helical or β conformations.

In some very recent studies sequential polypeptides have been used to model two other structural proteins. Tropoelastin, the precursor protein of native elastin, has been found to have several extensive runs of repeating oligopeptides, and it has been suggested that these form a series of β bends comprising β helices or 'oiled coils' [76,77] . These ideas have been investigated, and apparently supported, by proton NMR on synthetic polypeptides modeling the repeating (Val-Pro-Gly-Gly), (Val-Pro-Gly-Val-Gly), and (Ala-Pro-Gly-Val-Gly-Val) sequences found in the protein [78,79] .

Amino acid sequence analysis of the muscle protein tropomyosin has indicated repeating units of seven residues in which the second and fifth positions are almost invariably occupies by hydrophobic residues [80,81] . This is consistent with the geometrical requirements of a two-stranded rope of α helices, each having 3.6 residues per turn, which are held together through the juxtaposition of sides rich in hydrophobic residues [82,82] . Measurements of molecular weight and circular dichroism of solutions of the synthetic sequential polypeptides poly(Lys-Leu-Glu-Ser-Leu-Glu-Ser) and poly(Leu-Glu-Ser-Lys) have indicated that the polyheptapeptide, but not the polytetrapeptide, form dimers of helices [84] . Studies of the effects of varying polyheptapeptide sequences on the stability of the coiled-coil conformation have also been initiated [84,85] . However, there have as yet been no reports of X-ray structure analyses of either the tropomyosin or the tropoelastin model polypeptides. These could well provide more detailed knowledge of conformation than X-ray studies of the native proteins.

MODELS FOR GLOBULAR PROTEINS

Globular proteins have far less regular conformations than fibrous proteins. It so happened that the first globular protein structures to be determined, those of myoglobin and haemoglobin, contain more than 70% helix [86,87] , a fortunate circumstance that greatly facilitated their elucidation. However, with only a few exceptions, the more than 50 protein structures determined since contain appreciably less than 50% helix

[88,89] . Most of these helical regions resemble the right-handed α helix of poly(L-alanine), but there are also variations resembling the 3_{10} [86, 90-97] , α_{II} [91,95,98,99] , and π [86,90] right-handed helices, which have not been observed in synthetic polypeptides. Apart from two short segments in pancreatic trypsin inhibitor [100] which resemble the polyproline-II conformation, no left-handed helical regions have been observed in globular proteins. Considerable regions both of parallel and of antiparallel β pleated sheets have also been found in various proteins, and these generally have a right-handed twist when viewed along the polypeptide chain [101,102] . In addition, globular proteins contain many sequences which do not fold into regular conformations, but form various bends and loops which largely serve to determine the tertiary structure of the molecule. In a few cases residues with a cis configuration about the peptide bond are reportedly involved in such bends [97,99,103,104].

Much effort has been directed towards understanding the structures of globular proteins by calculating minimum-energy conformations for protein sequences and by seeking empirical correlations between sequence and conformation. From these studies there has emerged a body of rules as to what combinations of amino acid residues occur in helices [83,105,106] , pleated sheets [106] and the various bend conformations [107,108,109] . These rules have been used with considerable success to predict which regions of proteins conform to these three categories. Predictions of helices and, to some extent, of β bends are more accurate than of β sheets [110,111,112] . This may be a consequence of the methods' inability to take account of interactions between amino acid residues far apart in the linear sequence. The various prediction methods also do not distinguish between different kinds of helices or bends.

DISCUSSION

The great usefulness of polypeptide models in the study of proteins has derived largely from two factors. The first is their simplicity, the possibility of incorporating various chemical and structural features of proteins in amino acid sequences of minimal complexity. The second is their versatility, the possibility of systematically varying polypeptide models and seeing how particular properties of proteins depend on composition and sequence.

There is also a third advantage, the possibility of obtaining crystals of low molecular weight oligopeptide models, which can facilitate far more precise and reliable X-ray structure determinations than are obtainable from either fibrous or globular proteins. In fact, a protein-like bend conformation has been found in a tetrapeptide [113] and in a pentapeptide [114] . A series of such single-crystal studies might perhaps indicate whether bends in proteins are fixed by local sequences or appreciably affected by long-range interactions. The identification of stable locally determined conformations could greatly facilitate procedures for calculating the conformations of more extended regions [115] .

The prospects of obtaining single crystals of oligopeptides with 3_{10}-, α_{II}-, or π-helix conformations would appear to be far less promising. These conformations are generally confined to the final turn of an α helix and would probably not be stable if these sequences were isolated from the adjacent regions of the protein.

It may, however, be possible to perform single-crystal structure analyses of oligopeptides in more stable regular conformations. In recent years several collagen-like polytripeptides of specified molecular weight have been synthesised [116,117] , and it has been found possible to grow single crystals of $(\text{Pro-Pro-Gly})_{10}$ [118] and $(\text{Pro-Hyp-Gly})_9$ [119] in which the chains are associated to form triple-stranded structures [120] . Similar syntheses and crystallizations of oligopeptides of uniform lengths with sequences appropriate to β pleated sheets, α helices and α-helical coiled coils can be envisaged. For example, suitable tropomyosin-like di- or tripeptides might provide precise details of helical dimensions and of interactions between complementary residues.

The potential precision of an X-ray structure determination can be guaged from the number of observed reflections per atom in the asymmetric unit or, more correctly, per parameter to be determined. In the high resolution (1.5 Å) analysis of rubredoxin, a small protein molecule with only 400 atoms, some 5000 reflections were used [121] , and in the most precise polypeptide fibre structure analysis 61 reflections were used to locate 5 atoms in the residue of α poly-L-alanine [13,14] . These are the best cases, and generally protein fibre structure analyses utilise far fewer reflections per atom. In contrast, single-crystal structure analyses of small molecules, such as tetrapeptides, may utilize about 100 reflections per atom, and for the $(\text{Pro-Pro-Gly})_{10}$ crystals the ratio is 44 [120] . Thus, with suitable single crystals, it may be possible to achieve far more precise determinations of bend conformations as well as of torsion angles, hydrogen bond dimensions and the degree of planarity of the peptide group in α helices and β pleated sheets.

However, there are clearly limitations on the use of polypeptides as structural models for proteins. Their utility depends on their incorporating the essential structural features of the protein which one wishes to study in a form which is reasonably easy to prepare and analyse. Not all structural problems can be conveniently modeled, and, even if they can, it may be more informative to study the protein directly. For example, it has recently proved possible to renature and crystallize quite small fragments of native collagen chains [122] and these can provide a more interesting, but not necessarily more difficult, X-ray analytical problem than the $(Pro-Pro-Gly)_{10}$ crystals. Many important structural properties of proteins, including their interactions with substrates [123] , haptens [124] , and even common denaturants [125] are probably inherently too complex to be usefully simulated by simplified polypeptide models, though protein modification, possibly even involving total synthesis, may provide valuable information. Even the structural periodicities which detemine the assembly of collagen molecules to form fibrils, and involve repeats of 42 and 234 residues [126,127] , would seem to be beyond the scope of model studies.

Nevertheless, in spite of these limitations, the field of polypeptide studies is clearly active and developing in new directions [128] , and one might foresee that, with increased understanding of sequence-structure relationships, many new polypeptides may be designed not primarily as protein models, but for their own inherently interesting and useful properties.

REFERENCES

1 R.B. Corey and L. Pauling, Proc. Roy. Soc. (London), B141(1953)10.

2 R.E. Marsh and J. Donohue, Adv. Protein Chem., 22(1967)235.

3 G.N. Ramachandran and A.S. Kolaskar, Biochim. Biophys. Acta, 303(1973) 385.

4 L. Pauling, R.B. Corey, and H.R. Branson, Proc. Nat. Acad. Sci. U.S., 37(1951)205.

5 L. Pauling and R.B. Corey, Proc. Nat. Acad. Sci. U.S., 37(1951)235.

6 L. Pauling and R.B. Corey, Proc. Nat. Acad. Sci. U.S., 37(1951)729.

7 L. Pauling and R.B. Corey, Proc. Nat. Acad. Sci. U.S., 39(1953)253.

8 C.H. Bamford, A. Elliott, and W.E. Hanby, Synthetic Polypeptides, Academic Press, New York, 1956.

9 M.F. Perutz, Nature (London), 167(1951)1053.

10 C.H. Bamford, L. Brown, A. Elliott, W.E. Hanby, and I.F. Trotter, Nature (London), 169(1952)357.

11 W. Cochran, F.H.C. Crick, and V. Vand, Acta Cryst., 5(1952)581.

12 A. Elliott and B.R. Malcolm, Proc. Roy. Soc. (London), A249(1959)30.

13 S. Arnott and A.J. Wonacott, J. Mol. Biol., 21(1966)371.

14 S. Arnott and S.D. Dover, J. Mol. Biol., 30(1967)209.

15 A. Elliott, in Poly-α-Amino Acids, G.D. Fasman (Editor), Marcel Dekker, New York, 1967.

16 C.H. Bamford, L. Brown, A. Elliott, W.E. Hanby, and I.F. Trotter, Nature (London), 171(1953)1149.

17 S. Arnott, S.E. Dover, and A. Elliott, J. Mol. Biol., 30(1967)201.

18 R.H. Karlson, K.S. Norland, G.D. Fasman, and E.R. Blout, J. Am. Chem. Soc., 82(1960)2268.

19 E.M. Bradbury, A.R. Downie, A. Elliott, and W.E. Hanby, Proc. Roy. Soc. (London), A259(1960)110.

20 E.M. Bradbury, L. Brown, A.R. Downie, A. Elliott, R.D.B. Fraser, and W.E. Hanby, J. Mol. Biol., 5(1962)230.

21 R.D.B. Fraser, T.P. MacRae, and I.W. Stapleton, Nature (London), 193 (1962)573.

22 P.M. Cowan and S.McGavin, Nature (London), 176(1955)501.

23 V. Sasisekharan, Acta Cryst., 12(1959)897.

24 V. Sasisekharan, Acta Cryst., 12(1959)903.

25 F.H.C. Crick and A. Rich, Nature (London), 176(1955)780.

26 G.N. Ramachandran, V. Sasisekharan, and C. Ramakrishnan, Biochim. Biophys. Acta, 112(1966)168.

27 W. Traub and U. Shmueli, in Aspects of Protein Structure, G.N. Ramachandran (Editor), Academic Press, New York, 1963, p. 81.
28 W. Traub and U. Shmueli, Nature (London), 198(1963)1165.
29 D.A.D. Parry and A. Elliott, Nature (London), 206(1965)616.
30 P. Saludjian and V. Luzzati, in Poly-α-Amino Acids, G.D. Fasman (Editor), Marcel Dekker, New York, 1967, p. 157.
31 W. Traub, U. Shmueli, M. Suwalsky, and A. Yonath, in Conformation of Biopolymers, G.N. Ramachandran (Editor), Academic Press, New York, 1967, p. 449.
32 V. Sasisekharan, J. Polymer Sci., 47(1960)373.
33 U. Shmueli and W. Traub, J. Mol. Biol., 12(1965)205.
34 W. Traub and A. Yonath, J. Mol. Biol., 25(1967)351.
35 M. Suwalsky and W. Traub, Biopolymers, 11(1972)623.
36 M. Suwalsky and L. de la Hoz, Biopolymers, 12(1973)1997.
37 Y. Mitsui, Biopolymers, 12(1973)181.
38 W.T. Astbury, Trans. Faraday Soc., 34(1938)378.
39 F.H.C. Crick, Nature (London), 170(1952)882.
40 L. Pauling and R.B. Corey, Nature (London), 171(1953)59.
41 C. Cohen and K.C. Holmes, J. Mol. Biol., 6(1963)423.
42 A. Elliott, J. Lowy, D.A.D. Parry, and P.J. Vibert, Nature, 218(1968) 656,
43 R.D.B. Fraser and T.P. MacRae, J. Mol. Biol. 3(1961)640.
44 R.D.B. Fraser, T.P. MacRae, and A. Miller, J. Mol. Biol., 14(1965)432.
45 R.D.B. Fraser and T.P. MacRae, Polymer, 14(1973)61.
46 Y. Takeda, Y. Iitaka, and M. Tsuboi, J. Mol. Biol. 51(1970)101.
47 J.M. Squire and A. Elliott, J. Mol. Biol., 65(1972)291.
48 R. Marsh, R.B. Corey, and L. Pauling, Acta Cryst., 8(1955)710.
49 W.M. Schroeder and L.M. Kay, J. Am. Chem. Soc., 77(1955)3408.
50 F. Lucas, J.T.B. Shaw, and S.G. Smith, Biochem. J., 66(1957)468.
51 J.O. Warwicker, J. Mol. Biol., 2(1960)350.
52 R.D.B. Fraser, T.P. MacRae, F.H.C. Stewart, and E. Suzuki, J. Mol. Biol., 11(1965)706.
53 A. Brack and G. Spach, Biopolymers, 11(1972)563.
54 A. Brack, A. Caille, and G. Spach, Monatshefte für Chemie, 103(1972) 1604.
55 B.B. Doyle, W. Traub, G.P. Lorenzi, F.R. Brown, and E.R. Blout, J. Mol. Biol. 51 (1970)47.
56 R.D.B. Fraser, T.P. MacRae, and F.H.C. Stewart, J. Mol. Biol., 19(1966) 580.

57 W. Traub and K.A. Piez, Adv. Protein Chem., 25(1971)243.
58 G.N. Ramachandran and G. Kartha, Nature (London), 176(1955)593.
59 V.N. Rogulenkova, M.I. Millionova, and N.S. Andreeva, J. Mol. Biol., 9(1964)253.
60 W. Traub and A. Yonath, J. Mol. Biol., 16(1966)404.
61 N.S. Andreeva, N.G. Esipova, M.I. Millionova, V.N. Rogulenkova, and V.A. Shibnev, in Conformation of Biopolymers, G.N. Ramachandran Academic Press, New York, 1967, p. 469.
62 D.M. Segal, J. Mol. Biol., 43(1969)497.
63 A. Yonath and W. Traub, J. Mol. Biol., 43(1969)461.
64 D.M. Segal, W. Traub, and A. Yonath, J. Mol. Biol., 43(1969)519.
65 W. Traub, A. Yonath, and D.M. Segal, Nature (London), 221(1969)914.
66 W. Traub and G. Salem, Acta Cryst., A28(1972)S38.
67 G. Salem, M.Sc. Thesis, The Weizmann Institute of Science (1973).
68 J. Engel, J. Kurtz, E. Katchalski, and A. Berger, J. Mol. Biol., 17(1966)255.
69 W. Traub, Israel J. Chem., 12(1974)435.
70 G. Salem and W. Traub, FEBS Lett., 51(1975)94.
71 W. Traub and P.P. Fietzek, FEBS Lett., in press.
72 J.M. Anderson, W.B. Rippon, and A.G. Walton, Biochem. Biophys. Res. Commun., 39(1970)802.
73 B.B. Doyle, W. Traub, G.P. Lorenzi, and E.R. Blout, Biochemistry, 10(1971)3052.
74 E.R. Blout, C. de Loze, S.M. Bloom and G.D. Fasman, J. Amer. Chem. Soc., 82(1960)3787.
75 A.G. Walton and J. Blackwell, Biopolymers, Academic Press, New York, 1973, p. 400.
76 W.R. Gray, L.B. Sandberg, and J.A. Foster, Nature (London) 246(1973), 461.
77 D.W. Urry, Proc. Nat. Acad. Sci. U.S., 69(1972)1610.
78 D.W. Urry and T. Ohnishi, in Peptides, Polypeptides, and Proteins, E.R. Blout, F.A. Bovey, M. Goodman, and N. Lotan (Editors), Wiley (Interscience), New York, 1974, P. 230.
79 D.W. Urry, L.W. Mitchell, and T. Ohnishi, Pro. Nat. Acad. Sci. U.S., 71(1974)3265.
80 R.S. Hodges, J. Sodek, L.B. Smillie, and L. Jurasek, Cold Spring Harbor Symp. Quant. Biol., 37(1972)299.
81 D. Stone, J. Sodek, P. Johnson, and L.B. Smillie, Proc. 9th FEBS Meeting, Proteins of Contractile Systems, E.N.A. Biro (Editor), North

Holland, Amsterdam, 31(1975)125.

82 F.H.C. Crick, Acta Cryst., 6(1953)689.

83 M. Schiffer and A.B. Edmundson, Biophys, J., 7(1967)121.

84 S.A. St-Pierre and R.S. Hodges, Biochem. Biophys. Res. Commun. In press.

85 L.R. Treiber, W.M. Wong, and A.G. Walton.

86 H.C. Watson, Progr. in Stereochemistry, 4(1969)299.

87 M.F. Perutz, H. Muirhead, J.M. Cox, and L.C.G. Goaman, Nature (London), 219(1968)131.

88 A. Liljas and M.G. Rossmann, Ann. Rev. Biochem., 43(1974)475.

89 B.W. Matthews, Ann. Rev. Phys. Chem., 27(1976)

90 R.E. Dickerson and I. Geis, The Structure and Function of Proteins, Harper & Row, New York, 1969.

91 A. Liljas, K.K. Kannan, P.C. Bergsten, I. Waara, K. Fridborg, B. Strandberg, U. Carlborn, L. Jarup, S. Lovgren, and M. Petef, Nature New Biol., 235(1972)131.

92 J.J. Birktoft and D.M. Blow, J. Mol. Biol., 68(1972)187.

93 F.S. Mathew, M. Levine, and P. Argos, J. Mol. Biol., 64(1972)449.

94 T.L. Blundell, G. Dodson, D. Hodgkin, and D. Mercola, Adv. Protein Chem., 26(1972)279.

95 T. Imoto, L.N. Johnson, A.C.T. North, D.C. Phillips, and J.A. Rupley, in The Enzymes, Third Edition, Vol. 7, P.D. Boyer (Editor), Academic Press, New York, 1972, p. 665.

96 R.G. Kretsinger and C.E. Nockolds, J. Biol. Chem., 248(1973)3313.

97 H.W. Wyckoff, D. Tsernoglou, A.W. Hanson, J.R. Knox, B. Lee, and F.M. Richards, J. Biol. Chem., 245(1970)305.

98 G. Nemethy, D.C. Phillips, S.J. Leach, and H.A. Scheraga, Nature (London), 214(1967)363.

99 F.A. Quiocho and W.N. Lipscomb, Adv. Protein Chem. 25(1971)1.

100 R.H. Huber, D.Kukla, A. Ruhlmann, and W. Steigemann, Cold Spring Harbor Symp. of Quant. Biol., 36(1971)141.

101 G.N. Ramachandran and V. Sasisekharan, Adv. Protein Chem. 23(1968)283.

102 C. Chothia, J. Mol. Biol., 75(1973)295.

103 C.S. Wright, R.A. Alden, and J. Kraut, Nature(London), 221(1969)235.

104 R. Huber and W. Steigemann, FEBS Lett., 48(1974)235.

105 P.N. Lewis, N.Go, M. Go, D. Kotelchuck, and H.A. Scheraga, Proc. Nat. Acad. Sci. U.S., 65(1970)810.

106 P.Y. Chou and G.D. Fasman, Biochemistry, 13(1974)211.

107 C.M. Ventatachalam, Biopolymers, 6(1968)1425.

108 J.L. Crawford, W.N. Lipscomb, and C.G. Schellman, Proc. Nat. Acad. Sci. U.S., 70(1973)538.

109 P.N. Lewis, F.A. Momany, and H.A. Scheraga, Biochim. Biophys. Acta, 303(1973)211.

110 G.E. Schulz, C.D. Barry, J. Friedman, P.Y. Chou, G.D. Fasman, A.V. Finkelstein, V.I. Lim, O.B. Ptitsyn, R.A. Kabat, R.R. Wu, M. Levitt, B. Robson, and K. Nagano, Nature, 250(1974)140.

111 B.W. Matthews, Biochim. Biophys. Acta, 405(1975)442.

112 P. Argos, J. Schwarz, and J. Schwarz, Biochim. Biophys. Acta, 439(1976) 261.

113 T. Ueki, T. Ashida, M. Kakudo, Y. Sasada, and Y. Katsube, Acta Cryst., B25(1969)1840.

114 T. Ueki, S. Bando, T. Ashida, and M. Kakudo, Acta Cryst., B27(1971) 2219.

115 M. Levitt, J. Mol. Biol., 104(1976)59.

116 S. Sakakibara, Y. Kishida, Y. Kikuchi, R. Sakai, and K. Kakiuchi, Bull. Chem. Soc. Japan, 41(1968)1273.

117 S. Sakakibara, K. Inouye, K. Shudo, K. Kishida, Y. Kobayashi, and D.J. Prockop, Biochim. Biophys. Acta, 303(1973)198.

118 S. Sakakibara, Y. Kishida, K. Okuyama, N. Tanaka, T. Ashida, and M. Kakudo, J. Mol. Biol., 65(1972)371.

119 S. Sakakibara, personal communication.

120 K. Okuyama, N. Tanaka, T. Ashida, and M. Kakudo, Bull. Chem. Soc. Japan, 49(1976)59.

121 K.D. Watenpaugh, L.C. Sieker, J.R. Herriott, and L.H. Jensen, Cold Spring Harbor Symp. of Quant. Biol., 36(1971)359.

122 A. Yonath, W. Traub, and E.J. Miller, FEBS Lett., 57(197)93.

123 D.M. Blow and T.A. Steitz, Ann. Rev. Biochem., 39(1970)63.

124 D.R. Davies, E.A. Padlan, and D.M. Segal, Ann. Rev. Biochem., 44(1975) 639.

125 A. Yonath, A. Sielecki, J. Moult, A. Podjarny, and W. Traub, submitted for publication.

126 D. Hulmes, A. Miller, D. Parry, K.A. Piez, and J. Woodhead-Galloway, J. Mol. Biol., 79(1973)137.

127 W. Traub, submitted for publication.

128 E.R. Blout, F.A. Bovey, M. Goodman, and N. Lotan (Editors), Peptides, Polypeptides and Proteins, John Wiley & Sons, New York, 1974.

STRUCTURE OF POLY(D-L-PEPTIDES)

BERNARD LOTZ
Centre de Recherches sur les Macromolécules (C.N.R.S. et U.L.P.)
6, rue Boussingault - 67083 STRASBOURG (France)

FRÉDÉRIC HEITZ and GÉRARD SPACH
Centre de Biophysique Moléculaire (C.N.R.S.)
Avenue de la Recherche Scientifique - 45045 ORLEANS (France)

SYNOPSIS

The structure properties of regularly alternating poly(D-L-peptides) are examined, as revealed by predictions based on conformational energy analysis, and by experimental work on solution and solid state conformations of a syndiotactic polypeptide, poly(γ-benzyl D-L-glutamate) (PBD-LG). A variety of helical structures are observed for PBD-LG, including the α helix and five helices both single (π_{DL}) and double ($\pi\pi_{DL}$) stranded, which are specific to the L-D stereosequence.

Poly(D-L-peptides) are high molecular weight stereochemical analogs to the linear antiobiotic gramicidin A, which permits the passive diffusion of alkaline ions through artificial and natural lipid bilayer membranes. Channel structures which span the membrane thickness have been proposed to account for these properties, and are based on either π_{DL} or $\pi\pi_{DL}$ helices of dimerized Gramicidin A. The contribution of investigations of poly(D-L-peptides) to understanding the solution and biological properties of gramicidin A is analyzed and possible further investigations of the model polypeptides outlined.

INTRODUCTION

Investigations of the conformation and structure of polypeptides which contain only regular sequences of L and D residues (i.e. which include no glycine) have recently become possible, with the advent of racemization free methods of periodic polypeptide synthesis [1,2,3]. Such studies are still in infancy, especially when compared to similar ones performed on sequential polypeptides which include glycine residues. Results obtained so far indicate that poly(D-L-peptides) are of particular interest firstly for their novel structural possibilities, linked to their stereosequence and secondly as models for a transducting ionophore, the antiobiotic Gramicidin A [4,5].

1) The "classical" structures of fibrous proteins and poly(L-peptides) are all based on the repetition of only one (or of nearly the same) conformation of the residues in the chain, even for specific chemical sequences which include glycine residues (e.g. Bombyx mori silk or collagen). Due to this conformational equivalence (or similarity), the structure of these proteins involves conformations of the glycine residue which could be assumed by L residues.

A similar situation can hardly occur with poly(D-L-peptides). Indeed, conformations sterically acceptable for both L and D residues are confined to narrow zones close to left and right handed α helices. Any structure other than the α helices must therefore involve widely different conformations of the two residues. Some of these structures, which had first been proposed in connection with work on the antiobiotic Gramicidin A, have now been observed with poly(γ-benzyl D-L-glutamate) (PBD-LG) [6,7] and will be described first.

2) Poly(D-L-peptides) also appear to be appropriate models for investigating possible structures of Gramicidin A. This linear antiobiotic is made of 15 apolar residues with a strict and, so far, unique alternation of L and D residues [8]:

HCO-L-Val-Gly-L-Ala-D-Leu-L-Ala-D-Val-L-Val-D-Val-$(\text{L-Trp-L-Leu})_3$-
L-Trp-NH-CH_2-CH_2OH

Gramicidin is known to permit the passive diffusion of alkaline ions through biological and artificial membranes. In contrast to cyclic peptides and depsipetides (nonactin, valinomycin...) which are cage-carriers, Gramicidin A appears to form a pore through the membrane. Determination of the conformation of Gramicidin A in the membrane has not been possible, and only tentative models of its structure have been suggested so far, which are based on the solution behavior of the antiobiotic [9,10]: two molecules of Gramicidin would form a helix with a large central core which spans the membrane thickness. Two types of association of the antibiotic, giving rise to two different helices, both satisfactory on a geometrical basis are however possible [9,10], and the biologically active conformation(s) is not yet known.

In the absence of conclusive evidence from investigations on Gramicidin A itself, a number of tailor-made analogs have been synthesized which differ from the model by the absence, modification or addition of specific amino acids in the sequence. Investigations of such analogs have shown [11] that only those with a strict stereosequence of alternating L and D residues behave like the antibiotic in solution and, more importantly, share its ion-conductance properties when included in membranes.

An alternative route to help elucidate the three-dimensional structure of Gramicidin A is the investigation of poly(D-L-peptides). Such polymers, are, admittedly, very crude models for the antibiotic in view of their higher molecular weight and different sequence of side-chains. They share, however, its main structural feature, namely the sequence of alternating L and D residues, while the residues of the stereosequence (i.e. the nature of the side-chains) can be selected according to the desired properties (solubility, processability...) of the polymer. Further, well-known methods for investigation of polymers can now be used; in particular, the crystal structure can be determined by classical methods of fibre science, while that of Gramicidin A has, so far, eluded single crystal x-ray diffraction analysis [12] . The present contribution to understanding the structure and mode of action of the antibiotic by comparison with the model polypeptides is analyzed in the last part of this paper.

STRUCTURE OF POLY(D-L-PEPTIDES) IN THE SOLID STATE

The various structures of poly(D-L-peptides) or, more specifically, of PBD-LG are best described as derived from investigations of the solid-state conformation. Diffraction evidence, which helps define the geometrical parameters of the various structures is insufficient to locate the position of the atoms. Such investigations therefore rely strongly on conformational energy calculations which not only provide probable atomic coordinates of the polypeptide chain, but also reveal otherwise unaccessible conformational features.

The α_{DL} helix

As for irregularly alternating PBD-LG, the α helix is a stable structure of PBD-LG (Figure 1a). In fact, it is the native structure of this polymer, obtained upon drying of solutions (such as trifluoroacetic-acid-chloroform mixtures) in which PBD-LG exists in a random coil conformation. The α helix is therefore able to accomodate L and D residues, which should however result in fairly close contact between C_β atoms of L and D side

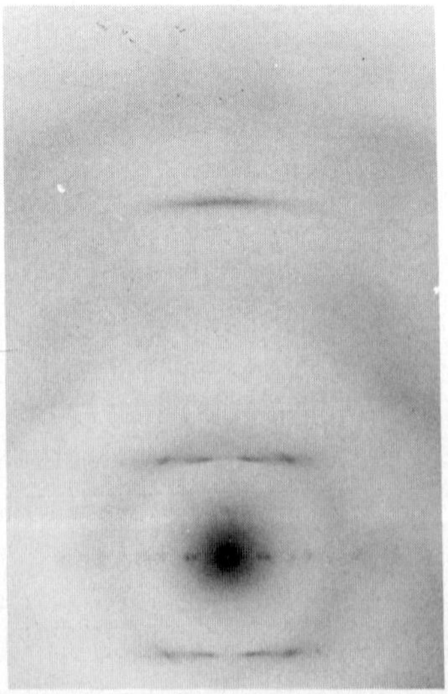

(a)

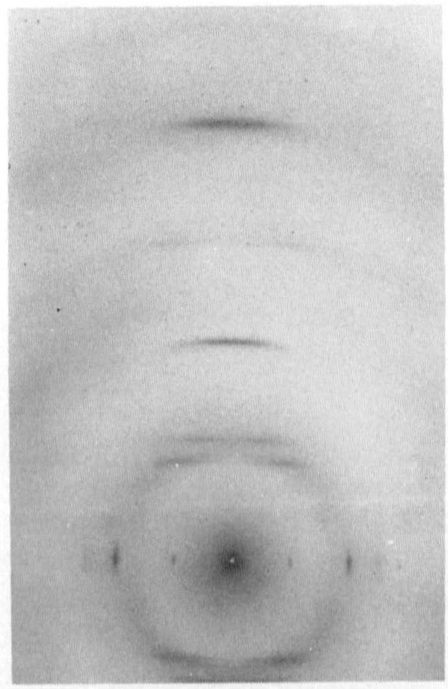

(b)

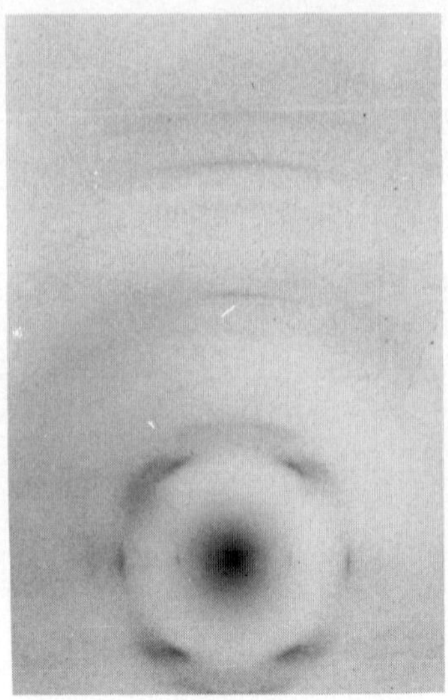

(c)

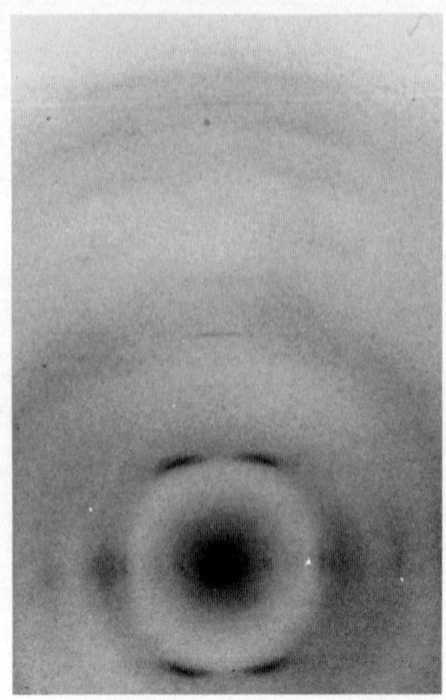

(d)

Fig. 1. Electron diffraction pattern of the (a) α_{DL} (b) $\pi_{DL}^{4.4}$ (c) $\pi\pi_{DL}^{5.6}$ and (d) $\pi\pi_{DL}^{9.0}$ helical structures of PBD-LG. Fiber axis vertical [7].

chains located in positions i and i + 3 of the helix. Such close contact had sometimes been considered as sterically prohibitive, but polypeptides with imperfect syndiotacticity have been found to exist in the α helical structure [13,14]. Indeed, recent energy calculations [15] have shown that the α helix is a stable conformation for poly(L-Ala-D-Ala). Further calculations [16] demonstrate that, while the contact between methyl groups is indeed acceptable, it results in a definite deformation of the helical backbone which tends to pull apart the C_α atoms bearing the two side-chains. These interactions do not appreciably modify the geometrical parameters of the α_{DL} helix as compared to the α_R one of poly(L-alanine) and result only in a slight increase in the total energy (- 11.16 versus - 11.70 kcal/mol-residue for α_{DL} and α_R helices, respectively).

The $\pi_{DL}^{4.4}$ helix

As already indicated, the α_{DL} helix of PBD-LG is thermally less stable than its α_R counterpart of PBLG. When heated in the solid state to 130 - 150°C (the transition temperature increases slightly with increasing molecular weight), a new structure is formed. Its diffraction pattern (Figure 1b) reveals a helical structure with parameters (n = 2.2 and b = 2.35 A, cf. Table 1) close to those expected for a helix specific to poly(D-L-peptides) first proposed by Urry [9] and by Ramachandran and Chandrasekaran [17] (Figure 2b). Following Urry's first nomenclature, this helix is designated $\pi_{DL}^{4.4}$ where the superscript is the number of peptide units per turn, i.e. 2 n.

Contrary to the α_{DL} helix which can be, to a good approximation, described in terms of a repeat unit reduced to one peptide group, the structural repeat unit of the $\pi_{DL}^{4.4}$ helix is now truly a dipeptide, in which L and D residues have conformations near those of their respective β structures. It follows that, unlike the α helix, carbonyl and NH groups point in both directions of the helical axis : inter-turn hydrogen bonding of the $\pi_{DL}^{4.4}$ helix is similar to the interchain one in the parallel β structure of poly(L-peptides).

Further, any hydrogen bond between a given residue and, say, the ith residue behind (or before) it, can similarly be made with residues i + 2, or i + 4, whose carbonyl and NH groups have the same orientation as residue i. This results in a family of helices with increasing number of residues per turn (and therefore larger diameter) and smaller h values. The $\pi_{DL}^{4.4}$ helix observed for PBD-LG is the tightest member of this family of helices,

Table 1

Main Characteristic of the Helical Structures of PBD-LG[(*)]

Structure	a(Å)	h(Å)	n	Dihedral Angles	Infrared Bands
α_{DL}	15.2	2.97	1.82	-60, -46, -44, -60	3290, 1665, 1550-1557
$\pi_{DL}^{4.4}$	17.0	2.33	2.22	-84, +101, +136, -95	3272, 1645, 1545
$\pi\pi_{DL}^{5.6}$	18.2	4.02	2.8	-116, +141, +159, -103	3280, 1632, 1536
$\pi\pi_{DL}^{7.2}$	22.6	2.95	3.6	-127, +146, +154, -126	3284, 1628-1631, 1525-1535
$\pi\pi_{DL}^{9.0}$	24.8	2.26	4.5	-138, +149 +157, -139	3285, 1628-1831 1525-1535
$\pi\pi_{DL}^{10.8}$	26.0	1.87[(1)]	5.4[(1)]	-145, +154, +159, -147	3285, 1628-1632, 1525-1535

(*) In this table, a is the average interhelical distance, h and n the rise and number of repeat (dipeptide) units per helical turn. Dihedral angles ϕ_L, ψ_L, ϕ_D, ψ_D are determined by energy analysis for right handed helices of poly(L-Ala-D-Ala) assuming ω and $\tau(NC_\alpha C')$ = 180° and 109°5, respectively, (for $\pi_{DL}^{4.4}$ helix, τ_L, τ_D, ω_L and ω_D = 106°3, 105°5, 187°1 and 174°1, respectively). For $\pi\pi_{DL}$ helices, the angles are for antiparallel strands. Wave numbers for Amide A, I and II vibrations are given and, when dichroism is observed, for orientations of the plane of polarization parallel and perpendicular to the fiber axis, respectively.

(1) calculated values.

while the $\pi_{DL}^{6.3}$ and $\pi_{DL}^{8.2}$ helices have not yet been observed. Conformational analysis reveals that the $\pi_{DL}^{4.4}$ helix can only be built with $\tau(NC_\alpha C')$ valence angles reduced to about 106°, yet its energy is only ∿ 0.5 kcal/mol-residue higher than that of the α_{DL} one (cf. Table 1). It is actually metastable at room temperature in the solid state. Like the α_{DL} helix however, it is thermally unstable as it gives rise to yet another structure at higher temperature.

<u>The family of $\pi\pi_{DL}$ helices</u>

*) $\pi\pi_{DL}^{5.6}$

At temperatures (220 - 230°C) close to the thermal decomposition

($\sim$ 260°C), PBD-LG in the solid state undergoes a second conformational transition. The new structure, which is also stable at room temperature, has a diffraction pattern (Figure 1c) again consistent with a helical structure, which is however double stranded (Figure 2c) as suggested in a model first proposed by Veatch et al [10].

This structure bears strong resemblance to the single stranded π_{DL} helices. It is made by the side by side association of two chains which are wrapped around the helical axis and is stabilized by extensive hydrogen bonding between each strand and its neighbour above and below it in adjacent helical turns. Double helices can of course be envisaged with parallel strands, but also with the more favourable association of antiparallel chains often encountered in β sheets of poly(L-peptides). These new structures can best be described as resulting from the intertwining of two π_{DL} helices (either parallel or antiparallel to each other) stretched out along their helical axes : they are therefore designated $\pi\pi_{DL}$ helices. They do not, actually, differ much from their parent structures in terms of the conformation at the residue level, which is still close to that of β structures.

The double helix obtained with PBD-LG is again rather tight, as judged from the small increase of the a parameter of the unit-cell, while in view of the steeper path of the chain, h increases to $\sim$ 4 Å, and 5.6 residues are needed to complete one helical turn. Conformational energy analysis of $\pi\pi_{DL}^{5.6}$ helices indicates that both antiparallel and parallel strand arrangements give rise to geometrically very similar structures, whose energies (- 10 and - 9.6 kca-/mol-residue, respectively) are only slightly higher than those of α_{DL} and $\pi_{DL}^{4.4}$ ones. Their diffraction patterns are expected to be very similar in view of their closely similar electron densities. Two off meridional reflections (at z $\sim$ 1.5 and 1.20 Å) are however specific to antiparallel stranded helices. Observation of these reflections for the $\pi\pi_{DL}^{5.6}$ structure of PBD-LG, together with an infrared spectrum typical of antiparallel β structures (with a shoulder at 1690 cm^{-1}) strongly suggests that antiparallel strand arrangement predominates in this structure.

*) Other $\pi\pi_{DL}$ helices

The $\pi\pi_{DL}^{5.6}$ structure of PBD-LG is the parent structure of a whole family of double helices. It is indeed stable only in the solid state and gives rise, when treated with various solvents, to helices whose

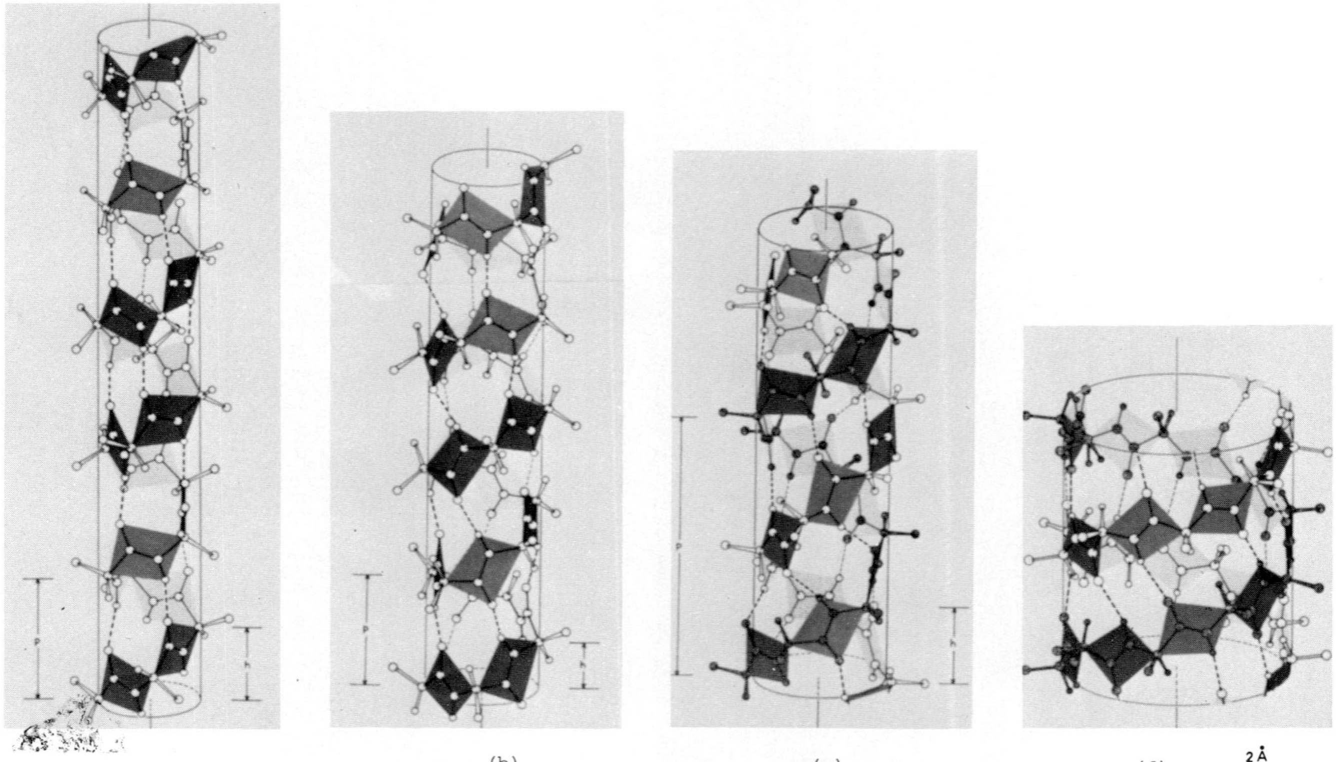

(a) (b) (c) (d)

Fig. 2. Schematic drawings of (a) α_{DL} (b) $\pi^{4.4}_{DL}$ (c) $\pi\pi^{5.6}_{DL}$ and (d) $\pi\pi^{9.0}_{DL}$ helices of a poly(D-L-peptide) chain. The planes of the peptide groups are shaded, and side-chains are reduced to C_β atoms. For the double helices, all atoms of one chain are shaded. The rise per repeat-unit (h) and pitch (p) are 2.97 and 5.40 Å, 2.33 and 5.12 Å, 4.02 and 11.24 Å, 2.26 and 10.10 Å for α_{DL}, $\pi^{4.4}_{DL}$, $\pi\pi^{5.6}_{DL}$ and $\pi\pi^{9.0}_{DL}$ helices, respectively [7].

structure depends specifically on the solvent used. These structures are derived from the $\pi\pi_{DL}^{5.6}$ helix in a manner very similar to that which yields the $\pi_{DL}^{6.3}$ and $\pi_{DL}^{8.2}$ helices from the $\pi_{DL}^{4.4}$ one. A similar sliding can indeed be performed with these double helices, provided that we consider now a pair of chains attached side by side by half the total number of hydrogen bonds. The second hydrogen bond which links a given chain to, say residue i of the other chain in the $\pi\pi_{CL}^{5.6}$ helix, can also be completed with residue i + 2, i + 4, etc... which are similarly oriented. The pattern of hydrogen bonds is therefore maintained, but these helices have more and more residues involved per helical turn, i.e. 5.6, 7.2, 9.0 and 10.8 residues. For these $\pi\pi_{DL}^{7.2}$, $\pi\pi_{DL}^{9.0}$ and $\pi\pi_{DL}^{10.8}$ helices, the diameter progressively increases and the rise per repeat unit decreases as the path of the chain becomes flatter. Further, the hydrogen bonds adopt positions more nearly parallel to the helical axis which results in increasing dichroic ratios of the IR spectra, which are otherwise similar to that of the $\pi\pi_{DL}^{5.6}$ structure.

Diffraction evidence obtained with PBD-LG helps determine the geometrical parameters of these helices (cf. Table 1), which are found by conformational analysis to correspond to stable structures of poly(D-Ala-L-Ala). This technique further reveals that larger helices are more stable, and therefore more likely, if made of antiparallel strands [16].

Helical structures with so many residues per turn would appear highly unlikely in view of their very large central cores. It turns out, however, that the latter are large enough to accomodate solvent molecules. More specifically, comparison of the core and solvent dimensions suggests that the size of the latter dictates what double helix is formed from the $\pi\pi_{DL}^{5.6}$ one. Thus methylene chloride, chloroform or dioxanne, and the larger collidine (trimethyl 2,4,6 pyridine) molecules induce formation of the $\pi\pi_{DL}^{7.2}$, $\pi\pi_{DL}^{9.0}$ and the larger $\pi\pi_{DL}^{10.8}$ helices, respectively (Figure 3). We may note that a similar behaviour has been previously observed for Amylose, which also forms molecular complexes with various slightly polar solvents (especially alcohols) and with iodine [18].

ome general remarks

To sum up therefore, it appears that PBD-LG, which is so far the poly(D-L-peptide) most thoroughly investigated, can exist in no less than six single or double stranded helical structures. Although determination of the structure is far from complete, the proposed conformations are satisfactory on geometrical and energetic grounds, and account for the main features of both diffraction and infrared evidence. Although more thorough analysis of the available experimental results may reveal finer details of these structures, their gross features are well established. The picture which emerges is that, due to the existence of a variety of almost iso-energetic structures. PBD-LG in the solid state has an unmatched structural versatility. This behaviour is not limited to the solid state but, as will be shown now, manifests itself in solution.

CONFORMATION OF PBD-LG IN SOLUTION

Solutions properties of PBD-LG have been investigated by various experimental techniques, including infrared, optical rotatory dispersion, circular dichroism and nuclear magnetic resonance. It has been possible to establish that most of the solid-state structures of PBD-LG also exist in solution, to establish fingerprints of physical properties which permit their unambiguous characterization and to analyze the transconformations between the various helical structures (and the disordered state). We briefly summarize results of extensive investigations [19, 20, 21, 22] on the solution properties of PBD-LG : its optical and spectroscopic properties will serve as guidelines or references in future work on other poly(D-L-peptides).

Recognition of different structures in solution

Best evidence for the existence of various structures of PBD-LG in solution is given by infrared techniques which also help relate solution and solid-state conformations. Indeed, PBD-LG may have three different IR spectra, both in solid-state and solution, characteristic of α_{DL}, π_{DL} and $\pi\pi_{DL}$ helical conformations. Discrimination between the various types of double helices must however rely on finer differences as measured by circular dichroism or ORD and, more interestingly, on typical NMR results. It appears indeed that the α_{DL} helix is characterized by α CH chemical shifts located at 3.65 and 3.83 ppm, while that of α_R helices of poly(γ-benzyl-L-glutamate) (PBLG) is at 3.92 ppm. $\pi\pi_{DL}^{4.4}$ and double helices have unusually low field α CH resonances at 4.45 and 5.5 ppm, respectively. Distinction of different double helices is possible through resonances related to the side-chain protons (for the $\pi\pi_{DL}^{9.0}$ structure, the aromatic hydrogens and benzyl CH_2 shifts are found at 7.0 ppm and 4.8 ppm compared

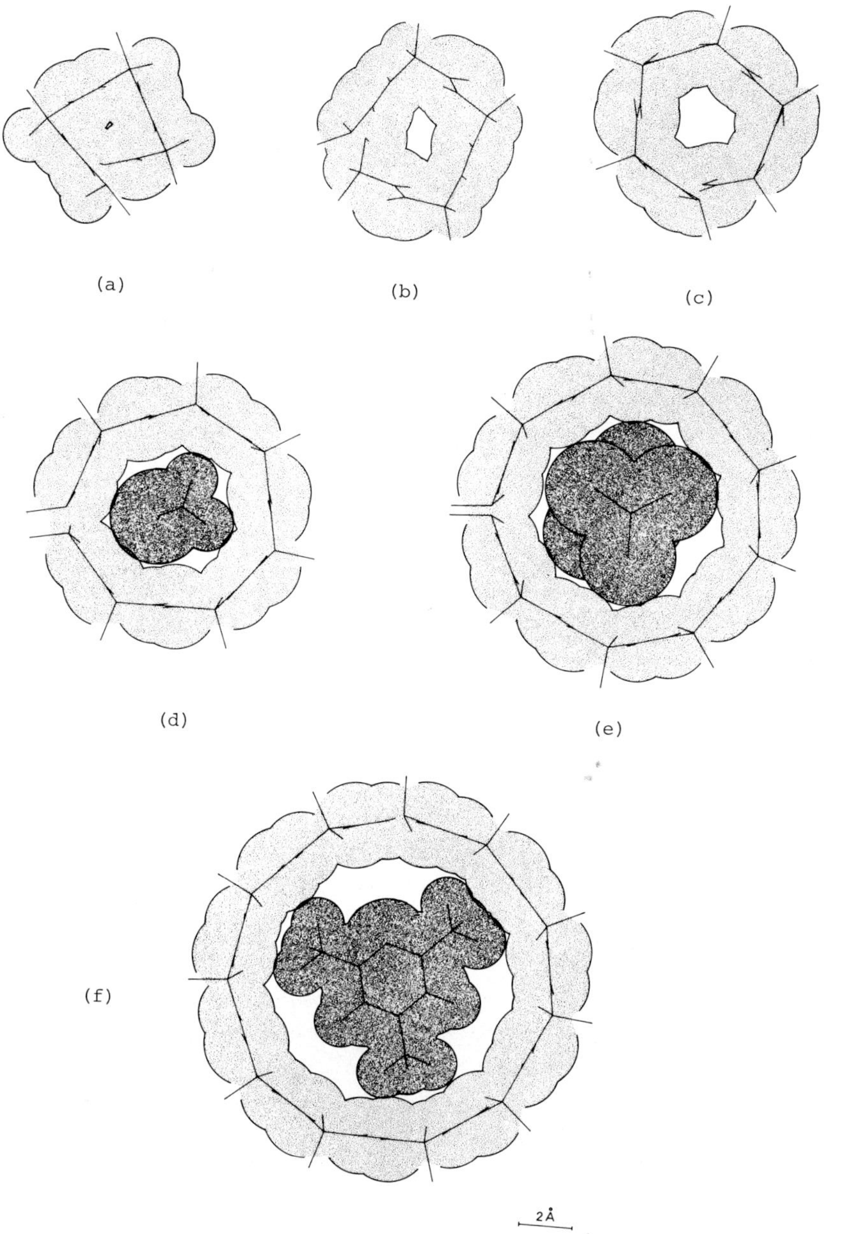

Fig. 3. End-on view of (a) α_{DL} (b) $\pi_{DL}^{4.4}$ (c) $\pi\pi_{DL}^{5.6}$ (d) $\pi\pi_{DL}^{7.2}$ (e) $\pi\pi_{DL}^{9.0}$ and (f) $\pi\pi_{DL}^{10.8}$ helices with van der Waals radii of the main chain atoms shaded in grey. Methylene chloride (d) pentachloroethane (e) and collidine (f) molecules (in dark grey) can be accomodated within the helical cores of the double helices, for which only one strand is shown [7].

to 7.3 and 5.0 ppm for α helical PBD-LG, respectively. $\pi\pi^{7.2}_{DL}$ and $\pi\pi^{9.0}_{DL}$ helices can also be distinguished by their NH resonance.

The various conformations of PBD-LG can therefore be characterized and distinguished one from the other. The transconformations between the various forms may thus be investigated with adequate experimental techniques.

Transconformations between the various structures of PBD-LG

Since PBD-LG exists in six helical structures (in addition to the random coil conformation), over thirty transconformations should be considered and, if they are possible, their experimental conditions should be defined. As an illustration of the conformational versatility of PBD-LG, the main characteristics of four families of transconformations will be outlined here.

* $\alpha_{DL} \rightarrow \pi_{DL}$

This transition, heat-induced in the solid state at 130°C, also takes place upon heating a dioxane solution of the polymer to 95°C. In contrast to the solid-state behaviour, the α_{DL} structure reappears again on recooling to room temperature. A conformational equilibrium between the two structures therefore exists, which is displaced in favour of the π_{DL} one by increased temperature.

* α_{DL} or $\pi_{DL} \rightarrow \pi\pi_{DL}$

As for the $\alpha_{DL} \rightarrow \pi_{DL}$ one, this transition is observed upon heating of dioxane solutions to ∿ 95°C (∿ 220°C in the solid state). The intramolecular character of this transition is suggested by the higher concentrations of polymer that are required; they do not convert back to α_{DL} once upon cooling of the solution. As already indicated, the $\pi\pi^{5.6}_{DL}$ helix is unlikely to be obtained (it is unstable in the presence of solvent), and the specific type of $\pi\pi_{DL}$ helix that is formed appears to depend on the solvent used. This statement is further illustrated by observation of transitions between different $\pi\pi_{DL}$ helices in mixtures of solvent.

* Transitions between $\pi\pi_{DL}$ helices

Such transitions are merely induced by exposure to (when starting from the $\pi\pi^{5.6}_{DL}$ helix) or changes of the solvent. This behaviour is not unexpected, since solvent molecules, located inside the helical core (except for the $\pi\pi_{DL}^{5.6}$ helix), are structural componants of this class of helices. The most characteristic, and best documented example involves $\pi\pi_{DL}^{7.2}$ and $\pi\pi_{DL}^{9.0}$ helices in mixtures of methylene chloride and dioxane or chloroform : transition to the $\pi\pi_{DL}^{9.0}$ structure occurs when the concentration of dioxane and chloroform exceeds $\sim$ 16 and 40%, respectively (the transition is of course reversible when the concentration of methylene chloride is increased again). The fact that widely different concentrations of dioxane and chloroform are needed to induce the transition demonstrates that the stability of these complexes not only depends on the geometrical fit of solvent and helical core dimensions, but that the polarity and/or possible interactions with the helical core walls (as observed in Amylose complexes [18] must occur.

* π_{DL} or $\pi\pi_{DL} \rightarrow \alpha_{DL}$

Such transitions are not observed in the solid state and only the $\pi_{DL} \rightarrow \alpha_{DL}$ one can occur in solution. The family of $\pi\pi_{DL}$ helices is a class on its own. The easiest route to the "native" α_{DL} helical conformation of PBD-LG involves casting from the random-coil conformation, obtained by use of strong acids such as triflouro- or dichloroacetic acid.

Influence of the history of the sample

The variety of structures observed for PBD-LG not only leads to a rather complicated pattern of transconformations, but also underlines the importance of the history of the sample and the care needed in its handling. For example, PBD-LG can exist, in the solid state, in any of the six different helical structures, which can be simply inferred from the previous history of the sample. Most of these difficulties are linked to the stability of both the α_{DL} and the double helices. Therefore either one or the other structure is obtained upon dissolution, depending on whether the starting material was in a single or double stranded conformation. Moreover, starting with α_{DL} helical PBD-LG, dissolution and casting from a dioxane solution yields either α_{DL} or $\pi\pi_{DL}^{9.0}$ structures, depending on whether dissolution is made at low temperature or 95°C, respectively.

This behaviour does not appear to be limited to PBD-LG, since influence of the history of the sample has also been reported for Gramicidin A. Investigations on PBD-LG however can help describe such a behaviour in simple structural terms.

STRUCTURAL INVESTIGATIONS OF OTHER POLY(D-L PEPTIDES) AND OF GRAMICIDIN A

This behaviour of PBD-LG may well differ from that of other poly(D-L-peptides) or Gramicidin A, as side-chains and/or molecular weight may be of importance. We now briefly review theoretical and experimental work performed on other poly(D-L-peptides) and on the antibiotic. A general picture of the behaviour of poly(D-L-peptides) however does not yet emerge as most of the work discussed is in progress and therefore fragmentary. Also, conclusions drawn from investigations performed on polypeptides with imperfect syndiotacticity may be suspect. Finally, some interpretations proposed in earlier works may need be re-examined in the light of the more recent results on PBD-LG.

Possible existence of other structures. Influence of side-chains.

No systematic search for stable structures of poly(D-L-peptides) by conformational energy analysis has been performed to date. This appraoch is actually rather unrealistic, as it would require the investigation of the four-dimensional map of ϕ_L ψ_L ϕ_D and ψ_D angles (if ω and τ are fixed). By this method, however, the $\pi_{DL}^{4.4}$ helix (which requires smaller τ angles) would be missed, as well as the family of double helices, since multi-stranded structures are not generated by this method. Conformational energy analysis has therefore mainly helped refine structures which intuitively or by model building, appeared reasonable. Apart from the α_{DL}, and various π_{DL} and $\pi\pi_{DL}$ helices, two sheet structures have been considered: the LD ribbon [17] based on a succession of β turns, and the α pleated sheet [23] based on alternating α_L and α_R conformations. Some support for the latter structure has been obtained from low molecular weight PBD-LG [24].

Conformational investigations of poly(D-L-peptides) have seldom examined the influence of side-chains other than the methyl group of alanine residues. Hesselink and Scheraga [15] however have shown that $\pi_{DL}^{6.3}$ helices of syndiotactic poly valine, polyphenylalanine and poly(α-aminoheptanoic acid) (a model for polylysine) are energetically acceptable. In view of the shortcomings of this technique in dealing with long side-chains, further investigations in this direction appear however unwarranted: the experimental approach, which is now possible, appears *a priori* more promising and should prove more fruitful. The available experimental evidence on model polypeptides other than PBD-LG, although fragmentary and limited to a few polymers often with uncertain stereoregularity, do support this view. Not only π_{DL} or $\pi\pi_{DL}$ helices other than those described, but also different structures may well be observed. For example, poly(L-Ala-D-Val) has been tentatively reported to exist in a $\pi_{DL}^{8.2}$ helix [25] and a structure with β type hydrogen bonding has been suggested for poly(D-L-methionine) [26]. We have observed, for poly(L-Ala-D-Ala), two structures, not yet elucidated, but different from those described so far. Also, preliminary investigations suggest that poly(D-Leu-L-Trp), which reproduces the CO terminal sequence of six residues of Gramicidin A has yet another conformation in solution, as judged from its infrared characteristics. The conformational versatility of poly(D-L-peptides) is, therefore, not limited to structures observed so far with PBD-LG.

Investigations on Gramicidin A

The extensive investigations of solution and solid state properties of Gramicidin A have suffered from the lack of a clear correlation between its spectroscopic properties and well characterized conformations, as determined for example by single crystal structure analysis. Numerous observations [27, 10] indicate that its behaviour in solution is unusually complicated. The conformation appears to depend on a number of factors, including solvent, concentration, temperature and even, under fixed conditions, time. The latter parameter, as well as influence of the history of the sample, appears to have often been overlooked in earlier studies on Gramicidin A.

The most thorough investigations of the solution conformation of Gramicidin A reported so far, led Blout and coworkers [10, 28, 29] to propose the existence of various double stranded helices. They observed that the antibiotic can be obtained, from dioxane-water (100 : 1) in four different species, denoted 1 to 4. Only the most prominent of these (species 3) has spectroscopic properties consistent with a double stranded, antiparallel helix, while parallel-β type hydrogen bonding is suggested for the other species, which could be either dimerized π_{DL} helices or, more probably according to the authors, parallel stranded double helices. In addition, the dissociation of the four species into monomers has been followed [28] which suggests that the antibiotic may also exist in a single stranded conformations. A π_{DL} single stranded helix has indeed been suggested for Gramicidin A on the basis of NMR investigations [30]. A comparative study, by infrared and ORD techniques, on Gramicidin A in DMSO and TFE was interpreted to support this assignment. However, the CD spectrum of hydrogenated Gramicidin A in TFE solution is, in view of its similarity with that of PBD-LG, equally well interpreted on the basis of a left-handed α_{DL} helix [20].

Correlations of this kind between spectroscopic properties of Gramicidin A and PBD-LG appear of course attractive. In attempting such correlations however, we have met with only limited success. Indeed, given the usual infrared, ORD, CD and NMR information on a structure of Gramicidin A, comparison with results on PBD-LG can only be used with fair confidence to eliminate possible structures rather than to define a given one. The situation will hopefully improve after polypeptides with sequences and molecular weight closer to that of the antibiotic have been investigated. The fact remains that Gramicidin A appears to have more conformations and easier transitions between them, than the model poly(D-L-peptides). These characteristics do not facilitate, of course, the analysis of the relationship between its structure and its biological activity.

THE RELATION BETWEEN STRUCTURE AND ION-CONDUCTANCE PROPERTIES OF GRAMICIDIN A

All the important work on Gramicidin A [4,5,31,32, etc] cannot be reviewed in the present context. We will only outline characteristic properties which give indications of, and need be explained by, the conformation of the antibiotic.

Physical data on membrane properties of Gramicidin A are difficult to obtain because of its very low concentration. Most information therefore results from conductance measurements performed on membranes doped with small quantities of the antibiotic (typically 10^{-9} molar solutions in 10^{-6} liter range are used). Under these conditions, step fluctuations of the membrane current under fixed applied potential (typically 100 mV) are observed, which can be associated to events affecting individual molecules or complexes. The flux of ions, about 10^{7} sec^{-1}, is much larger than that ($\sim 10^{4}$ sec^{-1}) of cage-carriers (e.g. valinomycin, nonactin), which strongly argues in favour of a fixed structure present in the membrane, i.e. a channel mechanism. Channels remain active (open) typically for about $\sim$ 0.2 seconds, which suggest that they are formed by transitions which involve inactive structures already in the membrane. Both the frequency of occurence and the average duration of the channels increases for thinner membranes (thicknesses ranging from 25 to $\sim$ 60 Å have been used). Activation energies of the channel conductance and termination have been estimated at $\sim$ 5 (for Na^+ and K^+ ions) and 19 kcal-mol^{-1}, respectively.

The approximate length and diameter of the channel can be inferred from the thickness of the membrane and the size of the permeating ions. Thus H^+, NH_4^+, C_A^+, Rb^+, K^+, Na^+, Li^+ and dimethyl ammonium permeate, while the larger tetramethylammonium (diameter $\sim$ 7 Å) does not (no polyvalent cations and no anions permeate). A channel length of 30 - 35 Å and diameter of $\sim$ 4 Å appear to be reasonable figures. These dimensions are incompatible with the existence of channels made of a single molecule. Indeed, from the relationship between the membrane conductance and gramicidin concentration, it has been suggested that two molecules are needed to form the conducting channel [33, 34]. Also, conductance versus concentration data of N,N-(dideformyl gramicidin A) malonamide, i.e. two gramicidin molecules linked chemically head-to-head, is of the first order [30].

Early conductance/concentration results were only approximate since the concentration of Gramicidin A in the membrane was not measured. Recently however, Veatch et. al. [35] have reported simultaneous conductance-concentration studies of dansyl-gramicidin A, a fully active analog of the antibiotic, whose concentration in the membrane can be determined (down to 10^{-14} mol cm^{-2}) by fluorescence techniques. They have confirmed that the active channel is a dimer of a non-conducting species already present in the membrane [31] and that most dimers are active channels. Further, the rate of dimerization depends on the type of lipid, and decreases with increasing thickness of the membrane.

Translation of these results into information concerning the conformation and structure of Gramicidin A can, at present, only remain tentative, in the absence of direct structural evidence. Geometrical, stereochemical, and electrical requirements however limit the number of satisfactory models.

On the basis of data then available, Urry [9,30] proposed that the conducting species of Gramicidin A adopts a $\pi_{DL}^{6.3}$ helical structure, and that two molecules, attached head-to-head, form the transmembrane channel 25 to 30 Å long. The cation permeation would occur along the helical axis, i.e. in the central core which has the right diameter ($\sim$ 4 Å) and is lined with the peptide oxygen moieties which can interact with the partly dehydrated, migrating ions. At the same time, the apolar side-chains characteristic of Gramicidin A are located on the outer walls of the helix, in contact with the hydrophobic inner portion of the bilayer membrane. From the various arguments in favour of this structure [9,30] we wish to point out that it provides a very elegant structural justification for the presence of the formyl moiety in the Gramicidin A chemical sequence.

An alternate channel model based on double helices would equally well provide the polar lining of the central core and the hydrophobic exterior of mostly apolar side-chains. A $\pi\pi^{7.2}_{DL}$ double helix made of two Gramicidin A molecules would have also a length of $\sim$ 25 Å and a central core of $\sim$ 4 Å. Another possibility might be the head-to-head association of two $\pi\pi^{9.0}_{DL}$ helices (total length of $\sim$ 33 Å). The channel would then be made of four molecules and the larger core would permit passage of hydrated ions. In the latter structure, the presence of formyl ends would have the same structural justification as for the $\pi^{6.3}_{DL}$ helix.

Interpretation of the opening and closure of transmembrane channels is at present even more speculative, since many mechanisms may be envisaged. Urry [9] suggested a conformational transition from the $\pi^{6.3}_{DL}$ helix to a structure in which two intra-turn hydrogen bonds (reminiscent of those of cyclo hexaglycine) would close the central channel. The more recent results of Veatch et. al. [35] suggest however that activation and desactivation of the channel is very closely related to the association and dissociation of non-conducting species. In this respect, one may note that the head-to-head association of gramicidin helices through hydrogen bonds may be rather labile. It is not known however whether the pairing and unpairing of non conducting species is induced or followed by structural modifications.

Comparison of the various models of gramicidin channels need not be pursued further in the absence of conclusive experimental data. In the search for such data, investigations of membranes doped with Gramicidin molecules with well established conformations are desirable. Similar experiments with models which do not have the structural versatility of the antibiotic would permit a clearer identification of the conducting structure- or structures.

CONCLUSION

Structural investigations of poly(D-L-peptides) contribute to a better understanding of the conformation of polypeptides and of the antibiotic activity of Gramicidin A.

Although our experience with poly(D-L-peptides) is still limited to merely one polymer, progress has already been made in understanding of their conformations in general : structures specific to the LD stereosequence have been observed, including a family of helices based on the same hydrogen bonding pattern. Only the $\pi_{DL}^{4.4}$ helix has been observed for PBD-LG, but other π_{DL} helices may well be found when investigating different poly(D-L-peptides).

The conformation of PBD-LG in solution is also well established and many transitions between the various structures have been observed. This behaviour may not be generalizable to poly(D-L-peptides) with other side-chains, but suggests a fair structural versatility for this class of polymers.

The contribution of poly(D-L-peptides) to understanding the structure of Gramicidin is also important. It has already confirmed the feasability of two structures previously proposed for this antibiotic. Although clear cut correlation between spectroscopic properties of Gramicidin and the model polypeptide in solution are not yet, in general, possible, this situation should also improve with investigations of other poly(D-L-peptides) with chemical sequence closer to that of the antibiotic.

The most exciting goal of these investigations is the elucidation of the channel conductance properties of Gramicidin. The models proposed so far explain these properties by the formation of pores of sufficient size to permit the diffusion of ions. Investigations of model poly(D-L-peptides) have confirmed the existence of structures whose geometry is compatible with the transduction properties of Gramicidin A. They do not, however, convincingly establish any of the possible models. They provide only a static picture of the pore : they neither explain the dynamics of the "opening" and "closing" of the channel nor give any indication of the conditions which activate these mechanisms. Further studies of the interaction of poly(D-L-peptides) and various ions, and of their membrane properties will help however elucidate the most intriguing behaviour of Gramicidin A.

REFERENCES

1. R. D. Cowell and J. H. Jones, J. Chem. Soc. (C) (1971), 1082-1090.
2. Y. Trudelle, J.C.S. Perkin I., (1973), 1001.
3. A. Caille, F. Heitz, and G. Spach, J.C.S. Perkin I., (1974), 1621-1625.
4. S. B. Hladky and D. A. Haydon, Biochim. Biophys. Acta, 274 (1972), 294-312.

5. V. B. Myers and D. A. Haydon, Biochim. Biophys. Acta, 274, (1972), 313-322.

6. F. Heitz, B. Lotz, and G. Spach, J. Mol. Biol., 92 (1975), 1-13.

7. B. Lotz, F. Colonna-Cesari, F. Heitz, and G. Spach, J. Mol. Biol. 106 (1976), in press.

8. R. Sarges and B. Witkop, Biochemistry, 4, (1965), 2491-2498.

9. D. W. Urry, Proc. Nat. Acad. Sci. USA, 69 (1972), 1610-1614.

10. W. R. Veatch, E. T. Fossel, and E. R. Blout, Biochemistry, 13 (1974), 5249-5256.

11. E. N. Shepel, St. Iordanov, I. D. Ryabova, A. I. Miroshnikov, V. T. Ivanov, and Yu. A. Ovchinnikov, Bioorgan. Chim. (USSR) 2, (1976), 581.

12. W. R. Veatch, Ph. D. Thesis, (1973), Harvard University.

13. F. Heitz and G. Spach, Macromolecules, 4 (1971), 429-432.

14. P. M. Hardy, J. C. Haylock, D. I. Marlborough, H. N. Rydon, H. T. Storey, and R. C. Thompson, Macromolecules 4 (1971), 435-440.

15. F. T. Hesselink and H. A. Scheraga, Macromolecules, 5, (1972), 455-463.

16. F. Collonna-Cesari et. al, to be published.

17. G. N. Ramachandran and R. Chandrasekaran, Ind. J. Biochem. Biophys., 9, (1972), 1-11.

18. A. Sarko, in "Structure of Fibrous Biopolymers", (1975), E.D.T. Atkins and A. Keller, eds. Butterworths, p. 335-351.

19. F. Heitz and G. Spach, Macromolecules, 8, (1975), 740-745.

20. F. Heitz and G. Spach, to be published.

21. F. Heitz, P. D. Cary, and C. Crane-Robinson, Macromolecules, 8, (1975), 745-750.

22. F. Heitz et. al, to be published.

23. P. DeSantis, S. Morosetti and R. Rizzo, Macromolecules, 7, (1974), 52-58.

24. B. Lotz, F. Heitz and G. Spach, C.R.H. Acad. Sci., 276C (1973), 1715-1718.

25. F. Ascoli, G. DeAngelis, F. DelBianco and P. DeSantis, Biopolymers, 14, (1975), 1109-1114.

26. E. Dellacherie, J. Neel and F. Colonna-Cesari, Biopolymers, 14, (1975) 1447-1463.

27. B. E. Isbell, C. Rice-Evans and G. H. Beaven, FEBS Lett., 25, (1972), 192-195.

28. W. R. Veatch and E. R. Blout, Biochemistry, 13 (1974), 5257-5264.

29. E. T. Fossel, W. R. Veatch, Y. A. Ovchinnikov and E. R. Blout, Biochemistry, 13 (1974), 5264-5275.

30. D. W. Urry, M. C. Goodall, J. D. Glickson and D. F. Mayers, Proc. Nat. Acad. Sci. USA, 68, (1971), 1907-1911.

31. E. Bamberg and P. Läuger, J. Membrane Biol., 11 (1973), 177-194.

32. E. Bamberg, K. Noda, E. Gross and P. Läuger, Biochim. Biophys. Acta, 419 (1976), 223-228.

33. D. C. Tosteson, T. E. Andreoli, M. Tieffenberg and P. Cook, J. Gen. Physiol. 51, (1968), 373S-384S.

34. M. C. Goodall, Biochim. Biophys. Acta, 219 (1970), 471-478.

35. W. R. Veatch. R. Mathies, M. Eisenberg and L. Stryer, J. Mol. Biol. 99, (1975), 75-92.

MOLECULAR ARCHITECTURE OF THE ANIMAL AND SOME MICROBIAL EXTRACELLULAR POLYSACCHARIDES

EDWARD D.T. ATKINS
H.H. WILLIS PHYSICS LABORATORY, UNIVERSITY OF BRISTOL, BRISTOL BS8 1TL (U.K.)

SUMMARY

The purpose of this article is to briefly review the conformations observed in the solid state for the animal polydisaccharides or glycosaminoglycans and in addition to show some recent results from the more complex microbial polysaccharides. Conformations for hyaluronate, the chondroitin sulphates, dermatan sulphate, keratan sulphate, heparan sulphate and heparin are discussed. The interaction of these molecules with proteins and their molecular architecture in cartilage is mentioned. Three examples of the complex bacterial capsular polysaccharides are briefly examined. In particular the three serotypes K5, K57 and K8 from the genus *Klebsiella* which exhibit two-fold, three-fold and four-fold helical symmetries respectively.

INTRODUCTION

X-ray diffraction analysis is an established technique for the elucidation of the three-dimensional shape and structure of molecules. The most detailed information is obtained when the material is crystalline and in the case of long chain polymeric substances with essentially regular repeating chemical sequences, such as the polysaccharides discussed below, the best that can usually be achieved is to induce the molecular chains to orient in approximately parallel register and to encourage disciplined molecular interconnections between neighbouring chains [1]. Such an array is analogous to the molecular morphology of many natural and synthetic fibres. Indeed the study of the x-ray diffraction of fibres and in particular the polysaccharides is almost as old as the x-ray diffraction of crystals [2].

Figures 4, 6, 10 and 14 are examples of x-ray diffraction patterns obtained from polysaccharide fibres or oriented films. The patterns exhibit diffracted intensity confined to (approximately) horiz lines which reflects the periodic nature along the chain dire measurement of the layer line spacing gives the pitch of the helix. Furthermore, for a helix with n units in one turn (i.e. per pitch (p)), diffracted intensity on the meridian (the vertical bisector of the pattern) only occurs on layer lines whose integer index ℓ is a multiple of n. The quotient p/n yields the axial projection (h) which corresponds to the structural repeat. Models of molecular helices incorporating the known stereochemistry of the various saccharide units, together with the glycosidic linkage geometries, can be computer generated while preserving the helix symmetry and dimensions as defined by the values of n and h. Additional information about the molecular structure involves the use of the full intensity data in a high resolution x-ray diffraction pattern together with more sophisticated chain packing procedures.

During the last five years notable advances have been reported concerning the molecular conformations of the animal extracellular polysaccharides. Progress in this area has emanated from the development of crystallization techniques, involving stress fields and annealing [1, 3], to provide oriented, semi-crystalline samples suitable for x-ray diffraction analyses.

In this contribution a brief review is given of the progress to date of the animal extracellular polysaccharides (glycosaminoglycans) together with some recent examples of the even more complex bacterial capsular polysaccharides from the *Klebsiella* serotypes.

ANIMAL POLYSACCHARIDES

For the purpose of this article these polysaccharides may be divided into two groups. The first group encompasses hyaluronate, the chondroitin sulphates, dermatan sulphate and keratan sulphate. They are all essentially linear polydisaccharides with alternating 1 → 3 and 1 → 4 glycosidic linkages and their chemical repeats are shown in Figure 1. Hyaluronic acid is high molecular weight glycosaminoglycan (about 10^6 Daltons) found in soft connective tissues such as umbilical cord, synovial fluid and vitreous humour. Unlike hyaluronic acid, the chondroitin, dermatan and keratan sulphate chains are of lower molecular weight (typically 50,000 Daltons for chondroitin and dermatan sulphate and 20,000 Daltons for keratan sulphate) and occur *in vivo* covalently linked to a protein core. The resulting proteoglycan is similar to a laboratory test tube brush. Recent biochemical studies [4] have established a specific interaction between hyaluronic acid and the proteoglycans in the intercellular matrix

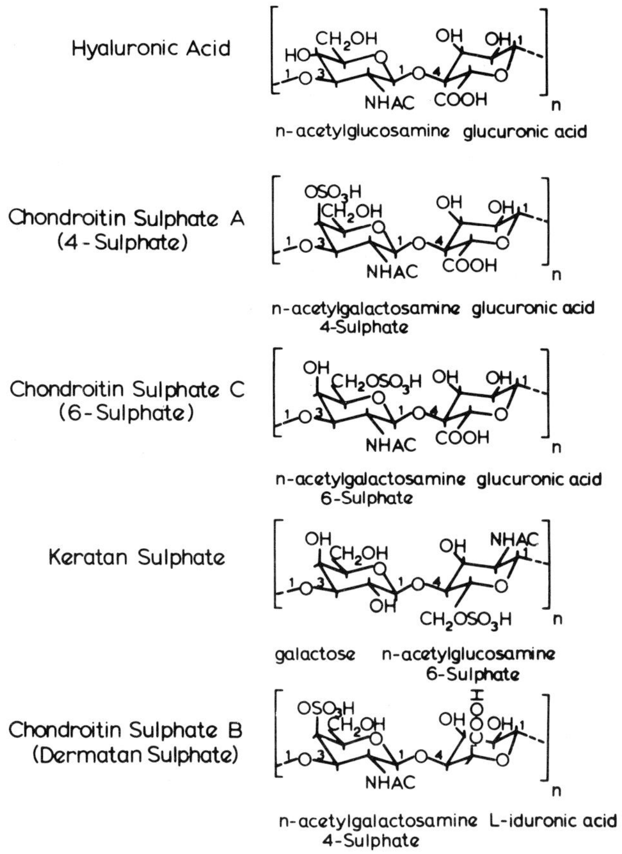

Fig. 1. Idealized chemical repeats of the connective tissue polysaccharides based on alternating 1 → 3 and 1 → 4 glycosidic linkages.

in cartilage. The essential feature of the proposed model (Figure 2a) is that many proteoglycans are able to bind along the entire length of a hyaluronic chain and that at saturation there is one proteoglycan bound to each region of hyaluronic acid of about twenty disaccharide units long (about 20 nm). Thus for a hyaluronic acid chain of molecular weight 500,000 Daltons there are about forty proteoglycans regularly distributed. Each proteoglycan can bind to only one hyaluronic acid chain so the system does not readily form a network or gel by an interaction of the form HA-PG-HA. These proteoglycan aggregates containing hyaluronic acid interact electrostatically through charges on the polysaccharide with collagen to form the molecular organization in cartilage (Figure 2b). X-ray diffraction patterns from such proteoglycan aggregates have been reported [5].

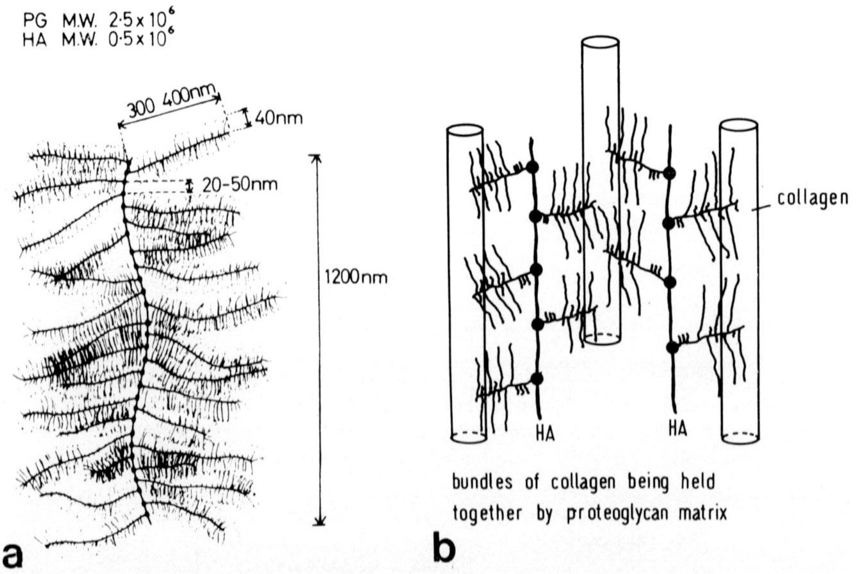

Fig. 2. (a) is a schematic representation of the proteoglycan-hyaluronic acid complex. A single hyaluronic acid chain associates with many proteoglycans. The fine hairs represent chondroitin and keratan sulphate chains. (b) shows the proposed arrangement of connective tissue in cartilage.

The second group consists of heparan sulphate and the blood anticoagulant heparin. Both have a polydisaccharide backbone but the glycosidic linkages are exclusively 1 → 4. Heparan sulphate (molecular weight typically 50,000 Daltons) occurs as a proteoglycan but heparin has a much lower molecular weight (typically 12,000 Daltons) and occurs as polymeric packets in mast cells. The idealized chemical repeats of these polysaccharides are shown in Figure 3.

HYALURONATES

Analyses of the x-ray diffraction patterns recorded from oriented, semi-crystalline specimens clearly show that this polyanion can exist in a number of regular helical conformations in the inspissated phase [3,6-13]. Extended conformations exhibiting helical symmetries 2_1, 3_2 and 4_3 have been reported [3,6,8,13]. Furthermore the latter four-fold helical conformation has also been observed to occur in a contracted posture, with a helix pitch approximately 10% lower than the corresponding extended conformation [7,8].

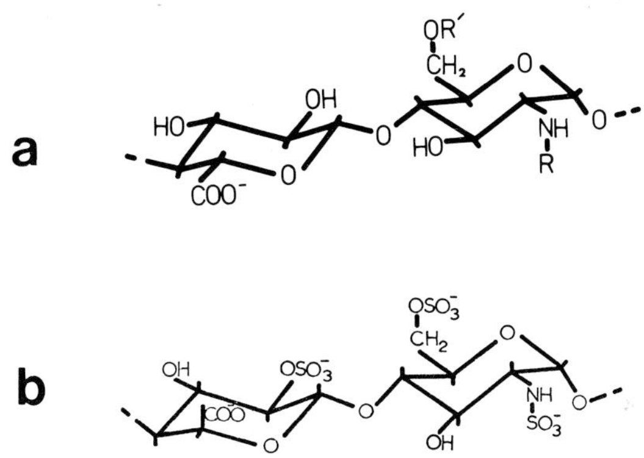

Fig. 3. Repeating sequences of: (a) heparan sulphate, and (b) heparin. R is either $COCH_3$ or SO_3^- and R' is usually H about sometimes SO_3^-.

Examples of the x-ray diffraction obtained from these four different conformations are illustrated in Figure 4. The layer line spacing in Figure 4a is 1.96 nm with meridional reflections occurring on even order layer lines. The axially projected repeat (h) of 0.98 nm correlates directly with the dimensions expected for the disaccharide repeat [3,6]. The maximum theoretical extension for such a disaccharide repeat, using standard geometries for the saccharide units, is 1.02 nm and so the simplest interpretation is a highly extended 2_1 helix. No detailed structure analyses have been undertaken relating to this conformation although a stereochemically satisfactory model has been computed (Figure 5a). This two-fold helix usually occurs at low pH although it is still uncertain whether it truly represents the free acid or alkali salt form of hyaluronate.

A number of salt forms, typically sodium, potassium and calcium readily provide x-ray diffraction patterns similar to Figure 4b [9]. In this case the layer line spacing is 2.85 nm and meridional reflections occur on every third layer line. Such a conformation was originally interpreted as an extended three-fold helix and preliminary model building favoured a left-handed (3_2) conformation [3,6]. A more recent x-ray structure refinement [13] of this model substantiates the earlier indications of a 3_2 helix and projections are illustrated in Figure 5b. In the particular structure refined the molecules pack in a trigonal lattice with dimensions $\underline{a} = \underline{b} =$ 1.17 nm, $\underline{c}$ (fibre axis) = 2.85 nm, $\gamma = 120^{\circ}$. The 3_2 helix is stabilized by O...O intramolecular hydrogen bonds across the two kinds of glycosidic

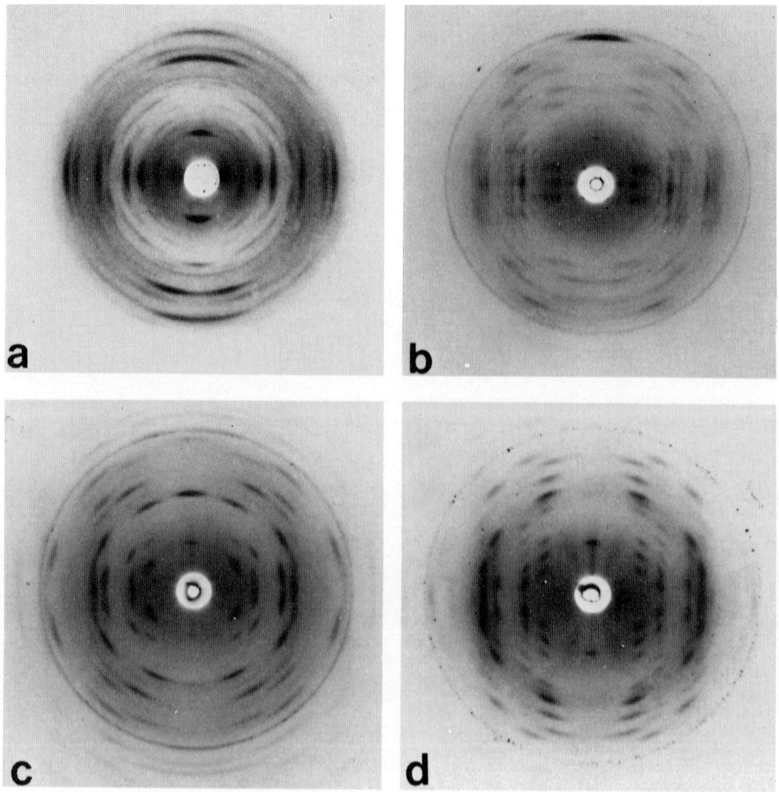

Fig. 4. X-ray diffraction patterns obtained from hyaluronate (a) 2_1 helix with $\underline{h}$ = 0.98 nm, (b) 3_2 helix with $\underline{h}$ = 0.95 nm, (c) 4_3 helix with $\underline{h}$ = 0.93 nm, and (d) 4_3 helix with $\underline{h}$ = 0.84 nm.

linkage. Two chains pass through the unit cell in opposite directions and are related by the diad axis symmetry of the space group $P3_221$ [13].

The extended 4_3 helical conformation [8] (Figure 4c) is currently being refined against the x-ray data. The evidence at this stage substantiates a left-handed conformation and a trial projection to give an indication of the overall shape of the molecule is shown in Figure 5c.

The contracted version of this conformation derived from the x-ray pattern shown in Figure 4d was initially interpreted as a novel, double helical model, with one chain draped around the other in an antiparallel fashion [7]. It was established from density calculations that two chains passed through the orthorhombic unit cell and computerized chain packing

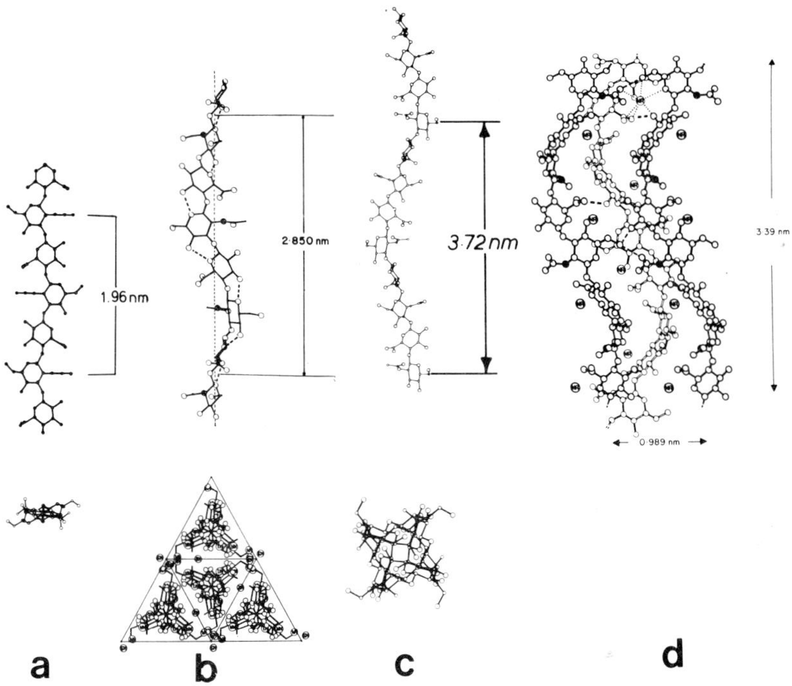

Fig. 5. Helical conformations relating to the x-ray patterns shown in Fig. 4(b) and (d) taken from [13] and [12] respectively.

procedures apparently ruled out the possibility of chains on the corners and another at the center. This result led to the double helix model which was argued offered an explanation for the rheological properties of hyaluronate solutions. However re-examination of these calculations coupled with x-ray refinement now favour single 4_3 helices [12], although no clear explanation is apparent for such a contraction. A projection of the refined structure is shown in Figure 5d.

Chondroitin 4-sulphate

Ordered conformations of both the sodium salt and free acid forms have been reported [14,15]. The crystallinity of the sodium salt form is extremely sensitive to humidity and Figure 6 illustrates a sequence of x-ray diffraction photographs taken with the sample maintained at various degrees of relative humidity during exposure. The layer line spacing remains constant, within experimental error, at a value of 2.88 nm and meridional reflections occur on every third layer line. These spacings

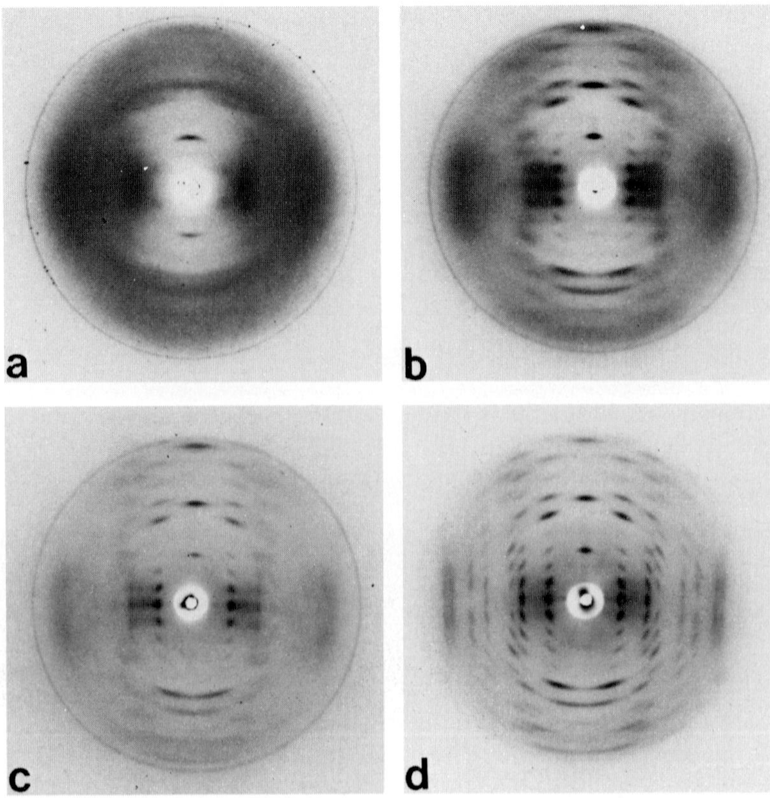

Fig. 6. Effect of relative humidity on the x-ray diffraction patterns from chondroitin 4-sulphate (a) taken at room humidity (30-40%), (b) at 86%, (c) at 90%, and (d) 93% relative humidity.

and the general distribution of intensities are sufficiently similar to those obtained for sodium hyaluronate [3,13] to suggest a 3_2 helix [10,16]. A stereochemically feasible conformation is shown in Figure 7a but no detailed structure refinement has yet been undertaken. In the most crystalline pattern shown in Figure 6d, the reflections index on a trigonal unit cell of side 1.45 nm. Density measurements and space group criteria favour two chains running through the unit cell and so each chain occupies approximately 50% more space than the corresponding sodium hyaluronate chain [6,13]. The 3_2 helix may be converted to a 2_1 helix with a value of $\underline{h}$ = 0.98 nm by immersion in an ethanolic acid bath [15]. No detailed structure refinements have yet been reported.

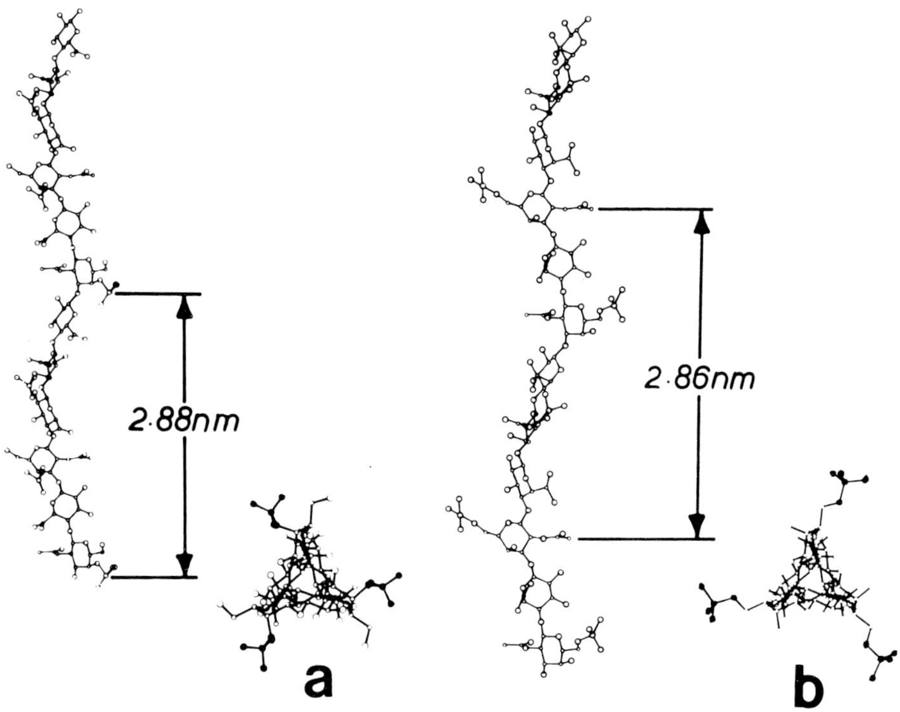

Fig. 7. Projections of the 3_2 helices for (a) chondroitin 4-sulphate, and (b) chondroitin 6-sulphate. The charged sulphate groups on the periphery of the molecule are marked with filled circles.

Chondroitin 6-sulphate

The sulphate group in chondroitin 6-sulphate is equatorially positioned (see Figure 1) and extends some 0.2 nm further from the helix axis than for the case of chondroitin 4-sulphate. Three conformations in the solid state have been reported [17 - 20] . A three-fold helix with $\underline{h}$ in the range 0.95 - 0.96 nm [17,18] has been examined using conformational analysis and is similar to the 3_2 helices reported for hyaluronate and chondroitin 4-sulphate. Figure 7b shows a trial model for comparison with chondroitin 4-sulphates and clearly shows how the charged sulphate groups extend on the periphery of the molecule. On conversion to the free acid form a 2_1 helical conformation is observed [17] with a value of $\underline{h}$ = 0.93 nm. In addition an eight-fold conformation ($\underline{h}$ = 0.98 nm) has been reported [18, 20] which was initially interpreted as an 8_1 helix [18] with a turn angle of 45° between successive disaccharide repeats. However, such a model involves considerable distortion and a more likely conformation is an

8_5 helix [20,16] analogous to that suggested for dermatan sulphate [21,19, 10]. No detailed structure determinations have yet been reported.

Keratan sulphate

Only one conformation, an extended 2_1 helix with a value of h = 0.98 nm, has been reported for this non-uronic acid containing polysaccharide (see Figure 1) [22]. The x-ray diffraction pattern is of poor quality with considerably layer line streaking. A model with no unacceptably short non-bonded contacts is illustrated in Figure 8a.

Dermatan sulphate

Of those connective tissue polyuronides based on the hyaluronic acid repeat, the molecular shape of dermatan sulphate has probably aroused most discussion. The chemical repeat is characteristically distinguished by replacement of the usual D-glucuronic acid by its C(5) epimer L-iduronic acid. A schematic representation of the shape and disposition of side groups for both uronic acids in the normal $^4C_1$ chair form are shown in Figure 9. Such a chair conformation for D-glucuronic acid appears quite satisfactory with all the side appendages favourably positioned equatorially (Figure 9a). On the other hand, L-iduronic acid has a large carboxyl group axially positioned (Figure 9b) and serious consideration must be given to the alternative $^1C_4$ chair conformation. The effect of converting the L-iduronic acid residue to the alternative $^1C_4$ chair conformation would introduce a diaxially linked unit substantially altering the shape of the backbone. X-ray diffraction patterns obtained from dermatan sulphate [14,21] yield values of h = 0.93 nm (8_5 helix) and 0.98 nm (2_1 helix). These spacings are not compatible with the L-iduronic acid residue in the alternative $^1C_4$ chair conformation [21,22]. Computer projections of the 8_5 helix are shown in Figure 8b. However, ^{13}C n.m.r. results [23] favour the alternative $^1C_4$ chair conformation for L-iduronic acid in dermatan sulphate or at least a closely associated skew boat conformation. It could be that the L-iduronic acid residue exists in the normal $^4C_1$ chair in the solid state but the axially positioned carboxyl perturbes the ring conformation in solution. Another possibility is that dermatan sulphate is more heterogeneous than we suspect and the x-ray diffraction results of course only relate to the parts of the macromolecule that crystallize. Such features have been observed in the case of heparan sulphate (see below).

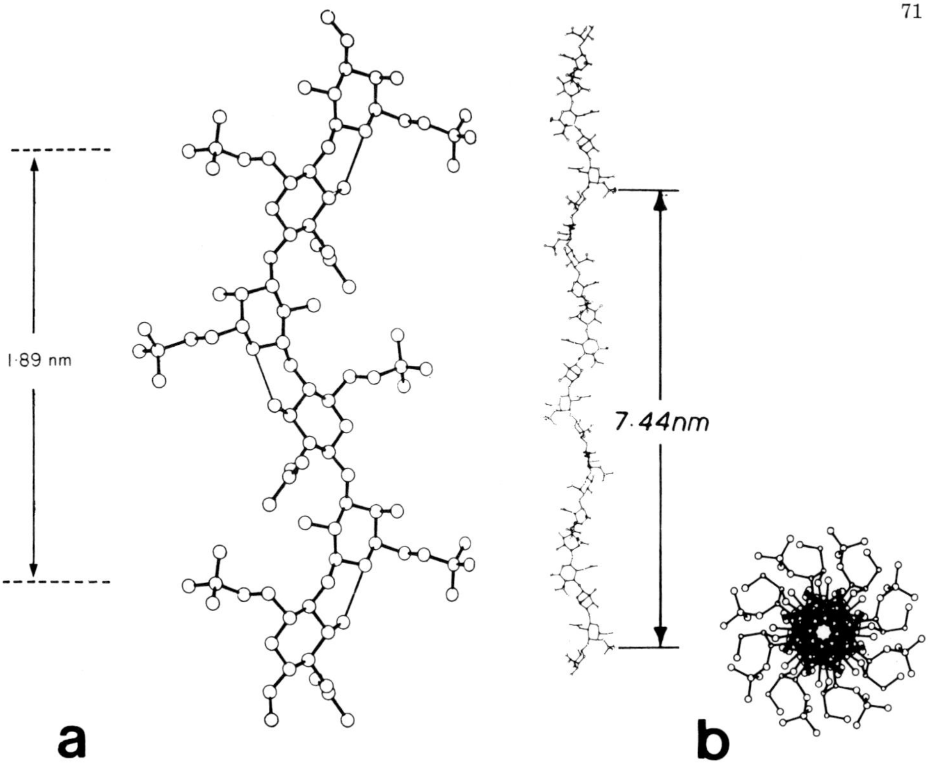

Fig. 8 (a) projections of the 2_1 helix after keratan sulphate (taken from [22], (b) 8_5 helix for dermatan sulphate.

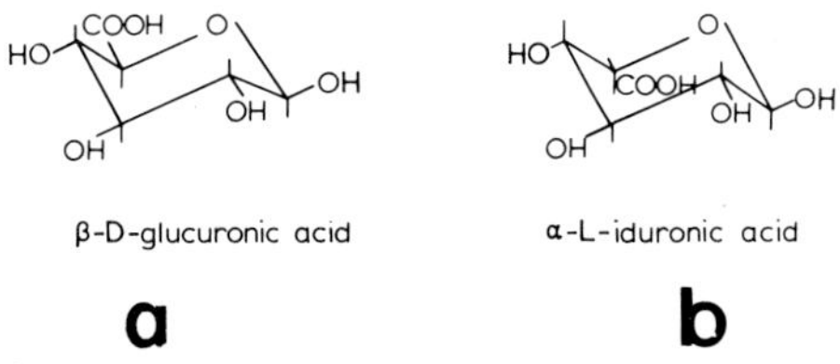

Fig. 9. Schematic representations of (a) β-D-glucuronic acid, and (b) α-L-iduronic acid.

In the analogous situation of the brown seaweed alginates the C(5) epimers D-mannuronic acid and L-guluronic acid exist in $^4C_1$ and $^1C_4$ chairs respectively in the solid state [24,25].

Heparan sulphate

Heparan sulphate consists of repeating disaccharide units of glucosamine and uronic acid. The uronic acid may either be D-glucuronic acid or its C(5) epimer L-iduronic acid. The glucosamine is partly N-sulphated and partly N-acetylated and also contains some sulphate ester groups. The D-glucosamine and L-iduronic acid residues have the α-1,4 configuration while recent chemical evidence favours the β-1,4 configuration for D-glucuronic acid [26]. Such a repeat is shown in Figure 3a.

The x-ray diffraction pattern obtained from ordered specimens in the condensed phase for sodium heparan sulphate [14] has layer lines of spacing 1.86 nm and with meridional reflections on even layer lines only, yielding a value for h = 0.93 nm [27-29] . These results are consistent with an extended two-fold helical structure as shown in Figure 11.

On formation of the calcium salt of the same preparation of heparan sulphate a new x-ray fibre pattern emerges (Figure 10b) similar to that obtained from calcium heparin (Figure 10c) [30]. These results suggest that heparan sulphate preparations can also contain regions formally similar to heparin. Independent evidence from chemical degradation methods also supports the presence of a trisulphated disaccharide that is apparently identical to the repeating sequence of heparin [31]. It would appear that heparan sulphate is a block copolymer containing a "heparan sulphate-like" phase and a "heparin-like" phase and that the former crystallites preferentially in the sodium salt form.

Heparin

Structural investigations of the blood anti-coagulant heparin using chemical degradation methods [32] currently favour a linear polydisaccharide of alternating 1,4 - linked 2-deoxy-2sulphamino-α-D-glucose-6-sulphate and 1,4-linked 2-sulphated-α-L-idopyranosyluronic acid to represent the major portion of the macromolecule. The covalent repeat (Figure 3b) is supported by nuclear magnetic resonance spectroscopy [33,34] and is consistent with the x-ray fibre diffraction results [35] and electron microscopical observations [36]. A typical x-ray diffraction pattern is shown in Figure 10a, which together with other x-ray results favor a two-fold helical structure [37-39] as shown in Figure 12.

BACTERIAL CAPSULAR POLYSACCHARIDES

In this section some examples have been taken from the capsular

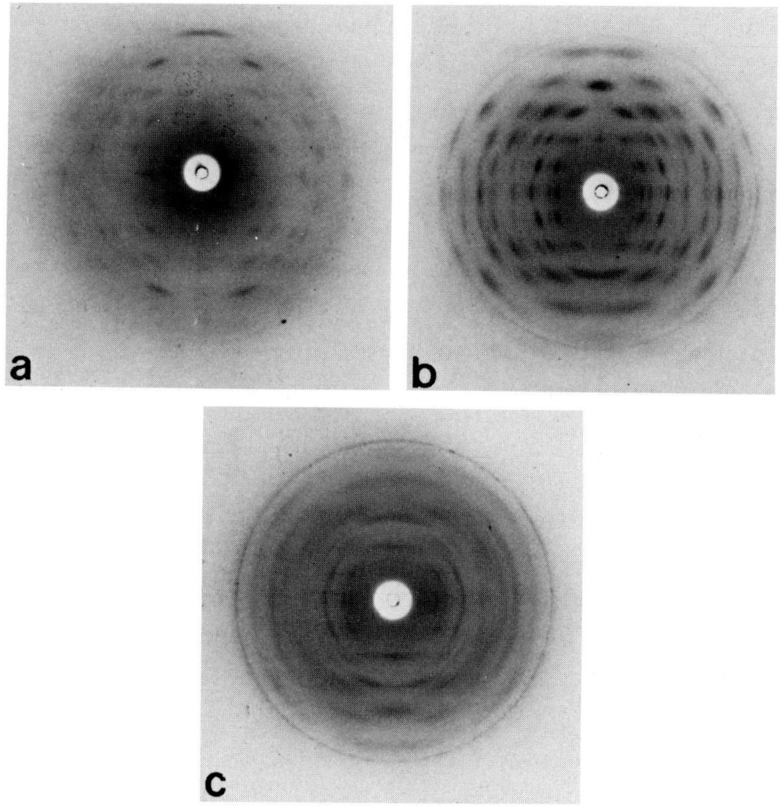

Fig. 10. X-ray diffraction patterns from (a) sodium heparin, (b) calcium heparan sulphate, and (c) calcium heparin.

polysaccharides of the genus Klebsiella. They have been chosen in order of increasing complexity and an overall model building procedure has been adopted described as follows.

Molecular models were generated using a procedure similar to the linked-atom description [40,41]. Bond lengths, bond angles and sugar residue conformations were held constant. Only the torsion angles at the various backbone glycosidic linkages were allowed to vary. Pendant groups were placed in stereochemically reasonable positions after the chain conformation had been determined. A conformation was considered possible when it simultaneously met two criteria; adherence to the appropriate helix pitch and symmetry while having no unacceptable non-bonded contacts. The presence (or absence) of unacceptable interatomic contacts was defined

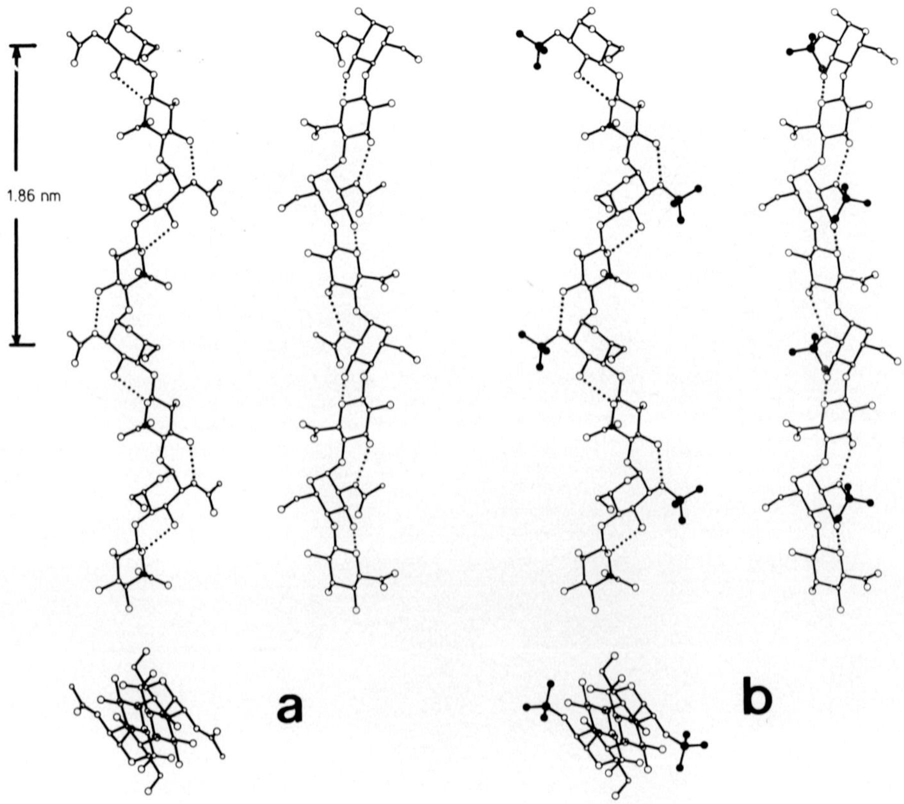

Fig. 11. Projections of the 2_1 helix for heparan sulphate. (a) with no sulphate groups included, and (b) with NSO_3^- groups on the glucosamine residues (filled circles).

using steric criteria [42]. In cases where alternative models were possible the model which contained the maximum number of intramolecular hydrogen bonds were favoured. Currently this methodology is being supplemented with energy minimization procedures although it appears to have little effect on the conformations outlined here.

Klebsiella serotypes

The genus *Klebsiella* belongs to the family of *Enterobacteriacae* and are found in the intestines of man and animals. In general *Klebsiella* bacteria produce extracellular polysaccharides which surround the bacterium as an additional outer layer. These capsules act as a viscoelastic barrier against mechanical damage and protect the cells against desiccation and phagocytosis. The capsular materials also exhibit specific antigenic

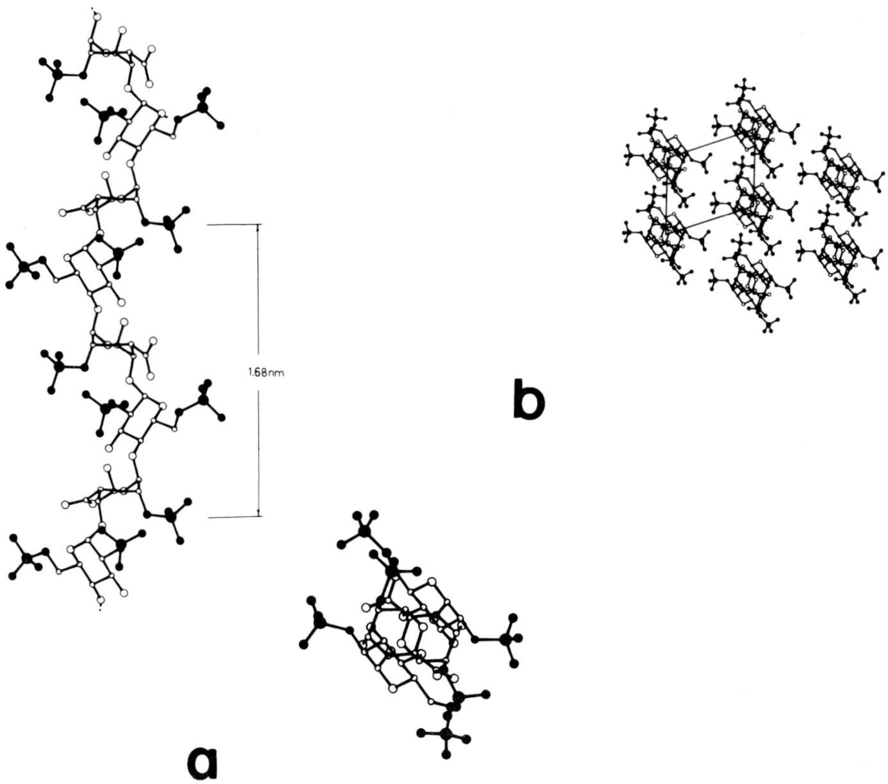

Fig. 12. Projections of (a) 2_1 helical conformation for heparin, and (b) packing of molecules in the unit cell. Note the charged sulphate groups (filled circles).

properties, and the serological classification of Klebsiella is based on immunological reactions of these capsular or K antigens [43]. To date about eighty serologically different strains have been recognized, the variety of which presumably has evolved due to the host defence mechanisms. In all cases known the isolated capsular material consists of heteropolysaccharides with a regular repeating unit of three to six saccharides, which are chemically different for each strain. The repeating unit is often branched being a mono or disaccharide as a side appendage. The presence of a uronic acid, pyruvate or both seems to be a characteristic feature of these glycans [44].

Serotype K5

Of the three Klebsiella serotypes to be discussed the K5 polysaccharide

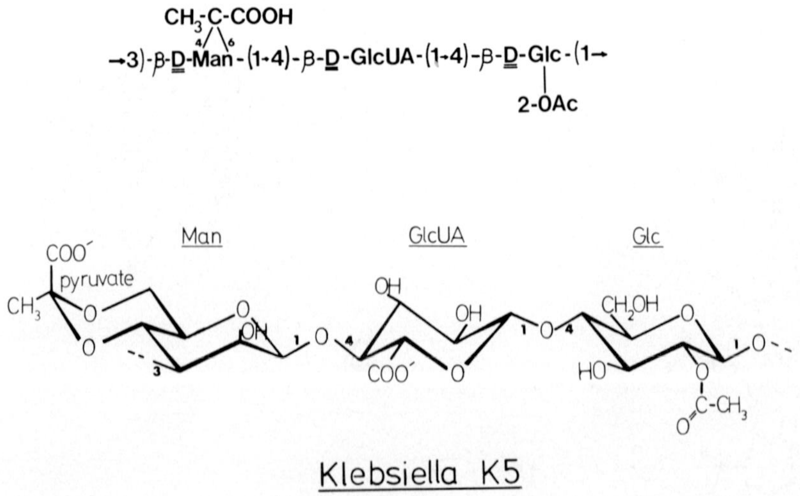

Fig. 13. Chemical repeat of serotype K5.

has the least complicated chemical covalent repeat. It is a linear poly-trisaccharide of the form $(-A-B-C)_n$, and the detailed chemical constitution given by Dutton et. al. [45] is shown schematically in Figure 13. The essential backbone structure consists of two neutral sugars, a 1,3-linked-β-D-mannose and a 1,4-linked-β-D-glucose, sandwiching a 1,4-linked-β-D-glucuronic acid residue. There are two appurtenances worth mentioning: pyruvic acid is linked to the D-mannopyranose as a 4,6 acetal, and an O-acetate group is attached to the 2-position of the glucopyranose ring. Thus the repeating sequence contains two charged carboxylate groups and the glycosidic linkage geometry is illustrated in Figure 13. All three saccharide units would be expected to exist in the normal $^4C_1$ chair conformation resulting in a pair of 1 → 4 diequatorial glycosidic linkages together with a single 1 → 3 diequatorial linkage. Both these linkage geometries are common to the simpler animal polysaccharides discussed above. If the vectors between each glycosidic oxygen atom were to align precisely the maximum theoretical extension per chemical repeat would be 1.56 nm. Of course it is extremely unlikely that a stereochemical acceptable model could exist with such a repeat but at least it indicates an upper limit. For the simpler polyuronides such as the alginates and connective tissue polyuronides the reported axially projected repeats in the

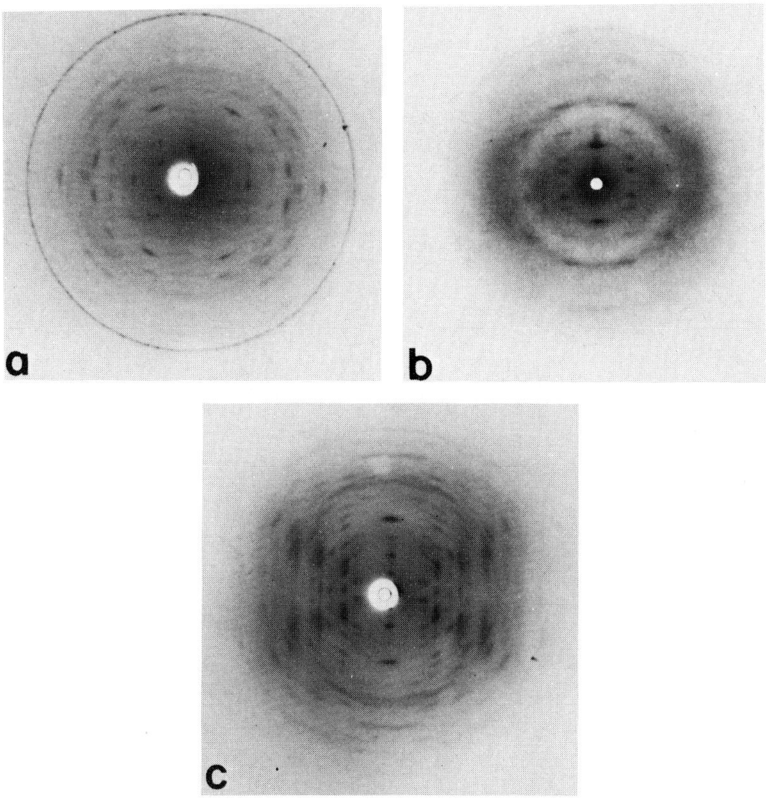

Fig. 14. X-ray diffraction patterns from Klebsiella serotypes (a) K5, (b) K57 and (c) K8.

condensed phase are all within 18% of the theoretical limit and typically highly extended conformations gave repeats centered around value 10% less than the maximum [10].

The x-ray diffraction results (Figure 14a) from the sodium salt of K5 [46] yielded a layer line spacing of 2.70 nm and with meridional reflections only occurring on even layer lines suggesting a two-fold helical conformation of the molecule. The value for $\underline{h}$ of 2.70/2 = 1.35 nm correlates well with the trisaccharide repeat and is only 14% below the theoretical limit suggesting an extended chain conformation.

Stereochemical acceptable structures with the required helical parameters have been computer generated and it has been found possible to form intra-chain hydrogen bonds across all three glycosidic linkages and fit

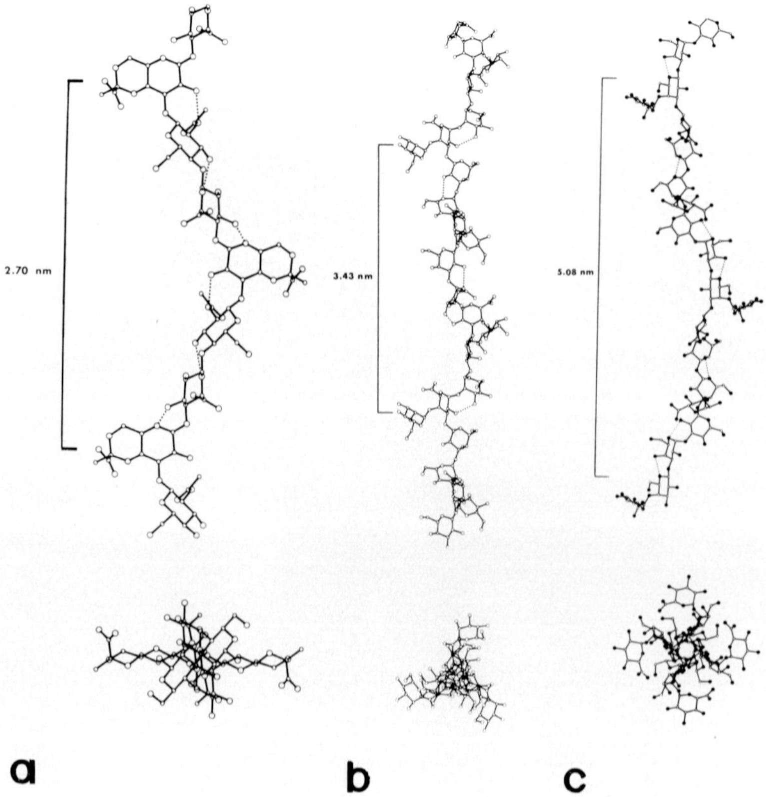

Fig. 15. Projections of the helical models for Klebsiella serotypes K5, K57 and K8. (a) two-fold helix for K5, (b) three-fold helix for K57, and (c) four-fold helix for K8.

the experimentally recorded helical symmetry and dimensions. Computer drawn projections are shown in Figure 15a. The structure incorporates an O(5)...H-O(3) hydrogen bond at both the β-D-Man-(1 → 4)-β-D-GlcUA and the β-D-GlcUA-(1 → 4) - β -D-Glc glycosidic linkages. This hydrogen bond is present in many homopolysaccharides such as cellulose [47] and the connective tissue polydisaccharide structures [12]. It is gratifying to find it also appearing in this more complex structure. At the third glycosidic linkage (β-D-Glc-(1 → 3) -β-D-Man) there is an O(2)...O(2) hydrogen bond.

Serotype K57

K57 is a polytetrasaccharide consisting of three saccharide units in

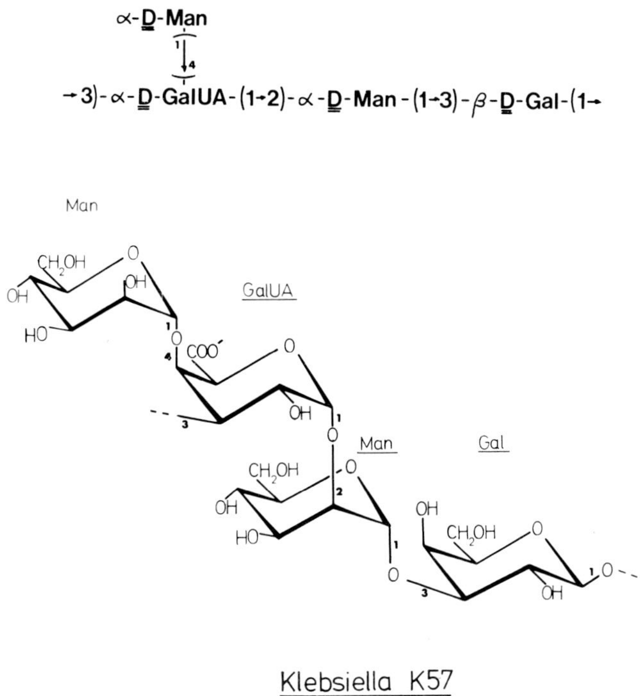

Fig. 16. Chemical repeat of serotype K57.

the backbone and with a single saccharide unit (α-D-mannose) in the side chain. The detailed chemical repeat has been established by Kamerling et. al. [48] and is given in Figure 16. As in the case of the K5 serotype the backbone consists of two neutral sugars and a uronic acid residue. This 1, 3 linked α-D-galacturonic acid residue is attached to a 1,2 linked α-D-mannose residue followed by a 1,3 β-D-galactose residue. The side appendage is attached to the uronic acid residue. Again it would be anticipated that all saccharide units exist in the normal $^4C_1$ chair conformation resulting in one 1 → 3 diequatorial glycosidic linkage, a 1 → 2 diaxial linkage and one lax → 3 eq linkage (Figure 16). In addition the mannopyranose side chain is 1 → 4 diaxially attached. This structure introduces some novel glycosidic linkage geometries together with the added complication of a side chain. The maximum theoretical extension for the chemical repeat, following the method described previously, is 1.27 nm.

The x-ray diffraction pattern (Figure 14b) shows layer lines with a measured spacing of 3.43 nm with meridional reflections present only on every third layer line [49]. The simplest interpretation of this diffraction pattern is that the polysaccharide backbone forms a three-fold helix with the value of $\underline{h}$ = 1.143 nm. This value, 10% less than the maximum permissible, correlates with the chemical repeat and indicates a highly extended conformation.

Trial models have been generated conforming to the helical symmetry and dimensions. Both left-handed and right-handed models have been generated using the techniques and criteria outlined above. Attempts were made to form the maximum number of intra-chain hydrogen bonds. It was found that no model could be constructed that included hydrogen bonds across all three backbone glycosidic linkages. Only a left-handed helix allowed the formation of two intra-chain hydrogen bonds in the backbone: α-D-GalUA-O(2) ...O(3)-α-D-Man and α-D-Man-O(5)...H-O(2)-β-D-Gal. Computer drawn projections of this conformation are illustrated in Figure 15b.

Serotype K8

The covalent chemical repeating sequence has been established by Sutherland [50] and is given in Figure 17. It is similar to the K57 polysaccharide having a polytrisaccharide backbone and a single saccharide unit for the side chain. Both structures have a uronic acid in the repeat, however, in K57 the uronic acid is incorporated in the backbone and the side chain is a neutral sugar. The converse is true in serotype K8 which has an uncharged backbone and an α-D-glucuronic acid side chain. All four saccharide units are expected to exist in the normal $^4C_1$ chair conformation as shown in Figure 17,

The theoretical maximum extension for the chemical repeat is 1.38 nm falling partway between the values for K5 and K57. The x-ray diffraction pattern for the sodium salt of the K8 polysaccharide is shown in Figure 14c. The material is highly oriented and crystalline and from the systematic absences of odd order meridional reflections it can be seen that the molecule is a 2_1 helix. However the layer line spacing of 5.078 nm is far too large for a repeat of two asymmetric units. In fact it is very close to the theoretical maximum extension for four complete covalent repeats, i.e. 4 x 1.38 = 5.52 nm. Thus the observed repeat is only some 10% less than the maximum permissible extension.

From preliminary model building it appears reasonable to assume that the structure of the isolated model is a perfect four-fold helix with an

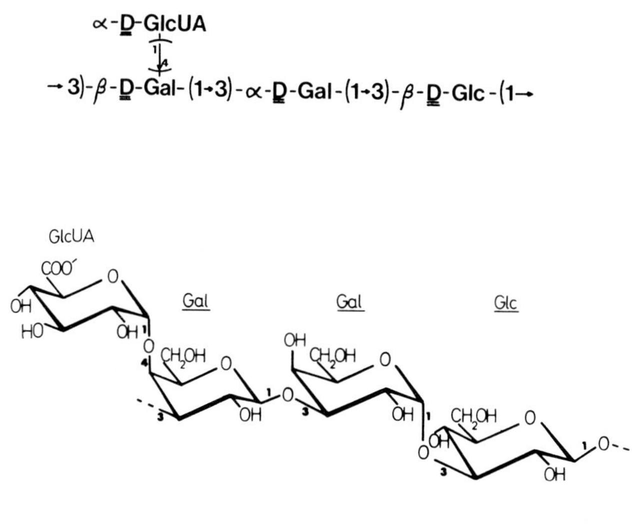

Klebsiella K8

Fig. 17. Chemical repeat of serotype K8.

axial advance per covalent repeat (h) = 1.27 nm. Perturbations from an idelized four-fold helix would be expected to result in a lower symmetry and consequently the packing of the molecular chains in an orthorhombic rather than a tetragonal unit cell [51]. The phenomenon has also been observed in hyaluronic acid [12].

Stereochemically feasible models were constructed to fit a four-fold helical symmetry and incorporating the maximum number of intra molecular hydrogen bonds. It was found impossible to construct a model which incorporated hydrogen bonds at all three backbone linkages. Only a left-handed helix allowed the formation of two hydrogen bonds in the backbone. Projections of this structure are shown in Figure 15c. The conformation contains hydrogen bonds O(5)...O(4) and O(5)...O(2) at the β-D-Gal-(1 → 3) - α-D-Gal and α-D-Gal-(1 → 3) -β-D-Glc glycosidic linkages respectively. It was found not possible to form a four-fold helix (with the observed extension) which contained a hydrogen bond at the β-D-Glc-(1 → 3) - β-D-Gal glycosidic linkage. This linkage is adjacent to the attachment site of the uronic acid side chain.

Lacking any information to define the position of the side chain in the isolated molecule, other than that it must be in a stereochemically allowed position, it was necessary to use x-ray intensity information to determine the orientation of the pendant residue [51].

DISCUSSION

Examples have been briefly reviewed covering the connective tissue polysaccharides and the more complex bacterial capsular polyuronides from Klebsiella. Conformations have been derived from the x-ray diffraction evidence in an attempt to provide a general outlook on the behaviour of these macromolecules in the condensed phase. This has recently become possible with the crystallization of these macromolecules in a form suitable for x-ray diffraction studies. It is only through these x-ray studies that the helical parameters necessary for meaningful model building can be obtained. One noticeable feature of the results presented is that all the conformations are highly extended, all that is except for the contracted four-fold of hyaluronic acid [12]. No simple explanation appears available to explain this puzzling phenomenon.

In the case of the connective tissue polydisaccharides based on the $1 \rightarrow 3$ and $1 \rightarrow 4$ glycosidic linkage geometry (see Figure 1) the three basic helical symmetries in the solid state are 2_1, 3_2 and 8_5. If we take the 2_1 helix as the standard conformation the 8_5 and 3_2 helices may be obtained by successive azimuthal rotations of 45° and 60° respectively per structural repeat. Such azimuthal rotations of course result from a series of rotations about the glycosidic linkage bonds. The relative similarity of these extended conformations clearly demonstrates the controlling influence of the backbone linkage geometry.

The quite different conformations for sodium heparan sulphate and sodium heparin may be correlated with the distinct change in the uronic acid chair conformation. The change from a $1 \rightarrow 4$ diequatorial to a $1 \rightarrow 4$ diaxial linkage substantially alters the polysaccharide conformation. A feature first noticed in the alignates [24,25].

As more conformations of the Klebsiella serotypes are examined it will be interesting to see if any kind of trend emerges and whether any rules can be formulated relating glycosidic linkage geometry, saccharide constition and chair conformation.

We have witnessed an avalanche of molecular information relevant to the animal and microbial polysaccharides which will naturally take time to be completely digested and fully understood, yet we have probably only scratched the surface. Much work is still to be done in refining the molecular conformations outlined in this report. However, one instinctively feels that this is an exciting period since not only has the crystallization of these complex substances been achieved, but a whole variety of molecular conformations are evident as a function of the chemical and thermodynamical environment. The ability to monitor detailed molecular geometries, even in the condensed phase, will greatly facilitate our general understanding of their properties and molecular biology.

ACKNOWLEDGEMENTS

I am indebted to my colleagues Drs. K. H. Gardner, D. H. Isaac and Ms. H. F. Elloway, for many valuable discussions and for allowing me to reproduce some of their results. My thanks also to the many collaborators who have generously supplied material which made much of this work possible. This work was financed by the Science Research Council and in part by the Arthritis and Rheumatism Research Council.

REFERENCES

1. E.D.T. Atkins and W. Mackie, Biopolymers, 11 (1972) 1685.
2. S. Nishikawa and S. Ono, Proc. Math. Phys. Soc., Tokyo, 7 (1913) 131.
3. E.D.T. Atkins and J. K. Sheehan, Biochem. J., 125 (1971) 92; Nature New Biology, 235 (1972) 253.
4. T. E. Hardingham and H. Muir, Biochem. J., 139 (1974) 565.
5. E.D.T. Atkins, T. E. Hardingham, D. H. Isaac and H. Muir, Biochem. J., 141 (1974) 919.
6. E.D.T. Atkins, C. F. Phelps and J. K. Sheehan, Biochem. J. 128 (1972) 1255.
7. I.C.M. Dea, R. Moorhouse, D. A. Rees, S. Arnott, J. M. Guss and E. A. Balazs, Science, 179 (1973) 560.
8. E.D.T. Atkins and J. K. Sheehan, Science, 179 (1973) 562.
9. E.D.T. Atkins, J. H. Brown, J. M. Landsdall, I. A. Nieduszynski and J. K. Sheehan, J. Poly. Sci., C42 (1973) 1513.
10. E.D.T. Atkins, D. H. Isaac, I. A. Nieduszynski, C. F. Phelps and J. K. Sheehan, Polymer, 15 (1974) 263.

11. S. Arnott, J. M. Guss and W. T. Winter, in Extracellular Matrix Influences on Gene Expression, edits., H. C. Slavkin and C. C. Grenlich, Academic Press, New York (1975) p. 399.

12. J. M. Guss, D.W.L. Hukins, D.J.C. Smith, W. T. Winter, S. Arnott, R. Moorhouse and D. A. Rees, J. Mol. Biol. 95 (1975) 359.

13. W. T. Winter, D.J.C. Smith and S. Arnott, J. Mol. Biol., 99 (1975) 219.

14. E.D.T. Atkins and T. C. Laurent, Biochem. J., 133 (1973) 603.

15. D. H. Isaac and E.D.T. Atkins, Nature New Biology, 244 (1973) 252.

16. E.D.T. Atkins, in Inborn Errors of Skin, Hair and Connective Tissue, edits., J. Holton and J. T. Ireland, Medical and Technical Pub. Co. (1975) p. 119.

17. E.D.T. Atkins, R. Gaussen, D. H. Isaac, V. Nandanwar and J. K. Sheehan, J. Poly. Sci. B10 (1972) 863.

18. S. Arnott, J. M. Guss, D.W.L. Hukins and M. B. Mathews, Science, 180 (1973) 743.

19. S. Arnott, J. M. Guss, D.W.L. Hukins and M. B. Mathews, Biochem. Biophys. Res. Comm. 54 (1973) 1377.

20. M. Fleming, M. Sc. Thesis, University of Bristol (1973).

21. E.D.T. Atkins and D. H. Isaac, J. Mol. Biol. 80 (1973) 773.

22. S. Arnott, J. M. Guss, D.W.L. Hukins, I.C.M. Dea and D. A. Rees, J. Mol. Biol. 88 (1974) 175.

23. G. K. Hamer and A. S. Perlin, Carbohyd. Res. 49 (1976) 37.

24. E.D.T. Atkins, W. Mackie and E. E. Smolko, Nature, 225 (1970) 626.

25. E.D.T. Atkins, I. A. Nieduszynski, W. Mackie, K. D. Parker and E. E. Smolko, Biopolymers, 12 (1973) 1865.

26. T. Helting and U. Lindahl, J. Biol. Chem., 246 (1971) 5442; M. Hook, Thesis, University of Uppsala (1974).

27. I. A. Nieduszysnki and E.D.T. Atkins, in Structure or Fibrous Biopolymers: Colston Papers No. 26., edits., E.D.T. Atkins and A. Keller, Butterworths, London (1975) p. 323.

28. E.D.T. Atkins and I. A. Nieduszynski, in Heparin: Structure, Function and Clinical Duplications, edits., R. A. Bradshaw and S. Wessler, Plenum Press, New York (1975) p. 19.

29. H. F. Elloway and E.D.T. Atkins, in press: Biochem. J.

30. E.D.T. Atkins and I. A. Nieduszynski, in Heparin: Chemistry and Clinical Usage, edits., V. V. Kakkar and D. P. Thomas, Academic Press, London (1976) p. 21.

31. P. Hovingh and A. Linkar, Carbohyd. Res., 37 (1974) 181.

32. R. W. Jeanloz, in Heparin: Structure, Function and Clinical Duplications, edits., R. A. Bradshaw and S. Wessler, Plenum Press, New York (1975) p. 3.
33. A. S. Perlin, B. Casu, G. R. Sanderson and L. F. Johnson, Can. J. Chem., 48 (1970) 2260.
34. A. S. Perlin, In Proceed. Int. Symp. Macromolecules, edit., E. B. Mano, Elsevier Scientific Pub. Co., Amsterdam (1975) p. 337.
35. E.D.T. Atkins and I. A. Nieduszynski, Fed. Proc., in press.
36. S. Hirano, Int. J. Biochem., 3 (1972) 677.
37. I. A. Nieduszynski and E.D.T. Atkins, Biochem. J., 135 (1973) 729.
38. E.D.T. Atkins, I. A. Nieduszynski and A. A. Horner, Biochem. J., 143 (1974) 251.
39. I. A. Nieduszynski, K. H. Gardner and E.D.T. Atkins, ACS Spring Meeting, New York, April 1976, in press.
40. S. Arnott and A. J. Wonacott, Polymer 7 (1966) 157.
41. S. Arnott, in Symp. on Fibrous Proteins, edit., W. G. Crewther, Butterworths, Australia, (1968) p. 26.
42. G. N. Ramachandran and V. Sasisekharan, Adv. Protein Chem., 23 (1968) 283.
43. F. Kauffmann, The Bacteriology of Enterbacteriaceae, Munksgaard, Copenhagen (1966).
44. W. Nimmich, Z. Med. Mikrobiol. Immunol; 154 (1968) 117.
45. G.G.S. Dutton and Mo-Tai Yang, Can. J. Chem., 51 (1973) 1826.
46. K. H. Gardner, D. H. Isaac, Ch. Wolf, E.D.T. Atkins and G.G.S. Dutton, submitted Carbohyd. Res.; E.D.T. Atkins, K. H. Gardner and D. H. Isaac, ACS Spring Meeting, New York, April 1976, in press.
47. K. H. Gardner and J. Blackwell, Biopolymers, 13 (1974) 1975.
48. J. P. Kamerling, B. Lindberg, J. Lonnengren and W. Nimmich, Acta. Chem. Scand., B29 (1975) 593.
49. D. H. Isaac, K. H. Gardner, E.D.T. Atkins, U. Elsasser-Baile and S. Stirm, submitted Carbohyd. Res.
50. I. W. Sutherland, Biochemistry 9 (1970) 2180.
51. Ch. Wolf, K. H. Gardner, E.D.T. Atkins, W. Burchard and S. Stirm; K. H. Gardner and E.D.T. Atkins, submitted J. Mol. Biol.

SECONDARY STRUCTURES OF POLYNUCLEOTIDES

STRUTHER ARNOTT
Department of Biological Sciences, Purdue University, West Lafayette, Indiana U.S.A. 47907

SUMMARY

DNA duplexes containing Watson-Crick base-pairs can assume different helical secondary structures characterized by the two parameters (h,t), where h is the axial translation and t the rotation per nucleotide residue. Seven secondary structures have been characterized accurately and precisely for native or synthetic DNAs spun into oriented and sometimes crystalline fibers. These fall into two conformational classes called A and B. Within each class the conformation angles of each nucleotide residue are qualitatively similar. Between the two classes the major conformational differences are confined to the furanose rings which are puckered C3'-endo in A and C3'-exo in B. Within either class, differences in h and t are manifested in different orientations and positions of the base-pairs with respect to the helix axes. These differences, in turn, give rise to differences in the relative widths and depths of the major and minor grooves of the gross double-helical structures. External factors such as hydration affect secondary structure as do internal factors such as nucleotide composition and sequence.

Analogous RNA duplexes and DNA-RNA hybrid duplexes show less variability of secondary structure in that all the examples so far observed are confined to the A family. Three of these have been refined extensively.

Even the helical polynucleotides and their complexes which either have no base-pairing, like polycytidylic acid, or base-pairs not of the Watson-Crick kind, like quadruple-stranded polyguanylic acid, all have conformations of the A or B type.

DETERMINING PRECISE AND ACCURATE SECONDARY STRUCTURES OF POLYNUCLEOTIDES

Analyses of x-ray diffraction patterns from oriented and (sometimes) crystalline fibers of native DNAs and RNAs and of synthetic polynucleotide complexes have provided details [1-12] of their many different helical secondary structures. In the majority of these systems two antiparallel,

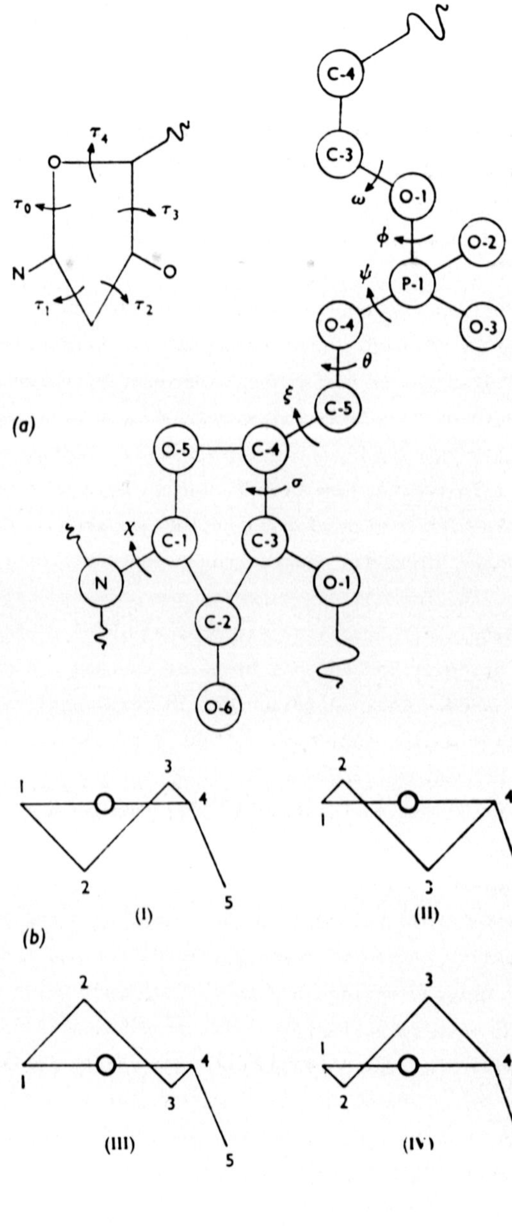

Fig. 1. (a) The conformation angles of a nucleotide residue. All angles except σ have a wide range of values. The discrete values of σ are determined by the sugar ring shapes in (b): (I) C2-endo; (II) C3-endo; (III) C2-exo; (IV) C3-exo. In the A genus of polynucleotides the rings are puckered C3-endo, in the B genus C3-exo (or perhaps C2-endo which is rather similar).

right-handed polynucleotide chains have complementary base sequences which are hydrogen bonded in the Watson-Crick [13] manner. However there are also triple-stranded complexes like poly(U)•poly(A)•poly(U) [6] and poly(dC)•poly(dI)•poly(dC) [12] where at least one of the base-pairings cannot be of the Watson-Crick kind, or like poly(I)•poly(A)•poly(I) [5] where none of the base-base interactions can be classical. The same is true of the isomorphous quadruple-stranded structures of poly(I) and poly(G) [8,14].

In most cases the asymmetric unit (from which the rest of the structure can be generated by the operation of symmetry elements) is a single nucleotide. In no case has it been necessary to consider an asymmetric unit of structure larger than three nucleotide residues. Despite the relative simplicity of these structures their analysis is not a trivial operation. Because the diffraction data from DNA fibers rarely extend to periodicities below 0.3nm, accurate determination of the structure of the asymmetric unit when the atomic coordinates are treated as independent variables is impossible. Since, however, the values of bond lengths and bond angles, and the shapes of purine, pyrimidine and desoxyribose rings are relatively invariant [15,16] it is possible to supplement the x-ray data from fibers with stereochemical data derived from accurate analyses of the structures of nucleosides and nucleotides in single crystals [17]. If the stereochemical features just listed are assumed to be known and fixed, determining the structure of a polynucleotide chain is reduced to defining the values of the six conformation angles ω, ϕ, ψ, θ, ξ, χ shown in Figure 1. The spacings of the layer lines and the intensity distribution in the fiber diffraction patterns permit the translation distance (h) and rotation about the helix axis (t) of successive nucleotides to be determined with high precision. Imposing these two constraints on the six conformation angles immediately reduces the number of degrees of implicit structural freedom to four. The variability of these parameters is further restrained by the need to form hydrogen bonds of standard lengths. Accurate structures can therefore be produced by optimizing the fit between calculated observed diffraction data even although these are of relatively low number and quality, and, since computer procedures have now replaced manipulation of wire models, structures can be presented with any desired degree of precision. Nor has the development of procedures for producing optimized molecular and crystal models with explicitly defined characteristics been of merely operational significance. It is not usually possible to use methods such as isomorphous replacement objectively to solve the x-ray phase problem in fiber diffraction analyses. Instead approximate phases are implicity provided by postulating some initial structural model and improved by the refinement of this model. However there is nothing in such a process which would automatically remedy an initial choice of the wrong kind of model. Where another model can be envisioned there is no remedy for the dilemma except to produce the optimum model of each kind then to compare the x-ray diffraction and stereochemical properties of each optimized model to determine whether one is clearly preferable. The importance of being able to do this conveniently for any fibrous system cannot be too strongly emphasized, especially when we consider that in the long -- 26 years old -- history of nucleic acid structure analysis no structures, except those produced by Maurice Wilkins' school, have survived reappraisal. For example, the original Crick-Watson [18] model for B-DNA contained C3-endo puckered furanose rings which we now know are associated with the A family of polynucleotide structures which have gross structural features quite different from those of the B family; the structure of reovirus RNA double helices is not a ten-fold double helix as suggested by Langridge and Gomatos [19] but an eleven-fold double helix [20,4] ; the structure of the eight-fold helical duplex of poly d(IC)•poly d(IC) is not the bizarre left-handed model proposed by Mitsui et. al. [21] but a right-handed, conformationally unexceptional member of the B family [7]; the multi-stranded complex formed with poly(I) or poly(G) is not triple-stranded as suggested by Rich [22] but quadruple-stranded [8,14] ; the poly(C) helices trapped in oriented, polycrystalline fibers by Langridge and Rich [23] are not duplexes with coaxial, base-paired, twelve-fold helical polynucleotide chains but single, six-fold helices with no base-base interactions [Arnott, Leslie, Chandrasekaran (unpublished)]!

It has been suggested that physical probes other than x-ray analyses of fibers might provide more accurate details of DNA structure. Ultraviolet circular dichroism spectra are indeed very sensitive to the secondary structures present but they are also sensitive to the base sequences present in a way that currently is being disputed [24,25]. Indeed it was an unwise interpretation of a circular dichroism spectrum which led directly to a structure of the wrong hand being proposed for poly d (IC)• poly d(IC) [21]. It is more likely, therefore, that this technique will prove more useful in deciding which (previously defined) structure is present and be a complementary probe to x-ray diffraction for studies of DNA in solution. Infra-red linear dichroism has been used to define the orientation of P-O and other bonds in DNA [26,27]. Nominally this technique should provide more accurate results than the x-ray fiber diffraction analyses. It turns out however that the conclusions from the IR studies are not generally compatible with the x-ray results. Moreover there is every reason to doubt some of the assumptions that are necessary for the IR calculations. As with UV circular dichroism it is therefore more likely that IR linear dichroism will prove more useful in monitoring the occurrence of structural transitions rather than in defining accurately features of the molecular structure being examined.

DESCRIBING POLYNUCLEOTIDE HELICAL SECONDARY STRUCTURES

The conformation angles which are the explicit variables in contemporary structural refinements are useful in defining the degree of relatedness among the different structures. Conformationally all the polynucleotide secondary structures (Tables 1,2,3) fall into two genera distinguished by the puckering of the sugar rings. The genera are called A and B after the eponymous A-DNA and B-DNA structures which contain respectively C3-endo and C3-exo (Figure 1) sugar rings. In monomers the angle χ, which describes the orientation of the base with respect to the sugar ring is strongly correlated with the ring puckering. The values of χ in A and B polynucleotides (Tables 1,2,3) show the same kind of differences as in monomers [28]. The other conformation angles are all qualitatively the same in different structural species even when these belong to different genera.

Even although structures within either genus are conformationally very similar, there are nevertheless profound differences in their gross features. These are most obviously described by the positions and orientations of the base pairs with respect to the helix axis. These in turn affect the widths and depths of the minor and major grooves of the Watson-Crick base-paired DNA and RNA duplexes (Tables 4,5). (The minor groove is the one in which the two glycosidic bonds of a base pair make an obtuse angle.) In the A genus the base pairs are invariably positioned "forward" so that the minor grooves are shallow and the major grooves deep. As h decreases the base pairs are increasingly tilted (away from being perpendicular to the helix axis) in a clockwise sense as one looks into the minor groove

Table 1

Comparison of conformation angles in Watson-Crick base-paired DNA duplexes and in triplexes where two strands are paired in the classical number

	Conformation angles	ω	φ	ψ	θ	ξ	σ	χ
Secondary Structure	B-DNA	155	-96	-46	-147	36	157	143
	B'-DNA	145	-87	-52	-136	39	157	145
	C-DNA	161	-106	-39	-160	37	157	143
	D-DNA	141	-101	-62	-152	69	157	144
	A"-DNA	-156	-74	-46	174	39	83	86
		-161	-68	-66	180	55	83	91
		-160	-77	-53	172	47	83	85
	A'-DNA	-155	-70	-61	176	51	83	83
		-151	-72	-82	173	72	83	84
		-153	-72	-65	178	54	83	84
	A-DNA	178	-47	-85	-152	45	83	86

The conformation angles (in degrees) are defined as in Arnott and Hukins [28] and illustrated in Figure 1. The values of σ correspond to those of fixed, standard furanose conformations (C3-exo in the B and C3-endo in the A genus). The conformation angles of A' and A"-DNA are from the triple-stranded complexes poly d(C)•poly d(I)•poly d(C) and poly d(T)•poly d(A)•poly d(T) respectively in which there is no diad axis relating the anti-parallel chains. The conformation angles of the Watson-Crick base-paired polypyrimidinenucleotide of these are given first, the polypurinenucleotide strands second, and the additional polypyrimidinenucleotide strand third.

[29]. The result is that the minor grooves maintain a fairly constant width, but the major grooves become narrower. In the B genus, in contrast, the minor grooves are deep, and as h decreases (or t increases) they become even deeper since the bases move "backwards". In this genus also, decreasing h results in increasingly tilted bases, but in an anticlockwise sense in this case. In B genus it is therefore the width of the minor groove that decreases with h, while that of the major groove remains relatively constant.

HELICAL SECONDARY STRUCTURES OF DNA

There are now five known species of B-type helices. The classical B-DNA helix [1,2] obtained with calf thymus and other "typical" duplex DNAs has h = 0.338 nm and t = 36°. The base pairs are sited centrally with

Table 2

Comparison of conformation angles in RNA duplexes and in triplexes where at least one RNA chain is paired in the Watson-Crick manner to a second polynucleotide chain.

	Conformation angles	ω	φ	ψ	θ	ξ	σ	χ
Secondary Structures	A'-RNA	-164	-63	-63	-171	45	83	79
	A-RNA	-147	-79	-60	175	49	83	73
	11_1 poly(U)•	-152	-74	-55	173	47	83	83
	poly(A)•	-156	-69	-76	176	64	83	89
	poly(U)	-149	-76	-63	174	53	83	83
	12_1 poly(U)•	-156	-75	-28	171	24	83	87
	poly(A)•	-163	-67	-66	-179	53	83	91
	poly(U)	-149	-83	-40	167	37	83	85
	poly(U)•	-152	-74	-57	172	49	83	83
	poly d(A)•	-160	-67	-71	178	58	83	92
	poly(U)	-149	-76	-62	174	53	83	84

In the duplex structures the two chains have the same conformations. For the triplex structures the conformation angles of the Watson-Crick base-paired polypyrimidinenucleotide strand are given first, those of the poly-purinenucleotide strand second, those of the second polypyrimidinenucleotide strand third.

respect to the helix axis and are approximately perpendicular to it. Both grooves are therefore wide and about equally deep (Table 4). B'-DNA is a distinct species observed with poly d(A)•poly d(T) [10] and poly d(I)• poly d(C) [unpublished work of Leslie, Tsivoglou, Arnott and Ratliff] and exists in crystalline forms different from B-DNA but, as its helical parameters (h = 0.325, t = 36°) suggest, it is a trivial variant of B-DNA. C-DNA (h = 0.331 nm, t = 38.6°) is another conformation observed with typical DNAs [30,31,11]. Initially [30] it was thought that there were notable conformational differences between the structures of C and B-DNA. More recently this has been shown to be untrue: in conformational and in other respects C-DNA is a close relative of B-DNA [11]. The fourth species of the B genus, obtained with poly d(AT)•poly d(AT) [7], poly d(GC)•poly d(GC) [7], poly d(ATT)•poly d(AAT) [31], poly d(AIC)• poly d(ICT) [unpublished work of Leslie, et. al.] etc. is conformationally

Table 3

Conformation angles of polynucleotide structures with no Watson-Crick base-pairs.

	Conformation angles	ω	ϕ	ψ	θ	ξ	σ	χ
Secondary Structure	poly(I)•	-152	-77	-57	171	52	83	84
	poly(A)•	-154	-75	-62	173	55	83	85
	poly(I)	-153	-75	-70	177	60	83	80
	B poly(I)	-175	-117	-73	-169	55	147	132
	A poly(I)	-150	-70	-104	170	96	83	74
	poly(C)	-129	-65	-57	173	47	83	82

Whether the four identical chains in poly(I or G) are B or A-type has not yet been resolved. The notable difference between the conformation of poly(C), the only single-stranded structure so far defined, and the multi-stranded polynucleotide complexes is in the value of ω, which more closely resembles the average value (-123°) observed in monomers.

unexceptional, but there are gross structural changes associated with the reduction of h to 0.303 nm and the increase of t to 45°. Most noticeable is the reduced width of the minor groove (Table 4). In this structure it would be impossible for an external agent to sense base-sequences except in the major groove. Recently Leslie, Arnott and Ratliff [unpublished] have observed yet another B-type DNA structure [with poly d(ITT)• poly d(ACC)] in which the value of h is low (0.32 nm) and the value of t is even higher (48°) than in D-DNA.

The three A-type DNA species [1,10 and unpublished work of Arnott, Bond, Selsing and Smith] all have C3'-endo sugar rings: consequently their base-pairs are forward from the helix axis, their minor grooves are shallow and their major grooves deep. The large differences in h (0.326 nm in A"-DNA, 0.316 in A'-DNA and 0.256 in A-DNA) result in striking differences in the widths of the major grooves. A'-DNA has been observed only with poly d(C)•poly d(I)•poly d(C) [12] and A"-DNA only with poly d(T) •poly d(A)•poly d(T) [10]. A-DNA is observed with typical DNAs but only in a highly crystalline form where some aspects of the molecular structure may be a consequence of lattice forces. If therefore A-type structures occur *in vivo* it is not clear which of the above species they would be most likely to resemble.

Table 4

Grooves in DNA double helices

Double Helix	$t(^o)$	h(nm)	m(nm)	M(nm)	d(nm)	D(nm)
B-DNA	36.0	0.337	0.57	1.17	0.75	0.85
B'-DNA	36.0	0.328	0.69	1.35	0.65	0.85
C-DNA	38.6	0.331	0.48	1.05	0.79	0.75
D-DNA	45.0	0.303	0.13	0.89	0.67	0.58
A"-DNA	30.0	0.326	1.09	1.00	0.42	1.26
A'-DNA	32.7	0.316	1.02	0.79	0.44	1.22
A-DNA	32.7	0.256	1.10	0.27	0.28	1.35

t = rotation per nucleotide; h = axial translation per nucleotide; m and M = widths of minor and major grooves and d and D = depths of minor and major grooves.

The groove widths are defined to be the perpendicular separations of the two helices which connect the phosphorus atoms of each strand, reduced by 0.58 nm (the diameter assumed for a phosphate group). For A' and A"-DNA, which occur in triple-stranded complexes, the cylindrical polar radii of the phosphorus atoms on different strands are not equal but for the purpose of these calculations it has been assumed that each is equal to the mean value.

Groove depths are defined below:

$$d = r_p - r_{N2G} + 0.14 \text{ for both the A and B genera;}$$

$$D = r_p - r_{N6A} + 0.14 \text{ for the B genus;}$$

$$D = r_p + r_{N6A} + 0.14 \text{ for the A genus.}$$

r_p, r_{N2G} and r_{N6A} are the cylindrical polar radii of the phosphorus, guanine N2 and adenine N6 atoms respectively. For A"-DNA, B'-DNA and D-DNA thymine O2 replaces guanine N2. For A'-DNA hypoxanthine O6 replaces adenine N6 and cytosine O2 guanine N2. The constant term (0.14 nm) represents the difference of the phosphate radius (0.29 nm) and the van der Waals radius of nitrogen (0.15 nm).

HELICAL SECONDARY STRUCTURES OF RNA CHAINS INVOLVED IN WATSON-CRICK BASE-PAIRING

No B-type conformations have been observed for RNA chains involved in Watson-Crick base-pairing. This may be because ribose rings have a slight preference for the C3-endo puckered form and because the C2 position which carries the additional -OH group is sterically restricted in B-type duplexes [33]. Two A-type species have been defined for the RNA duplexes

Table 5

Grooves in RNA double helices or in the Watson-Crick paired part of triple helices involving at least one RNA chain.

Double Helix	t(°)	h(nm)	m(nm)	M(nm)	d(nm)	D(nm)
A'-RNA	30	0.300	1.10	0.81	0.31	1.38
A-RNA	32.7	0.282	1.12	0.42	0.27	1.31
11_1 poly(U)• poly(A)• poly(U)	32.7	0.306	1.04	0.68	0.43	1.20
12_1 poly(U)• poly(A)• poly(U)	30	0.304	1.08	0.85	0.40	1.27
poly(U)• poly d(A)• poly(U)	32.7	0.304	1.04	0.67	0.44	1.17

which can be obtained from reovirus [20], wound tumor [34] virus and rice stunt virus [35], or with the synthetic duplexes poly(A)•poly(U) and poly(I)•poly(C) [33,4]. A-RNA has h = 0.282 nm and t = 32.7°. A'-RNA (h = 0.300 nm, t = 30°) is obtained in circumstances in which an A→B transition would occur in a typical DNA [33]. Both structures are conformationally similar to one another and to the A form of DNA [4]. Since for both RNAs h is larger than for A-DNA, the major grooves in A and A'-RNA are somewhat wider (Table 5) than in A-DNA but apparently not wide enough to accomodate a third (polypyrimidinenucleotide) strand. In the similar polynucleotide complexes where such a strand has been added to an RNA Watson-Crick duplex (Table 5) the values of h are somewhat larger (0.304 and 0.306 nm) and therefore the widths of the major grooves of the duplex parts of structures somewhat greater.

In hybrid duplexes where an RNA chain is base-paired in the classical complementary fashion with a DNA chain B-type structures are apparently also forbidden: the two structures of this type which have been studied are isomorphous with either A-DNA [36] or A'-RNA [37].

POLYNUCLEOTIDE SECONDARY STRUCTURES WITH BASE-BASE INTERACTIONS NOT OF THE WATSON-CRICK TYPE

Table 6

DNA secondary structures (shown underlined) so far observed in fibers of synthetic DNAs with repeated di- and trinucleotide sequences.

AA	AT	AC	AG	GC	GG
TT	AT	GT	CT	GC	CC
(B,B')[1]	(B,D)	(B,A)	(B)	(B,D)	(B,A)

AAA	AAT	AAC	AAG
TTT	ATT	GTT	CTT
(B,B')[1]	(B,D)[1]	(A,B)	
	AGT	AGC	
	ACT	GCT	
	(A)	(C,A)	
	GAT	GAC	
	ATC	GTC	
	(A)		
AGG	GGT	GGC	GGG
CCT	ACC	GCC	CCC
	(B,A)		(B,A)

[1] Apparently I:C pairs are equivalent to A:T and not to G:C pairs as far as conformation determining properties are concerned since poly d(I)•poly d(C) mimics poly d(A)•poly d(T) and not poly d(G)•poly d(C) and poly d(AIC)• poly d(ICT) mimics poly d(AAT)•poly d(ATT).

With the possible exception of the isomorphous quadruple-stranded complexes of poly(I) and poly(G) [8,14] all the multi-stranded polynucleotide systems have helical chains of the A-type whether all the chains are RNA as in poly(U)•poly(A)•poly(U) [6] or DNA as in poly d(T)•poly d(A)• poly d(T) [10]. Neither does the presence of Hoogsteen type AU(T) pairs where the interglycosylic link distance C1...C1 is 0.2nm shorter than for a Watson-Crick pair nor the presence of IA pairs with C1...C1 0.2 nm longer preclude the chains having unexceptional A-type conformations

(Tables 1,2,3). The structure of poly(I)•poly(A)•poly(I) [5] (h = 0.329 nm, t = 30°) where there can be no pairs of the Watson-Crick kind is the most striking examples of this.

Independently Arnott et. al. [8] and Zimmerman et. al. [14] determined that the multistranded structures of poly(I) and poly(G) contained four identical, parallel polynucleotide helices rather than three as had been proposed earlier [22]. However Arnott et. al. [8] proposed that the chains had B-type conformations while Zimmerman et. al. [14] produced a model with A-type chains. There are stereochemical infelicities in the Zimmerman et. al. model but Smith and Arnott [unpublished] find that these can be remedied (Table 3) and have further concluded that in this case the x-ray data are too poor to enable a choice to be made between the competing models.

SINGLE-STRANDED POLYNUCLEOTIDE HELICES

It has long been recognized that the hydrogen bonds are not the main factors stabilizing polynucleotide secondary structures: favorable conformations at each single bond and (hydrophobic) base stacking interactions are much more important. It would not be surprising, therefore, to find stable polynucleotide structures with no hydrogen-bonded base-base interactions. So far, however, there are only two examples of such structures. The rhombohedral form of poly(C) contains six-fold poly(C) helices (h = 0.311, t = 60°) and the only interactions between the non-coaxial chains involve the free -OH groups of the ribose rings [Arnott, Chandrasekharan and Leslie (unpublished)]. Interestingly Leslie and Arnott [unpublished] observe the same molecular conformation in an orthorhombic crystal form of poly(2'-OMe-C) where there can be no -OH... -OH hydrogen bonds.

EXTERNAL FACTORS AFFECTING STRUCTURAL TRANSITIONS IN DNA AND RNA DUPLEXES

The A→B transition was first observed in fibers of typical DNAs when the ambient relative humidity was increased from 75% to 92% [38]. Later it was recognized that the presence of extraneous ions (e.g. about 3% NaCl) was also necessary and that with fibers containing twice as much salt only B-DNA would be observed [39]. The C-DNA conformation is sometimes observed under intermediate conditions of salt content and relative humidity [30,31,11]. The type of cation present can also affect the transition points so that with NaDNA A, C, and B forms are all observed but with LiDNA only C and B forms. Transitions analogous to the B→C→A structural changes in fibers apparently occur in aqueous solutions to which alcohols have been added [40,41]. We may therefore conclude that the B conformation is favored when the environment is most polar, the other conformations as it becomes less so.

By contrast RNA duplexes and DNA-RNA hybrid duplexes are relatively little affected by changes in the polarity of the environment. The minor transition A→A' can be obtained in the same circumstances which would cause a A→B or D→B transition in a DNA duplex [33].

INTERNAL FACTORS AFFECTING DNA DUPLEX STRUCTURES

Internal factors are defined as nucleotide composition and sequence, and repetitiveness of sequence. It would appear that it is nucleotide sequence rather than composition or repetitiveness of sequence which is the important factor in determining the structural behaviour. It is true that the most A,T-rich DNAs are alike in favoring B forms under all conditions, but poly d(AGT)•poly d(ACT), which is very A,T-rich behaves like poly d(G)•poly d(C) [9], whereas poly d(GC)•poly d(GC) behaves not like its composition isomer, poly d(G)•poly d(C), but like its sequence analogue, poly d(AT)•poly d(AT) [7]. Before it became clear that sequence rather than composition might be an important parameter influencing the structural behaviour two sets of experiments were carried out which suggested that genomic DNAs rich in adenine and thymine might be structurally peculiar. First there were the x-ray fiber diffraction experiments with the A,T-rich DNAs like Clostridium perfringens, Cytophaga johnsonii etc. by Bram and Tougard [42] who suggested not only that the B→A transition was difficult for such DNAs but also that the secondary structures adopted were distinct from the classical forms. Infra-red linear dichroism experiments by Brahms et. al. [26,27] strongly supported at least the first proposition. However more recent x-ray experiments by Selsing and Arnott [unpublished] with the same DNAs have produced only classical conformations. Selsing and Arnott did obtain peculiar diffraction patterns from DNAs from Cl. perfringens etc. before final purification by Sepharose chromatography. This procedure removes a teichoic acid contaminant which is common in such preparations. It should therefore be concluded that the results of Bram and Tougard [42] are examples of an external effect (from teichoic acids, etc.) on DNA rather than examples of an internal effect (from sequence). On the other hand the results of Brahms et. al. [26,27] and Selsing and Arnott are probably reconcilable with the proposition that A,T-richness predisposes a DNA against A type conformations because -- as shown in Table 6 -- such DNAs must contain the kind of doublet sequences which favour B forms.

The question of whether it is the repetitiveness of a sequence (rather than the nature of the sequence itself) which leads to non-classical behaviour has been answered negatively by x-ray studies of the satellite DNAs from Drosophila virilis, Gecarcinus lateralis and Mus musculus [by Selsing, Leslie, Arnott, Gall, Skinner, Souther and Spencer (unpublished)]. These results were in apparent discord with UV circular dichroism studies of solutions of the three satellite DNAs of D. virilis and the satellite DNA of D. melanogaster [43] if one accepts the theory of Gray and Tinoco [24] that the CD spectrum of a DNA depends on the kind and frequency of doublet sequences present. However Arnott [25] has now shown that it is possible to predict rather accurately the experimental spectra for the satellite DNAs if one assumes that it is the kind and frequency of triplet sequences which are important. It is unnecessary to postulate peculiar molecular structures for these special DNAs.

Systematic x-ray studies are now being made [by Leslie, Tsivoglou, Arnott and Ratliff] of synthetic DNAs with repeated dinucleotide and trinucleotide sequences. These studies are not complete but the results so far (Table 6) are not incompatible with the following rules. Classical DNA structural behaviour -- that is to say, assumption of the B form in more polar conditions and the A form in the less polar -- is shown by DNAs with as few as one sixth of their intra-strand stacks GG, AG, or GA. Unusual behaviour is observed when at least two thirds of the stacks are the symmetrical, base-paired doublets AT:AT or GC:GC and any PuPu stacks are AA. Then, the structure preferred in less polar situations is not A-DNA but D-DNA which belongs (see Table 1) to the B conformational genus. A third type of behaviour is unique to poly d(A)•poly d(T) and poly d(I)• poly d(C) where the form stable in less polar conditions is neither A-DNA nor D-DNA but B'-DNA [10]. [Since A,T-rich DNAs must be rich in tracts like poly d(AT)•poly d(AT) and poly d(A)•poly d(T) it is not surprising that they favour B forms and not A.]

CONCLUSIONS

1) Pure DNAs in aqueous buffers will usually have the same secondary structure, identical, or very similar, to the B-DNA conformation observed at high relative humidities in fibers containing a small percentage of trapped salts. In the same circumstances RNA duplexes and DNA-RNA hybrid duplexes will be similar to A'-RNA.

2) Under less polar conditions (when alcohols are added to aqueous solutions, or when salt-free fibers are exposed to lower relative humidities or _in vivo_ when a tract of DNA is enclosed within a relatively hydrophobic binding site) the secondary structure of a DNA may change to C-DNA then to A-DNA if a sufficient number of PuPu stacks containing G are present, or to D-DNA when at least 67% of the stacks are AT:AT or GC:GC and the PuPu stacks are AA, or to B'-DNA in poly d(A)•poly d(T). Analogous changes will not occur in RNA-containing duplexes except that there may be a relatively minor change from A'→A.

3) Although one can envision DNA having regions with different secondary structures, the evidence is in favour of it having a relatively monotonous secondary structure. A local sequence on a DNA would be able to display its special structural preference only when isolated from the effects of other sequences and when in a less polar environment such as might be provided _in vivo_ by protein binding sites.

4) The inclination of oligo d(G)•oligo d(C) tracts to be A could be enhanced and the inclination of oligo d(A)•oligo d(T) tracts to be B-like could be reversed by triple-stranded complex formation with an appropriate oligopyrmidine-(DNA or RNA)-nucleotide since all triple-stranded polynucleotides are apparently required to have A-type chains.

POSSIBLE BIOLOGICAL ROLES FOR DIFFERENT DNA SECONDARY STRUCTURES

It would be surprising if the B→A transition and the sequence-dependent transitions within the B genus were not exploited in biological systems although it is difficult to obtain direct evidence of this.

One could envisage that the property whereby oligo(purine-nucleotide)• oligo(pyrimidinenucleotide) tracts on a DNA can remain intact but form a triple-stranded complex with a second complementary oligo(pyrimidinenucleotide) might be used in two ways. Firstly, two separated tracts of this kind could be brought together through complexing with a single-stranded oligo-RNA (or DNA) of appropriate sequence. Alternatively, since an A-type conformation is obligatory with triple-stranded complexes, such RNAs or DNAs might be used to produce B→A transitions which might be propogated some considerable distance along the uncomplexed DNA duplex. This could be mechanism for readying a distant sequence for recognition if an A conformation was prerequisite.

If, as has been suggested, there is an initial, but temporary, RNA tract in DNA replication fragments, this may be because some DNA polymerases require an A-type duplex for their action and as we know an RNA chain in a Watson-Crick base-paired duplex imposes A-type geometry [36, 37]. One would expect this A-type geometry to be maintained even after DNA (rather than RNA) nucleotides were being added to the growing chain. However as the length of the new DNA duplex increased relative to the DNA-RNA tract there would be an increasing probability of the structure returning to the B form. This would cause cessation of replication, disruption of the DNA-RNA tract, thus making endonuclease activity possible which would remove the RNA and prepare the site previously occupied for ligase activity. For this B-type geometry might be necessary. Note that the length of the DNA duplex tract required to overturn the A-producing effect of the RNA initiator would be less for sequences rich in AA:TT, AT:AT and GC:GC doublets. It is interesting that the same sequences which disfavour an A conformation in a DNA duplex are the ones which can be induced to form the peculiar B-type D conformation. This structure is in many respects -- shallow major groove, narrow minor groove -- the antithesis of an A structure. It may be that sequences recognized by proteins designed to maintain or restore B conformations are of this type.

ACKNOWLEDGEMENTS

Much of the work discussed here was supported by U. S. Public Health Services Grant GM 17371. I am grateful to my colleagues Andrew Leslie, Bob Ratliff, Peter and Stephanie Campbell Smith who provided some of the unpublished data.

REFERENCES

1. S. Arnott and D.W.L. Hukins, B.B.R.C., 47 (1972) 1504.
2. S. Arnott and D.W.L. Hukins, J. Mol. Biol. 81 (1973) 93.

3. S. Arnott, D.W.L. Hukins and S. D. Dover, B.B.R.C., 48 (1972) 1392.
4. S. Arnott, D.W.L. Hukins, S. D. Dover, W. Fuller and A. R. Hodgson, J. Mol. Biol., 81 (1973) 107.
5. S. Arnott and P. J. Bond, Science, 181 (1973) 58.
6. S. Arnott and P. J. Bond, Nature New Biol., 244 (1973) 99.
7. S. Arnott, R. Chandrasekaran, D.W.L. Hukins, P.J.C. Smith and L. Watts, J. Mol. Biol., 88 (1974) 523.
8. S. Arnott, R. Chandrasekaran and C. Marttila, Biochem. J., 141 (1974) 537.
9. S. Arnott and E. Selsing, J. Mol. Biol., 88 (1974) 551.
10. S. Arnott and E. Selsing, J. Mol. Biol., 88 (1974) 509.
11. S. Arnott and E. Selsing, J. Mol. Biol., 98 (1975) 265.
12. S. Arnott, in V. G. Allfrey, E.K.F. Bautz, B. J. McCarthy, R. T. Schmike, and A. Tissierès, (editors), Organization and Expression of Chromosomes, Berlin: Dahlem Konferenzen, 1976.
13. J. D. Watson and F.H.C. Crick, Nature, 171 (1953) 737.
14. S. R. Zimmerman, G. H. Cohen and D. R. Davies, J. Mol. Biol., 92 (1975) 181.
15. D. Voet and A. Rich, Progress Nucl. Acid Res. Mol. Biol., 10 (1970) 183.
16. S. Arnott and D.W.L. Hukins, Biochem. J., 130 (1972) 453.
17. S. Arnott, S. D. Dover and A. J. Wonacott, Acta Cryst., B25 (1969) 2192.
18. F.H.C. Crick and J. D. Watson, Proc. Roy. Soc. A., 223 (1954) 80.
19. R. Langridge and P. J. Gamatos, Science, 141 (1963) 694.
20. S. Arnott, M.H.F. Wilkins, W. Fuller and R. Langridge, J. Mol. Biol., 27 (1967) 535.
21. Y. Mitsui, R. Langridge, B. E. Shortle, C. R. Cantor, R. C. Grant, M. Kodama and R. D. Wells, Nature, 228 (1970) 1166.
22. A. Rich, Biochim. Biophys. Acta, 29 (1958) 502.
23. R. Langridge and A. Rich, Nature, 198 (1963) 725.
24. D. M. Gray and I. Tinoco., Jr., Biopolymers, 9 (1970) 223.
25. S. Arnott, Nucleic Acids Research, 2 (1975) 1493.
26. J. Pilet, J. Blicharski and J. Brahms, Biochemistry, 14 (1975) 1869.
27. J. Pilet and J. Brahms, Nature New Biol., 236 (1972) 99.
28. S. Arnott and D.W.L. Hukins, Nature, 224 (1969) 886.
29. S. Arnott, R. Chandrasekaran and E. Selsing, in M. Sundaralingam and S. T. Rao (editors), Proc. Fourth Harry Steenbock Symposium, University

Park Press, Baltimore, 1975, p. 577.

30. D. A. Marvin, M. Spencer, M.H.F. Wilkins and L. D. Hamilton, J. Mol. Biol., 3 (1961) 547.

31. J. Brahms, J. Pilet, T.-T. P. Lan and L. R. Hill, Proc. Nat. Acad. Sci., 70 (1973) 3352.

32. E. Selsing, S. Arnott and R. L. Ratliff, J. Mol. Biol., 98 (1975) 243.

33. S. Arnott, W. Fuller, A. Hodgson and I. Prutton, Nature, 220 (1968) 561.

34. K. I. Tomita and A. Rich, Nature, 201 (1964) 1160.

35. T. Sato, Y. Kyogoku, S. Higuchi, Y. Mitsui, Y. Iitaka, M. Tsuboi and K. Miura, J. Mol. Biol., 16 (1966) 180.

36. G. Milman, R. Langridge, and M. J. Chamberlain, Proc. Nat. Acad. Sci. (Washington), 57 (1966) 1804.

37. E. J. O'Brien and A. W. McEwan, J. Mol. Biol., 48 (1970) 243.

38. R. E. Franklin and R. G. Gosling, Nature, 171 (1953) 740.

39. P. J. Cooper and L. D. Hamilton, J. Mol. Biol., 16 (1966) 562.

40. J. Brahms and W.F.H. Mommaerts, J. Mol. Biol., 10 (1964) 73.

41. G. Malenkov, L. Minchenkova, E. Minyat, A. Schyolkina and V. Ivanov, FEBS Letters, 51 (1975) 38.

42. S. Bram and P. Tougard, Nature New Biol., 239 (1972) 128.

43. D. M. Gray and J. G. Gall, J. Mol. Biol., 85 (1974) 665.

STRUCTURE, MORPHOLOGY AND MECHANICAL PROPERTIES OF BIOPLASTICS

P. H. Geil, S. Barenberg and W. M. Wong

Case Western Reserve University, Cleveland, Ohio

SUMMARY

The wide variation in chemical structure (side chain type, sequence, configuration, crosslinking, etc.) and conformation (back-bone and side chain, ordered and disordered) inherent in polypeptides, combined with their potential biocompatibility, has resulted in increasing research on the physical structure and properties of representative materials. In addition to summarizing structure-property relationships for homopolymer polypeptides, using poly-γ-benzyl-L-glutamate as an example, primary consideration in this paper is given to random and block copolymers of this peptide with e.g. leucine and butadiene. Conventional methods of polypeptide and solid state polymer analysis have been used to characterize the secondary structure (chain conformation), tertiary structure (chain association) and bulk properties. The evidence indicates that both types of block copolymers form phase separated domain morphologies. Morphological and attendent property changes could be affected by modifying the method of film preparation. Emphasis is directed toward the need for adequate characterization and to the potential for control of physical properties in these systems by control of the structure and morphology.

INTRODUCTION

Over and above their use as models for biological macromolecules, more and more attention is being paid to potential practical uses of polypeptides as materials, both inside and out of the body, as biomaterials and as plastics. As materials we need to go beyond the primary structure and characterize the relationship of secondary and tertiary structure to their properties, physical and biological. It is believed that polypeptides have an even greater potential for controlled variation in properties than do normal synthetic polymers because of the diversity of conformations that are readily obtainable. Although research on polypeptides as materials is just beginning, the complexity is rapidly increasing as not only homopolymers, but also random and block copolymers, are both becoming available and are being shown to be of useful potential.

At the present time detailed conformation-morphology-property relationships are only just beginning to be worked out, with all of the problems of reproducibility (purity, molecular weight, etc.) and thus it is not feasible to fully review reported results. As will be shown, even for poly-γ-benzyl-L-glutamate (PBLG), which has probably been most studied, there is a lack of satisfactory, useable characterization of the type needed for material's design. Rather, in this paper we wish to briefly indicate the type of information obtainable and the value of obtaining it, using synthetic polymers as an example, and then consider the present status of the comparable information for polypeptides. For this purpose we will restrict ourselves to benzyl-L-glutamate based polymers.

The types of secondary structure or conformation possible in polypeptides are well known (α-helix, β sheets of various types, "distorted" helices, random coil, etc.). The tertiary structures possible and their role in determining properties are much less well understood. In fact, even for normal synthetic polymers (where typically one is concerned with only one conformation in the crystal, compared with a random conformation in amorphous regions, and the properties depend to a first approximation on the relative amounts of the two conformations (the percent crystallinity)) control of the morphology for desired properties is only now being obtained. For instance, polyethylene fibers can now be obtained with moduli varying between that of a rubber and steel wire, for the same degree of molecular orientation as determined by x-ray diffraction and degree of crystallinity; the difference is dependent on the morphology. Thus, we will briefly consider here first appropriate characterization techniques and results for synthetic homopolymers, their application to the homopolymer PBLG, followed by similar considerations for synthetic copolymers and corresponding results for benzyl-L-glutamate based copolymers. Most of the results for the copolymers will be drawn from research at Case Western Reserve University.

For synthetic polymers with a regular or tactic conformation, the basic morphological structure that develops during solidification from a random conformation in solution or the melt is the folded chain, lamellar crystal, a platelet on the order of 100 Å thickness in which the molecular axes are oriented more or less normal to the lamellar surface; the fold period is related to the lamellar thickness and is a function of crystallization temperature (Fig. 1) [1]. For some polymers, in particular polyethylene, considerable detail is known concerning the internal structure, growth kinetics, melting and annealing behavior, and physical properties of the individual crystal, as formed from dilute solution and of their aggregates (termed spherulites, Fig. 2) as usually formed from concentrated solution and the melt [1,2]. However, although similar lamellar crystals have been grown from dilute solution for a number of polypeptides, particularly by Keith and co-workers [3], and spherulitic structures are also obtainable from concentrated solutions (e.g. polyalanine, Fig. 3) essentially no similar detail is known. In fact, for no polypeptide has even the thickness of the crystal as a function of crystallization temperature nor the arrangement of the lamellae (if present) in the spherulites been reported.

In addition to the lamellae, which in some polymer-solvent systems are lath-like or ribbon-like but still have a folded chain structure, fibrillar structures have been observed during crystallization of some synthetic polymers from solution (Fig. 4). Although possibly directly related to the morphology of many polypeptides which also yield fibrillar structures when precipitated from solution, unfortunately there is as yet no description of molecular arrangement or effect of growth conditions.

Both the low and high strain properties of the bulk material, i.e. the modulus or stiffness, the yield and tensile strength, and the extensibility, are strongly influenced by this tertiary structure. For many practical purposes advantage is taken of the anisotropic bonding forces inherent in a polymer chain; the molecules are oriented, usually by stretching an already solidified polymer, so that the molecular axes are aligned to support the applied stresses (e.g. uniaxially in fibers, biaxially in films and hoop-wise in pipes). However, even in uniaxially oriented fibers or films the resultant properties depend strongly on the morphology; as indicated above, for the same high density polyethylene resin the properties can vary from an elastic material of relatively low modulus [6], to a typical fiber with a modulus of $5GN/m^2$ to recently developed high modulus fibers [7] which on a weight basis are stronger than steel and are approaching (ca. 1/3) the theoretical strength of the covalent band; potentially we should be able to approach this value even closer in the near future when the proper tertiary

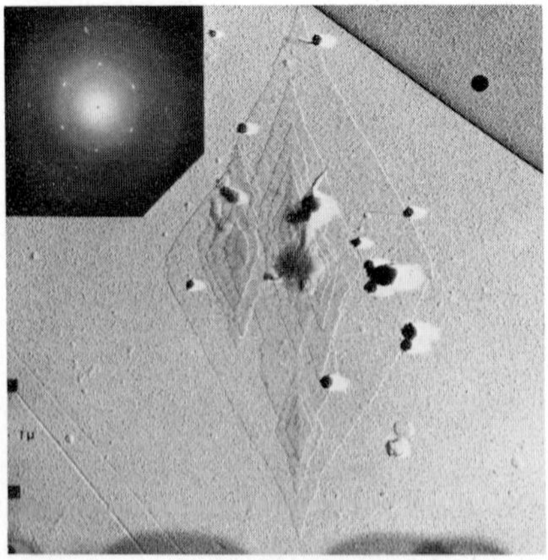

Figure 1. Single crystal of nylon 6 crystallized from dilute (0.005% wt/v) glycerine solution by slow cooling. An electron diffraction pattern from a similar crystal (inset) indicates the molecular axes are normal to the plane of the platelet. Several spiral growths can be seen on the crystal [1].

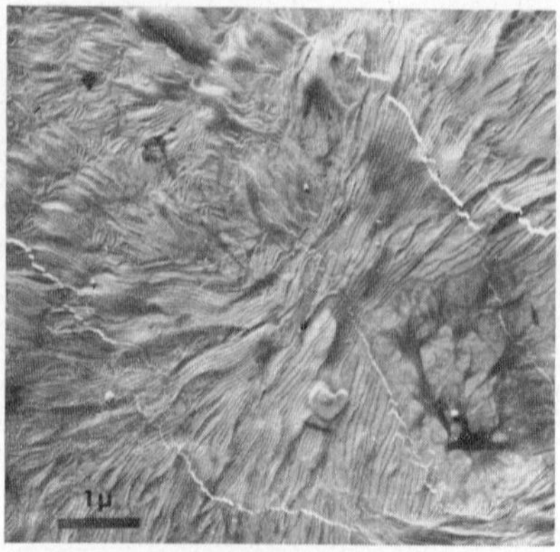

Figure 2. Spherulites of low density (ca. 50% crystalline) polyethylene; a) observed between crossed polaroids, b) replica observed in the electron microscope [1].

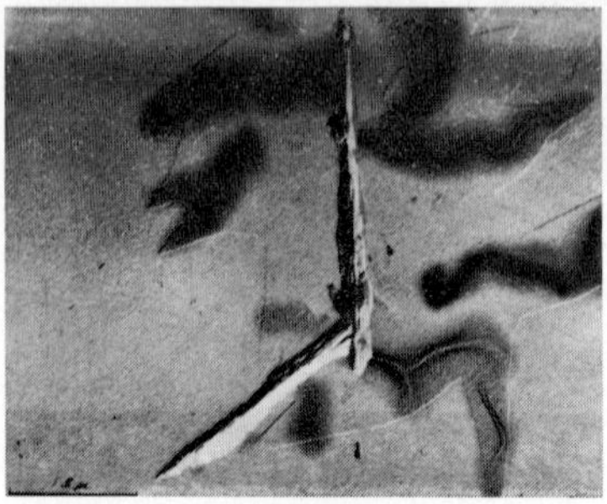

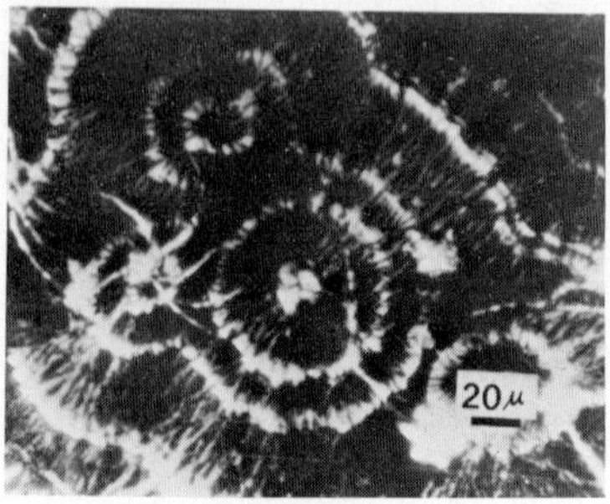

Figure 3. Polyalanine crystallized from dilute solution in a) hexafluoroacetone and b) trifluoroacetic acid-chloroform (1:99) by solvent evaporation [4].

structure is attained. As suggested by Figure 5, the typical polymer fiber we are all familiar with consists of a stack of lamellar crystals, interconnected by numerous (but yet uncountable) "tie" molecules. Needed for maximum modulus and tensile strength would be one in which linear molecules are both fully extended and staggered so that chain end defects are minimized, i.e. a structure basically reminiscent of that which is proposed for cellulose and collagen.

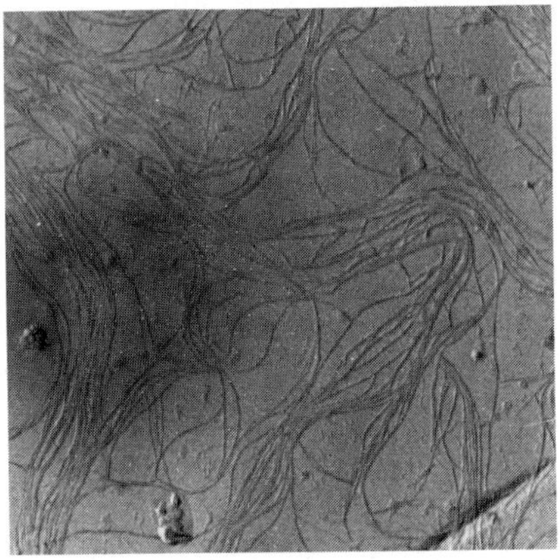

Figure 4. Fibrils of nylon 6 precipitated from glycerine solution only after storage at room temperature for 2 days. Single crystals similar to those in Figure 1 are also seen in this preparation, presumably forming during cooling to room temperature [5].

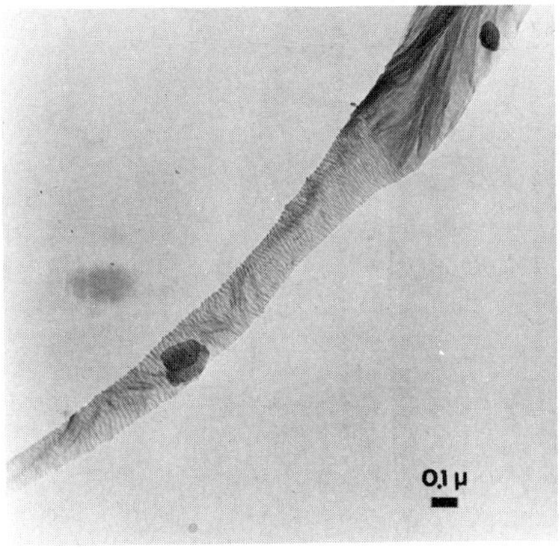

Figure 5. Defocus electron micrograph of a drawn thin film of high density polyethylene. The striations in the drawn region, to the left, correspond to lamellae, the contrast arising from phase effects due to the difference in mass thickness in the alternating crystalline and amorphous regions. The film to the right consists of lamellae more or less parallel to the film surface, in the undrawn portion of the original spherulite [8].

The low strain-properties of synthetic polymers are often characterized by dynamic mechanical spectroscopy, i.e. the determination of the real and imaginary (loss) parts of the modulus, as a function of temperature. Relaxation or decrease in the modulus, accompanied by mechanical loss in the sample, occurs in temperature ranges where particular types of molecular motion become activated at a frequency corresponding to that of the exciting mechanical force [9]. Significant changes in the magnitude of the loss are seen with changes in tertiary structure.

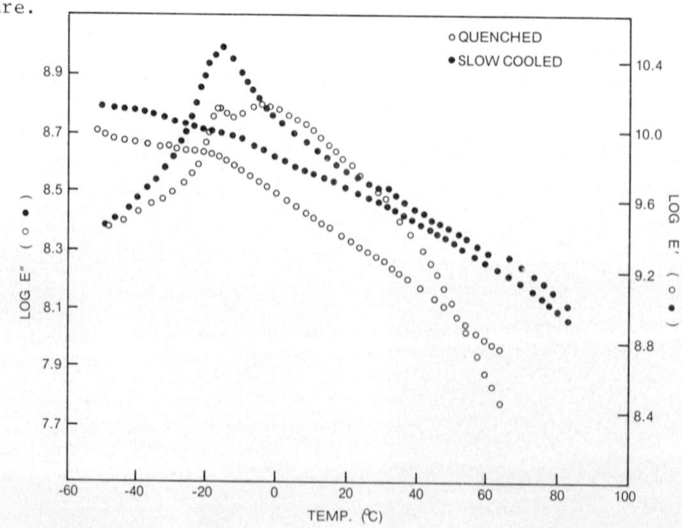

Figure 6. Dynamic mechanical spectra for low density polyethylene. In the slow cooled sample the α relaxation is at about 100°C and the β at -20°C. Plotted are the real (E') and imaginary (E", loss) parts of the modulus and tan δ = E"/E' [8].

For instance, in Figure 6, the magnitude of the α relaxation in low density polyethylene, attributed to the onset of segmental rotation and translation in the crystals and of the β relaxation, attributed to motion of segments in the disordered regions, are seen to depend on inverse manner on the relative amount of order, i.e. crystallinity; furthermore, a decrease in stiffness is seen to occur in the temperature region of the relaxations. This technique, as well as x-ray, infrared spectroscopy and electron microscopy, turns out to be of particular value in characterizing the tertiary structure of the copolymers; the temperature positions of the relaxations depend on the degree of intermixing of the 2 components.

For PBLG and a number of other polypeptides, the situation is perhaps even more complicated. Although single crystals can be grown for PBLG, (see Figure 15) this does not seem to be its preferred form of precipitation from solution. Instead fibrils of various size, shape and, apparently, molecular arrangement are formed [10,11]. Representative examples, as a function of solvent, are shown in Figure 7. As we will see, this fibril morphology is carried through to the random

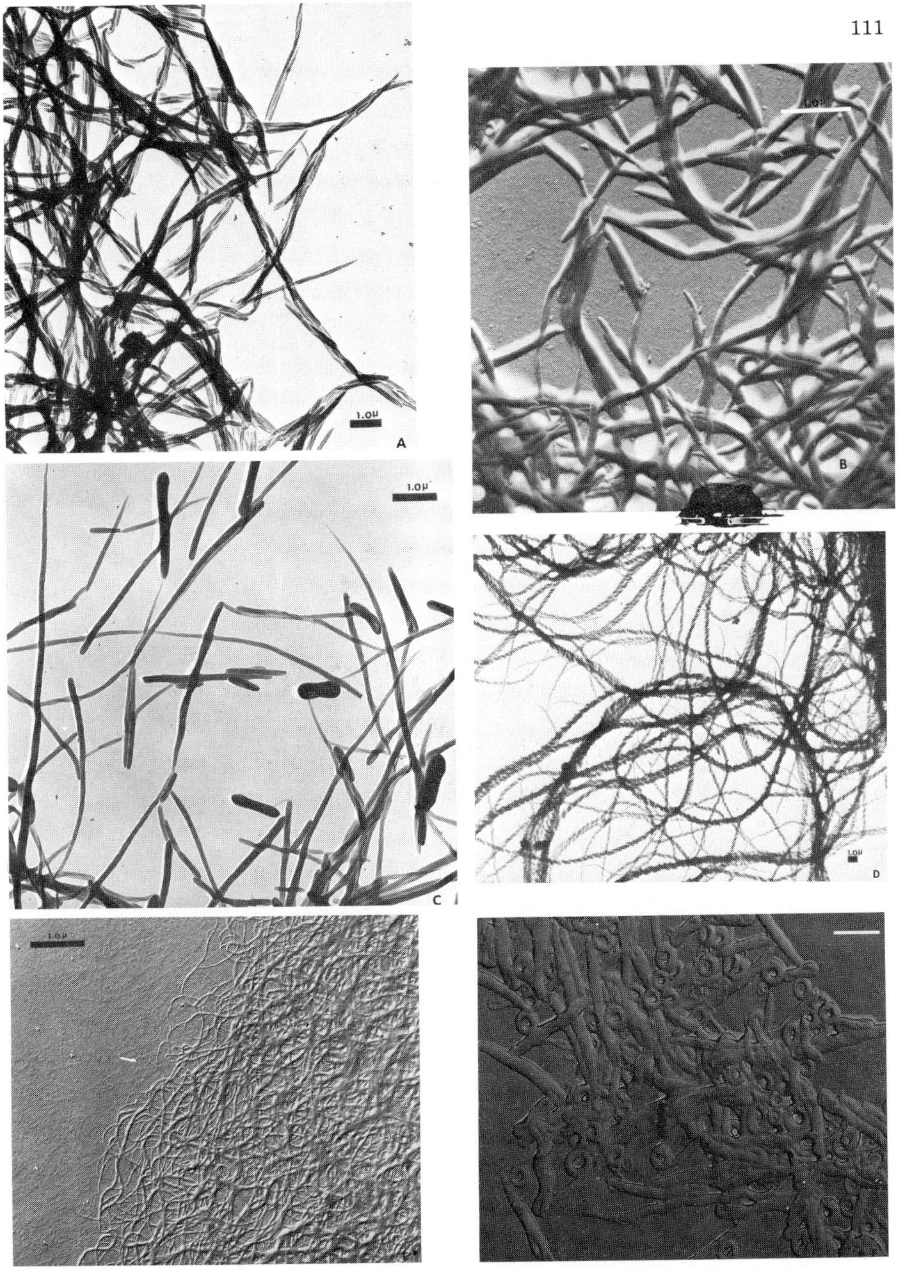

Figure 7. PBLG precipitated from 7.10^{-4} g/ml butoxyethanol-trimethyl benzene (BOE-TMB) solutions at 70°C: a) BOE, b) BOE:TMB-4:1, c) BOE:TMB-2:1, d) BEO:TMB-1:1, e) BOE:TMB-1:4, f) TMB [10].

and block copolymers. Although electron diffraction shows parallel alignment of α-helical molecular axes with the fibril axes [11], details of the molecular packing and arrangement in even the PBLG fibrils are still essentially unknown with reports from our own laboratory suggesting a basic 50–70 Å diameter fibril in some cases [10] but not in others [11].

Over the years, there have been numerous publications describing x-ray diffraction from PBLG, as well as from mixtures of PBLG & PBGD. A number of the results for PBLG are tabulated in Table I. Unfortunately, it is extremely difficult to compare the results; wide variations in spacings are reported with variations in solvent and precipitation conditions. Results of Tobolsky and coworkers probably are most representative of the range of variation possible and of reasonable classification of "crystal" forms [12,13,16]. They have suggested the use of the spacing of the innermost reflection to characterize 3 different molecular packings, A - 14.9 A, B - 14.5 A and C - 13.2 A. It is obvious intermediate forms are also possible and, to date, in none of even the "standard" forms is, for instance, the arrangement of the side chains fully clarified. Information of this type, however, is what is needed to correlate with the dynamic mechanical spectroscopy-solvent relationships that have been observed. X-ray diffractometer scans for films cast from the 3 solvents used for most of the results to be described below are shown in Figure 8. That cast from chloroform corresponds to Tobolsky's form C, a relatively crystalline form, whereas those from dioxane and benzene are form A, one which is relatively poorly ordered.

Typical dynamic mechanical spectra for PBLG, for temperatures above room temperature, are shown in Figure 9 [22]. There is an obvious change in the relative height of the two peaks, the high T or α peak decreasing with increasing crystallinity, while the β peak increases [20, 22]. This is in agreement with the suggested assignment of the α peak to the motion of disordered backbone segments [23] and the β process to disordering or melting of the side chains. As the crystallinity decreases there will be fewer ordered side chains to "melt", they, presumably, already being mobile at this temperature in disordered regions of the solid. However, it is noted, this does not agree with DSC [24] and dielectric [25] measurements which have been interpreted as suggesting the β relaxation resembles a glass-rubber transition, i.e. the onset of motion of already disordered side chains. Needed are

Table I

TECHNIQUE	X-RAY	X-RAY	X-RAY	E.D.	X-RAY	X-RAY	X-RAY	X-RAY	X-RAY	X-RAY	X-RAY	X-RAY	X-RAY	E.D.
SOLVENT	DMF	DMF	CHCl 3	CHCl 3	CHCl 3	CHCl 3	DMF	Ø	DIOXAN	EDC	EDC	DMF	CHCl 3	HFIP
REFERENCE	12	12	13	14	15	16	17	18	19	18	20	18	20	21
FORM	A	B	C	C	C									
INTERCHAIN SPACINGS												15.3		
	14.9	14.5						14.9				14.7		
			13.2							13.4		13.3	13.26	13.6
	12.8			12.62	12.59			12.8	12,7		12.8			
		11.9												
						10.35	10.3							
	9.7						9.6	9.7				9.8		
	8.8							8.8						
		7.79	7.58	7.28	7.35				7.4					
		7.27												
		6.50	6.68	6.40	6.29				6.4					
	5.05	5.37	5.26			5.2	5.06	5.15				5.15		5.35
							5.18		4.9					

E.D. = Electron Diffraction
DMF = Dimethylformamide
Ø = Benzene
CHCl 3 = Chloroform
HFIP = Hexafluoroisopropanol
EDC = Ethylenedichloride

Table I. Summary of x-ray diffraction results for PBLG [21].

x-ray (and possibly electron diffraction from single crystal) studies of the molecular packing as a function of temperature as well as solvent, etc. Interpretation of these peaks is considered further below in connection with the various copolymers. Regardless of the interpretation, however, we point out the potential of varying properties by control of the secondary and tertiary structure.

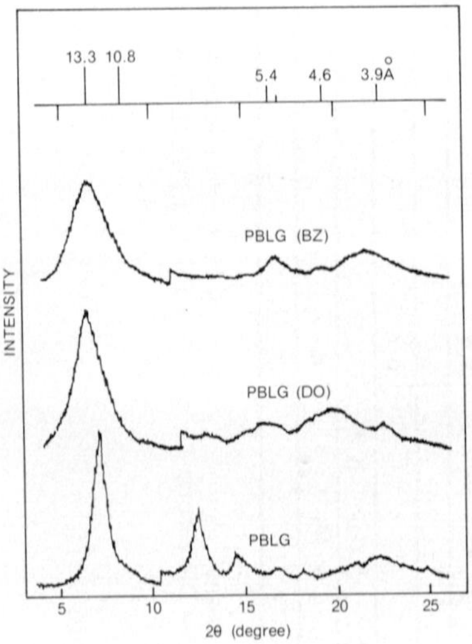

Figure 8. X-ray diffractometer scans for PBLG cast from chloroform, dioxane and benzene [data from reference 12].

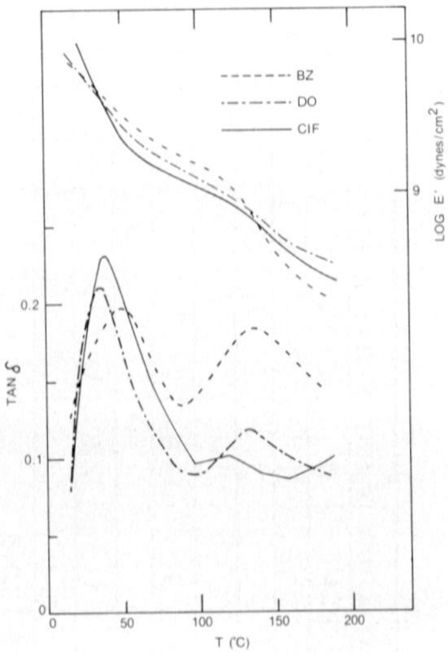

Figure 9. Dynamic mechanical spectra (modulus, E' and tan δ, E"/E1) for PBLG cast from chloroform, dioxane and benzene [data from reference 12].

In contrast to random copolymers of normal synthetic polymers, which are usually wholly amorphous or disordered, random copolymers of 2 or more amino acids frequently are ordered. X-ray patterns from copolymers of varying relative amounts of benzyl-L-glutamate (G) and leucine (L), are shown in Figure 10. Although in all of these polymers there still remains the fundamental question of the meaning of the term crystallinity, the results indicate substantial portions of the molecules have an α-helical conformation and have reasonable lateral regularity, the amount decreasing with increasing leucine content. The α-helical conformation has been confirmed by infrared (IR) and circular dichroism (CD) measurements [26,27]. If the x-ray data is interpreted in terms of crystallinity, the crystallinity is decreasing with increasing leucine content. In G-valine (V), random copolymers, however, IR and CD indicate the

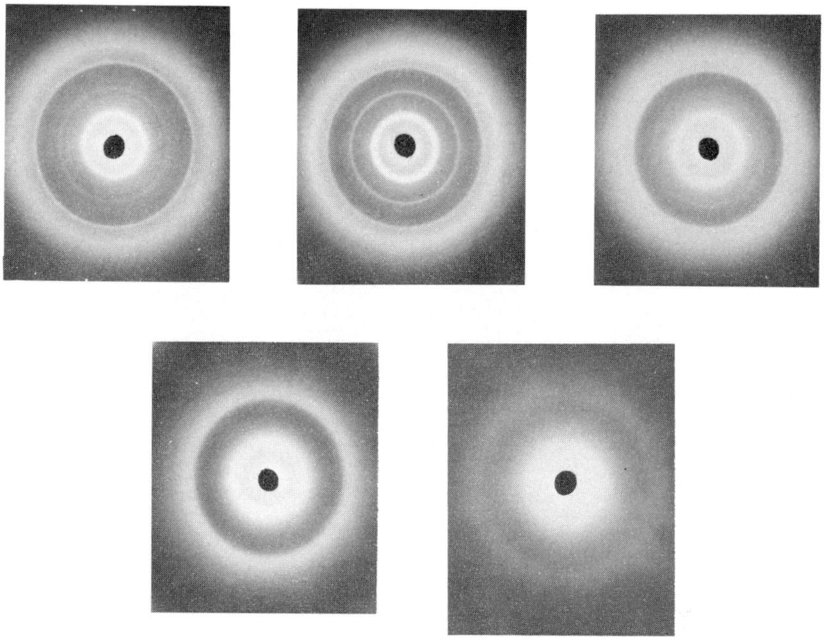

Figure 10. X-ray diffraction patterns of PBLG-PL random copolymers cast from dioxane: a) PBLG, b)GL-41, c) GL-11, d) GL-14 and e) PL [26].

presence of both α and β conformations, the β increasing with increasing valine content (Figure 11) [27]. However, although of fundamental interest, nothing is known as to the minimum segment length required for either the α or β conformations, although the short segments of valine in these random copolymers do result in a slightly higher amide I frequency than in pure or block valine polypeptides, in agreement with its position in globular proteins [27]. Electron micrographs and diffraction patterns of random GL copolymers cast from solution are shown in Figure 12. As in the case of the homopolymer, PBLG, fibrils are formed, with the diffraction patterns indicating the molecules are parallel to the fibril axes.

Dynamic mechanical spectra for GL random copolymers are shown in Figure 13. One notes that the temperature of the 50°C (B) PBLG relaxation is nearly independent of the amount of L copolymerized, but that its intensity decreases (in a linear fashion [28]) with decreasing G. As indicated above, this relaxation is attributed to side chain motion of the G, the results indicating this motion is not affected by the random inclusion of a comonomer. On the other hand, the PBLG

α relaxations were greatly shifted in temperature, even for the copolymer containing the smallest proportion (50%) of L, with the data suggesting an even larger shift in the other copolymers.

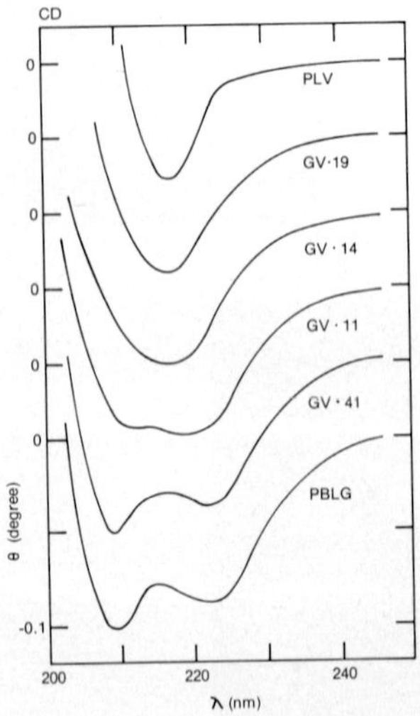

Figure 11. Circular dichroism spectra for random GV copolymers as a function of composition. The films were cast from trifluoroacetic acid on a thin film of polyethylene [27].

This is in agreement with the suggestion this peak is due to motion of small segments of the backbone [23], the added L tending to stabilize the α-helix [28]. Considering the γ (120°K) relaxation in PL, one notes that a small amount of G results in a drastic reduction in peak height, much larger than the corresponding effect on the β PBLG peak. Hiltner et al [28] suggest this is due to a coupling effect between neighboring side chains; i.e. that the relaxation involves, as in the β PBLG peak, side chain motion, but that motion of any one side chain is coupled to that of its neighbors. The added G side chains, being bulkier than the L side chains, would interfere with this coupling.

Because of the possibility of greater variation in mechanical properties for the same relative peptide concentration, block copolymers in general would be expected to have greater potential for practical use than random copolymers, although each may have unique properties. This potential, as well as examples of expected morphology, can be exemplified by styrene (S), isoprene (I), or styrene-butadiene (B) block copolymers [29]. For EM observations, the B segments can be stained with OsO_4. Solvents can be chosen which are preferential for one or the other component, or neither, the resulting morphology and properties depending on the solvent and relative block lengths. Examples for SIS cast from a mutual solvent are shown in Figure 14, the morphology varying from spheres of S in a continuous matrix of B, to rods of S, followed by alternating lamellae of S and B as the S content increases toward 50%, followed by a reversal as the S content increases still further.

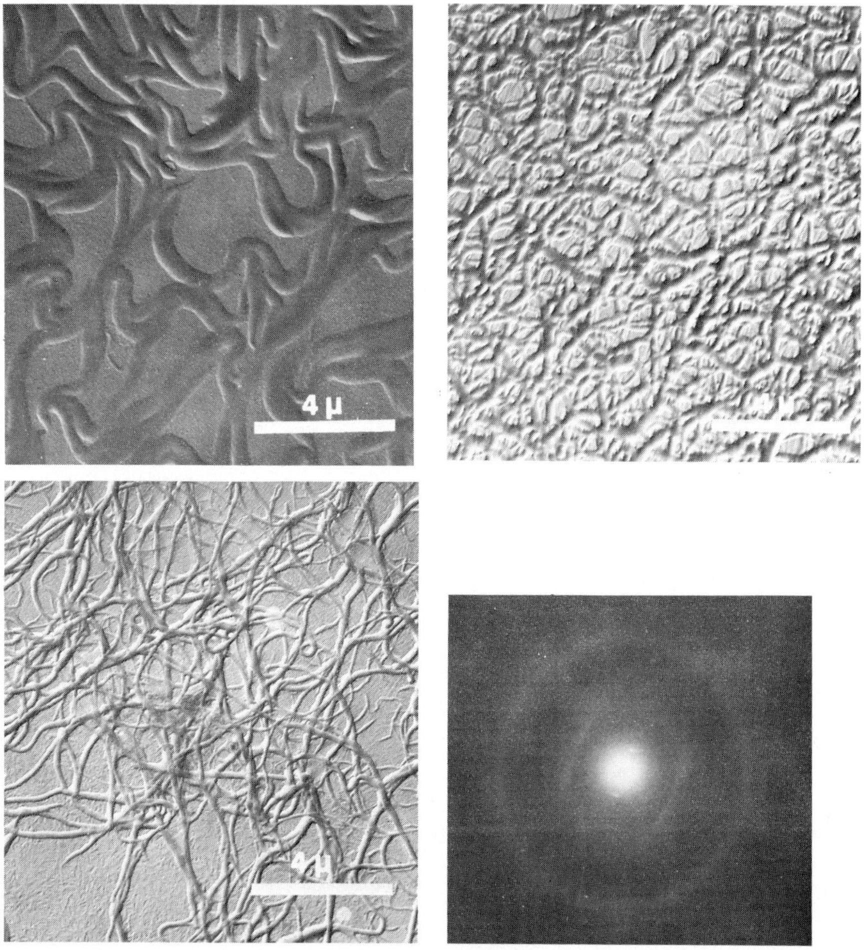

Figure 12. Electron micrographs of random GL (a. GL 41, b. GL 11, c. GL 14) copolymers cast from dioxane. Also included is an electron diffraction pattern from the GL-41 copolymer [26].

For a given polymer and a change in solvent, the morphology is shifted toward the form of greater concentration of the comonomer for which the solvent is preferential. Some variation in morphology is also produced for the same composition depending on whether one has an AB diblock, or

ABA or BAB triblock copolymers. With further division into smaller blocks, domains are still formed but the morphology becomes less well defined.

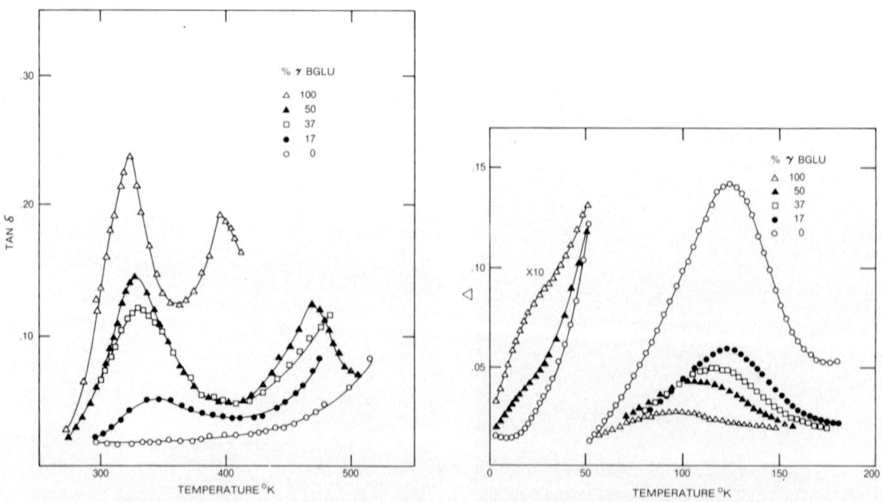

Figure 13. Dynamic mechanical spectra for GL random copolymers measured at 110 ps [28].

Of concern in all the materials, and still not thoroughly characterized, is the degree of intermixing at the domain boundaries; it would be expected to depend on the distribution in segment lengths, solidification conditions and subsequent annealing treatments. Corresponding variations in mechanical properties, for the same overall composition, as well as for variations in composition, are tabulated in Table II [31].

Recently we have completed an extensive study of benzyl glutamate-butadiene/acrylonitrile-benzyl glutamate (GBG) block copolymer [21], one which indicates similar features occur in these hybrid polymers, as well as related work on a number of G-L and G-V block polymers [22,26,27]. The latter work is still in progress and will be described in a poster paper at this meeting and is summarized later. Unfortunately, an electron microscope characterization of the domain structure, shown to be present by other techniques, has not yet been possible, preventing development of a full morphology-property correlation. It is the GBG polymer, instead, that we consider in somewhat more detail since not only have such correlations been determined but the material also shows potential as a biomaterial.

The polymerization of the GBG copolymer is described elsewhere [21].

TABLE II

Samples	ST/BD mole ratios (charged)	Elongn. to break	Light*** transmitt.	Impact strength	Heat distort. temp.****
S	100/0	7%	90%	2.1*	97°C
SB	60/40	2	62	2.5*	83
SBS	60/40	70	71	4.4*	67
BSB	60/40	2	51	2.4*	78
SBSB	60/40	23	70	3.1*	55
SBS—1	80/20	12	75	0.5**	100
SBS—2	70/30	25	81	0.4**	87
SBS—3	60/40	49	72	0.8**	66
SBS—4	50/50	17	68	>2.9**	54
SBS—5	40/60	>300	66	>2.9**	5
Commercial HI-PS	—	30	0	7.0*	75
Polyblend	60/40	2	0	0.5**	—

* Charpy impact strength (kg·cm/cm^2)

** Dynstatt impact value (kg·cm)

*** Average values of the light transmittances of 1 mm thick plates measured at the wave lengths of 400, 500, 600 and 700 mμ.

**** Clash & Berg method (JIS K-6745)

Table II. Physical properties of S-B block copolymers [31].

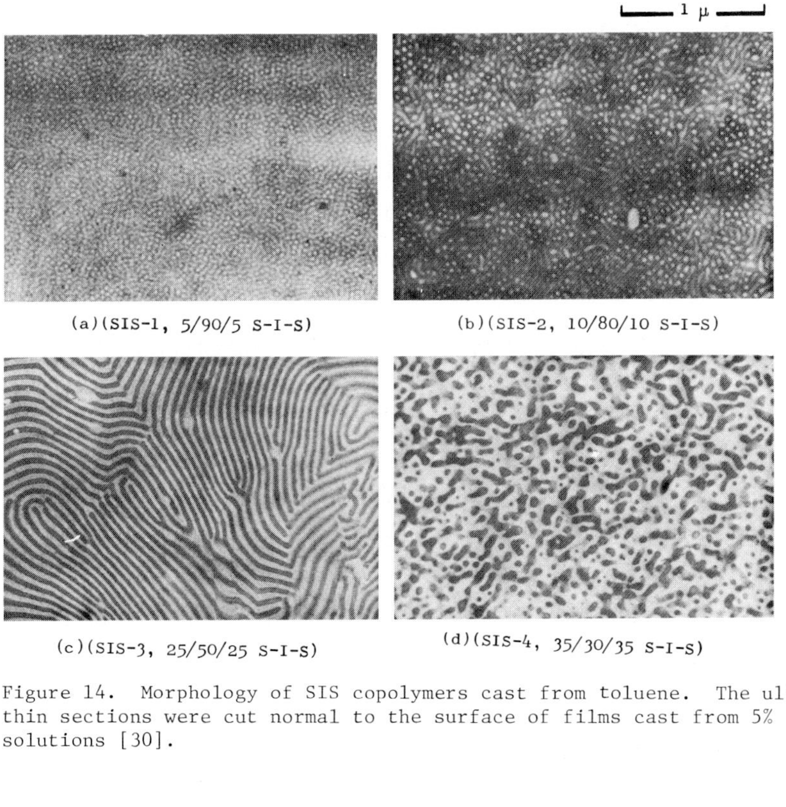

(a)(SIS-1, 5/90/5 S-I-S) (b)(SIS-2, 10/80/10 S-I-S)

(c)(SIS-3, 25/50/25 S-I-S) (d)(SIS-4, 35/30/35 S-I-S)

Figure 14. Morphology of SIS copolymers cast from toluene. The ultra-thin sections were cut normal to the surface of films cast from 5% solutions [30].

The center block, a primary amine capped random butadiene-acrylonitrile (83%-17%) copolymer (ATBN) with a molecular weight of 3400 (corresponding to an extended length of 280 A) was supplied by Dr. K. Riew of the B. F. Goodrich Co. Two block polymers were prepared, with PBLG α-helix lengths of ca 120 A (I) and 240 A (II). As solvents, dioxane and chloroform were used primarily, the dioxane being preferential for PBLG and chloroform being essentially non-preferential. Morphological studies were also carried out using xylene, slightly preferential for the B segments, but property studies were limited since it was found the morphology depended on the time the solution was stored at room temperature prior to casting of the samples.

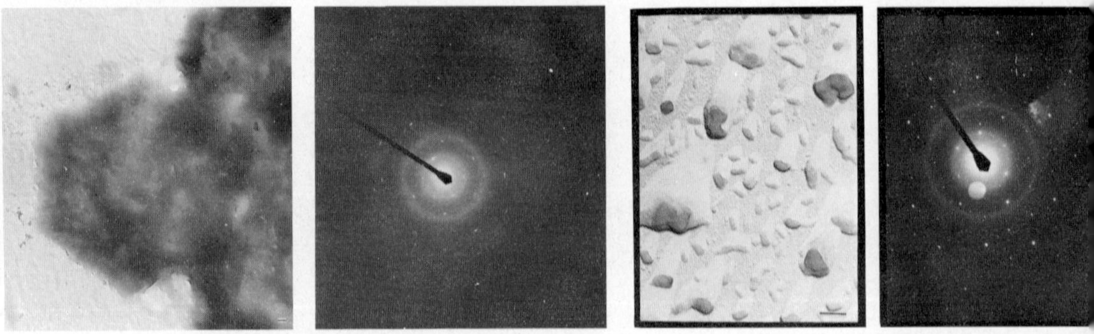

Figure 15. Single crystals and corresponding electron diffraction patterns for a, b, PBLG and c, d, GLG-II copolymer precipitated from hexafluoroisopropanol [21].

Before considering the normal mode of precipitation, we ask first whether it is possible to form lamellar crystals from a polymer such as this? Figure 15 shows crystals and corresponding electron diffraction patterns of pure PBLG and of copolymer II, the one with the longer PBLG segments, precipitated from hexafluoroisopropanol. The diffraction pattern indicates the G segments, only, have crystallized, in an α-helical conformation. Based on results for block copolymers of styrene and ethylene oxide [32] it is presumed the B segments are located on the surface of the GBG crystals. Unfortunately the GBG crystals are irregular enough that it was not possible to measure their thickness to determine if the G segments are folded; they are folded in the homopolymer PBLG crystals.

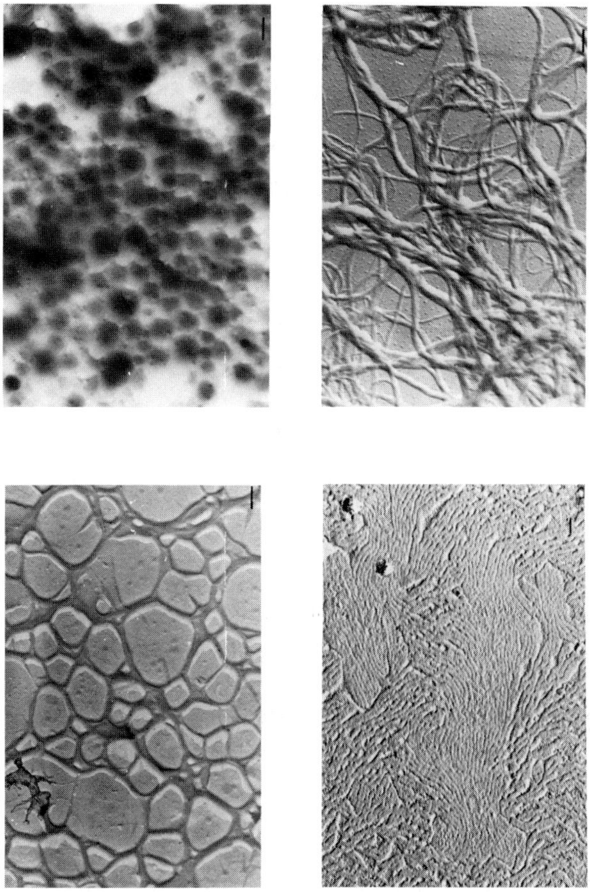

Figure 16. Thin films of ATBN, PBLG, a 50-50 mixture of ATBN and PBLG and the GBG-II copolymer cast from xylene and shadowed [21].

Figure 16 shows samples of ATBN, PBLG, a solution blend of the two homopolymers and of GBG-II cast from xylene. The micrograph of the solution blend is interpreted as indicating that phase separation has occured on precipitation, the PBLG precipitating as fibers and being coated with an amorphous layer of ATBN. In the copolymer, however, a coherent film is formed, with a large scale grain structure, each grain containing a series of nearly parallel striations.

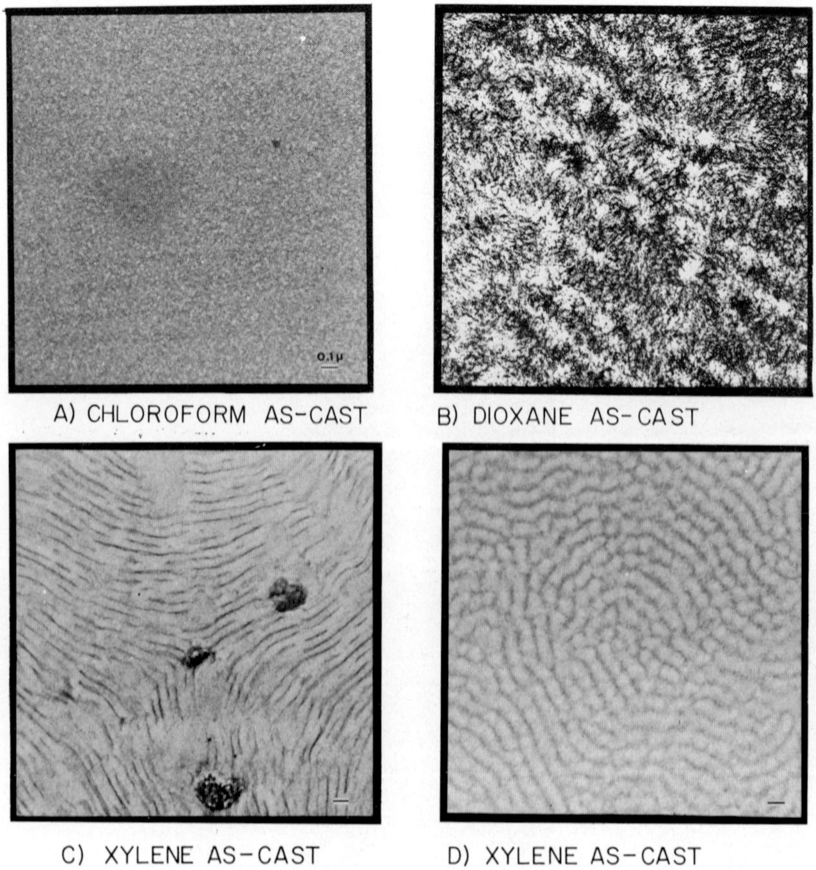

Figure 17. OsO_4 stained films of GBG-II cast from, a) chloroform, b) dioxane and c, d xylene. The film in d was cast after the solution had remained at room temperature for 2 days whereas that in c was cast immediatley after preparation [21].

One of the reasons for preparing the GBG polymers was that it would be possible to stain the B segments with OsO_4. Films cast from 0.1% (w/v) solution of copolymer II are shown in Figure 17. As in the case of the SBS copolymers, there is an obvious difference in morphology with variation in solvent. The non-preferential chloroform yielded a pseudo-homogeneous structure, the ATBN domains being on the order 150 Å in diameter in both copolymers.

On the other hand, the film cast from a fresh xylene solution consists of grains of alternating lamellae of B and G segments, the lamellae being oriented normal to the film surface and giving rise to the striations seen on the shadowed film. Obviously phase separation has occured. Again the B domains were 150 Å thick, i.e. about 1/2 the length of the extended chain segment, whereas the G lamellae were 425 Å thick. When cast from a solution stored at room temperature for two days, a lamellar structure is still seen, but the domain boundaries are less clearly defined. Related to this is an increase in their thickness, the overall periodicity now being about 1000 Å. The B lamellae are equal in thickness to their extended chain length and the G lamellae are about 750 A thick, i.e. about 3 times their α-helical length. Further study of this storage time effect is needed, related observations also having been made with nylon 6 solutions [5].

The morphology of the dioxane cast samples is intermediate between the other two; although difficult to measure the ATBN domains were about 150 A thick while the BLG layers were ca 300 and 500 A thick, respectively, for copolymers I and II. It is noted that in both the xylene and dioxane cast samples, not only must the BLG segments be fully extended helices, packed end to end, but probably are extended or deformed over and beyond the α-helical conformation.

TABLE III

WIDE ANGLE X-RAY*

PBLG	I	I	II	II	II	ATBN**
DIOXANE	$CHCL_3$	DIOXANE	$CHCL_3$	DIOXANE	XYLENE	NASCENT
13.6A°	13.6 A°	13.6 A°	14.0 A°	14.0 A°	14.9 A°	20.9 - 8.0 A°
10.3	10.3	10.3	10.3	10.3		
9.3		8.6			9.3	
7.9	7.8	7.8	7.9	7.9		
5.2	5.4	5.4	5.5	5.5	5.2	
4.5	4.6	4.6	4.7	4.7	4.5	

* Calibrated with CaF

** Amorphous scattering range

Table III. X-ray diffraction data for GBG copolymers [21].

X-ray diffraction data from thick films cast from 1% (wt/v) are tabulated in Table III. As can be seen, the innermost reflection of the dioxane and chloroform cast samples does not agree with any of the "standard" forms defined by McKinnon and Tobolsky [12,13], (form A - 14.9 Å, form B - 14.5 Å and form C - 13.2 Å), lying in between forms B and C whereas the G segments in the xylene cast sample are clearly in form A. We also note the presence of a weak 10.3 A reflection in the dioxane and chloroform cast samples, a spacing that has been attributed to, for instance, side chain distortion [13], a particular stacking of the side chains [33] or the presence of a 7_2 rather than 18_5 helix [16].

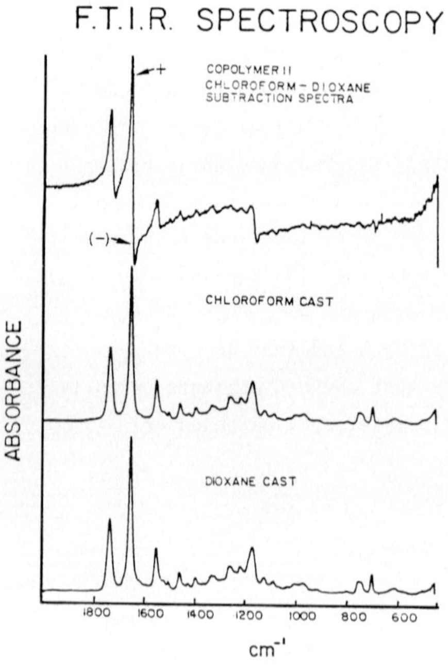

Figure 18. FT IR spectra of GBG-II cast from dioxane and chloroform and the difference spectra [21].

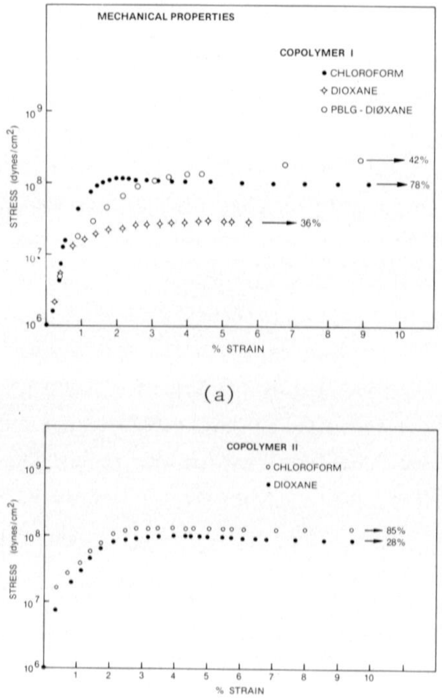

(b)

Figure 19. Stress strain curves for, a) PBLG and GBG-I and b) GBG-II [21].

Infrared spectroscopy indicated the G segments were in an α-helical form in all of the samples, with no unique bands seen for the copolymers. However, using a Fourier Transform spectrometer, subtraction of the dioxane cast spectra from the chloroform cast spectra (Figure 18) indicated a band shift in the 1650 cm^{-1} amide I region which, it was suggested, was due to more of the G segments being disordered (random coil, disordered α-helix, 7_2 helix, etc.) in the chloroform than in the dioxane cast sample.

Consider now the mechanical properties of the films. Stress-strain curves are shown in Figure 19, with the data summarized in Table IV.

TABLE IV

MATERIAL SOLVENT	MODULUS dynes/cm^2	% STRAIN at break	U.T.S.* dynes/cm^2
COPOLYMER I CHLOROFORM	3.22×10^9	78.0	9.12×10^7
COPOLYMER I DIOXANE	1.98×10^9	36.0	4.87×10^7
COPOLYMER II CHLOROFORM	5.13×10^9	85.0	1.46×10^8
COPOLYMER II DIOXANE	5.25×10^9	28.0	1.0×10^8
COPOLYMER II XYLENE	6.42×10^9	0.8	1.72×10^8
PBLG (69)	4.34×10^9	42.0	1.38×10^8

* U.T.S. ULTIMATE TENSILE STRENGTH

Table IV. Stress-strain data for GBG copolymers [21].

The modulus is seen to depend on molecular weight, being lower than homopolymer PBLG for copolymer I, having the largest relative amount of low modulus ATBN, and higher than PBLG for copolymer II, with the latter effect probably being due to the morphology and conditions of measurement used for the PBLG. The data suggests, in fact, that the PBLG is the continuous phase in all of the samples, supporting the initial stress.

The elongation or strain at break, on the other hand, is clearly dependent on the morphology, as is, to a lesser extent (depending also on molecular weight) the ultimate tensile strength, the stress at break. It is in this area, control of the mechanical properties by control of the morphology, that block copolymers have the greatest potential.

The appropriate dynamic mechanical spectra are shown in Figure 20. In addition to the peak at ca 300°K, PBLG cast from dioxane has a small relaxation just below 100°K and, as shown in Figure 9, higher temperature relaxations. The 100°K peak has been attributed to the onset of phenyl group motion in side chains [28] while the 300°K peak, as previously discussed, is attributed to large scale motion of the entire side chain (presumably initially disordered). The ATBN, on the other hand, has a broad peak centered about 130°K, a small peak at 185°K and a glass transition (Tg) peak at 220°k. The 130°K and 185°K peaks are attributed to small segment motion in trans and cis regions of the chain, respectively [34,35]. It is only the Tg peak that is seen in the copolymers, lying at a somewhat lower temperature in the chloroform cast sample than in the dioxane cast one.

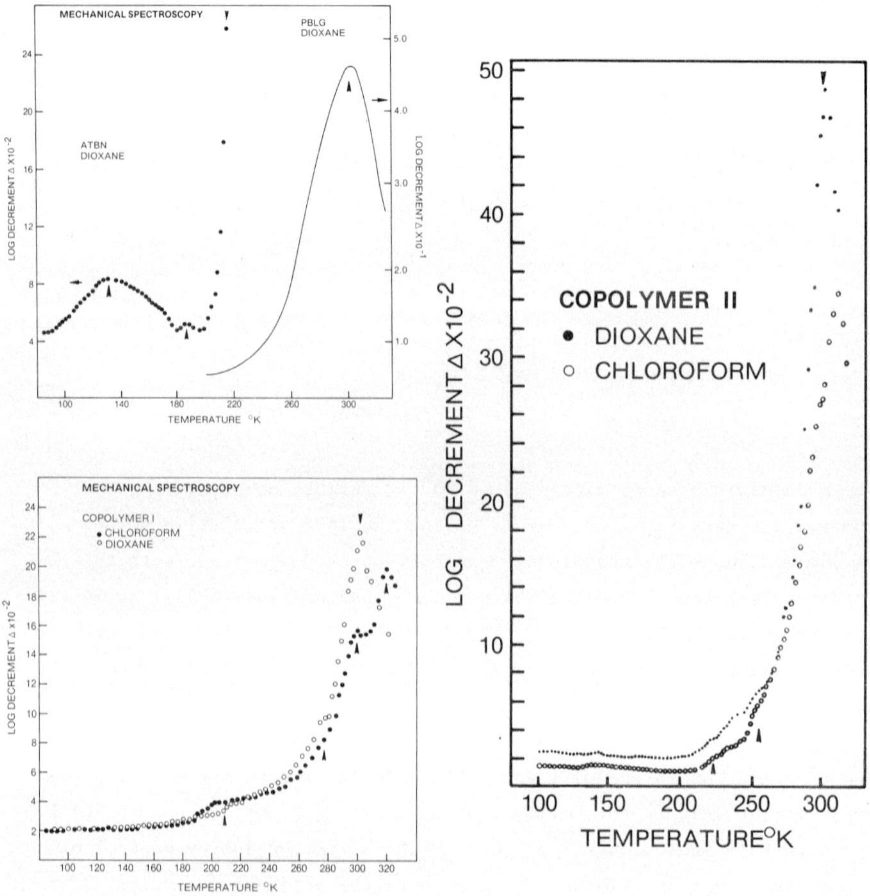

Figure 20. Dynamic mechanical spectra for ATBN, PBLG and GBG copolymers [21].

The presence of relaxation peaks in the copolymers at temperatures similar to those found in the homopolymers is indicative of phase separation in all of the samples; if homogeneously mixed on the scale of the molecular segments the peaks should have moved considerably on the temperature scale, the presence of one type of chain affecting the mobility of the other. The presence of some interaction, however, probably at the phase boundaries, is seen in the presence of a so-called interfacial peak [36] at approximately 260-280° (as a shoulder) in all of the copolymers.

Of most interest, however, is the apparent spliting of the 300°K peak in both chloroform cast samples, a new peak being observed at a higher temperature. This spliting for the chloroform cast samples was confirmed using dielectric spectrometry.

We presently attribute this new, higher temperature peak to the presence of a "distorted" or different helical structure for part of the G segments in the chloroform cast samples; the presence of this distorted helical form is consistent with the IR subtraction spectra as well as with the x-ray observations of a 10.3 Å reflection and a 100 reflection intermediate between form C and form B. However, it is noted that a xylene cast sample, presumably in form A, also showed a similar spliting of the 300°K peak. Annealing of the chloroform samples at 60°C, above the peak, followed by slow cooling, results in a broadening of both peaks. The presence of the peak at a higher temperature suggests that the associated motion is "more difficult" than that corresponding to the pure PBLG side chain peak with the annealing resulting in a disordering or randomizing of the side chain packing in the regions contributing to both peaks. The latter is reasonable since the side chain would be moving freely above the relaxation, repacking more irregularly during cooling than during solvent removal. However, although the distortion may be related in some way to interaction with the ATBN, full clarification of the details of the origin of the new peak is still required. X-ray or electron diffraction studies, as well as IR, as a function of temperature, should be useful.

In order to test the potential of these copolymers as biomaterials, 2 mil thick films, as cast from dioxane and chloroform, were implanted subcutaneously in rats for two and six weeks. Histological examination indicated minimal inflamatory response (Figure 21), independent of casting solvent and suggests further study of this type of material is warrented both scientifically and practically. It is also noted that changes in surface morphology and chemical reactivity can be induced after fabrication of appropriate physical forms, permitting preparation of samples with suitable physical and biological properties.

We consider now the application of similar characterization methods to block copolymers in which both blocks are polypeptides. A poster paper describing detailed results for several purified benzyl glutamate-leucine and benzyl glutamate-valine copolymers is being presented at this conference by Dr. Hiltner (see reference 22). We will, therefore, only summarize some initial results in which the morphology, x-ray diffraction, IR and dynamic mechanical spectra were examined [26]. The results point out the need for using more than one technique for characterizing these materials.

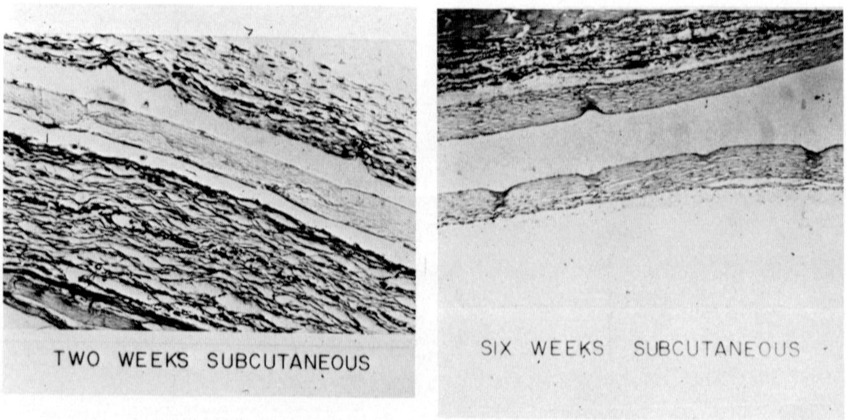

Figure 21. Histological section of GBG-II implanted subcutaneously in rats [21].

The blocks were prepared sequentially, (i.e. for GLG, a G segment to which was added L followed by more G) [37]. The theoretical lengths of the blocks, for the polymers used, in terms of numbers of monomer units and based on the monomer feed concentrations, are listed in Table V along with the x-ray diffraction data for samples cast from dioxane ($L_{30}G_{60}L_{30}$ was cast from a benzene:dioxane (3.2:1) mixture and the polyleucine was used as prepared, being insoluble). The x-ray results suggest that the G segments in all of the GLG copolymers are in the α-helical form C with essentially no x-ray evidence for the L component, even in those containing the greatest quantity of L. An apparent difference in structures between the $L_{30}G_{60}L_{30}$ and $G_{30}L_{60}G_{30}$ copolymers is seen, the LGL being much less ordered.

Typical electron micrographs are shown in Figure 22. All except the LGL are basically fibrillar in texture, with striations being seen perpendicular to the fiber axis in the $G_{50}L_{50}G_{50}$ sample and parallel to the fiber axis in the $G_{120}L_{60}G_{120}$ and $G_{60}V_{30}G_{60}$ samples. The latter

TABLE V

Polymer	Spacings (A°)							
PBLG	3.9 ↔		4.6_D	5.4_S	7.7_M		14.3_{VS}	
Leu			4.6 ↔	5.2_D		11.4_D		
$G_{60}L_{30}G_{60}$			← 4.6_D	5.4_S	7.7_M		13.2_{VS}	
$G_{50}L_{50}G_{50}$			← 4.6_D	5.4_S	7.7_M		12.9_{vs}	
$G_{30}L_{60}G_{30}$			← 4.6_D	5.4_S	7.7_M		13.3_{VS}	
$L_{30}G_{60}L_{30}$				← 5.4_D			13.3_D	
$G_{60}V_{30}G_{60}$	3.7_D	4.4_D	4.8_{VS}	5.4_W	7.2_W	7.9_W	14.2_{VS}	16.6_W
$G_{30}V_{60}G_{30}$	3.7_D	4.4_D	4.8_{VS}	5.4_M	6.7_W	7.7_M	14.3_{VS}	15.5_{VW}

D - Diffuse
S - Strong
M - Medium
W - Weak
V - Very
↔ - Spread

Table V. X-ray diffraction data for GLG and GVG copolymers [26].

are possibly related to the twisted fibrils seen in homopolymer PBLG fibers (see Figure 7) while the former may be related to a domain structure. The $L_{30}G_{60}L_{30}$ sample, in agreement with the x-ray results, shows no apparent morphology.

FTIR spectra showed that all of the G-L copolymers were α-helical in nature, whereas the $G_{60}V_{30}G_{60}$ copolymer (and a $G_{60}S_{30}G_{60}$ copolymer, S being serine) had both α and β conformations [38,39]. We assume phase separation has occured even though the EM observations do not permit its detection (in stroked, oriented films of the GSG and GVG copolymers, IR and electron diffraction show that the α-helical G segments were parallel to the fiber axis, whereas teh β conformation S or V segments are perpendicular to the fiber axis. The relationship of the cross-β structure to the morphology is now yet known).

Dynamic mechanical measurements for the various copolymers are shown in Figure 23. All of the GLG, and the $G_{30}V_{60}G_{30}$ block copolymers showed

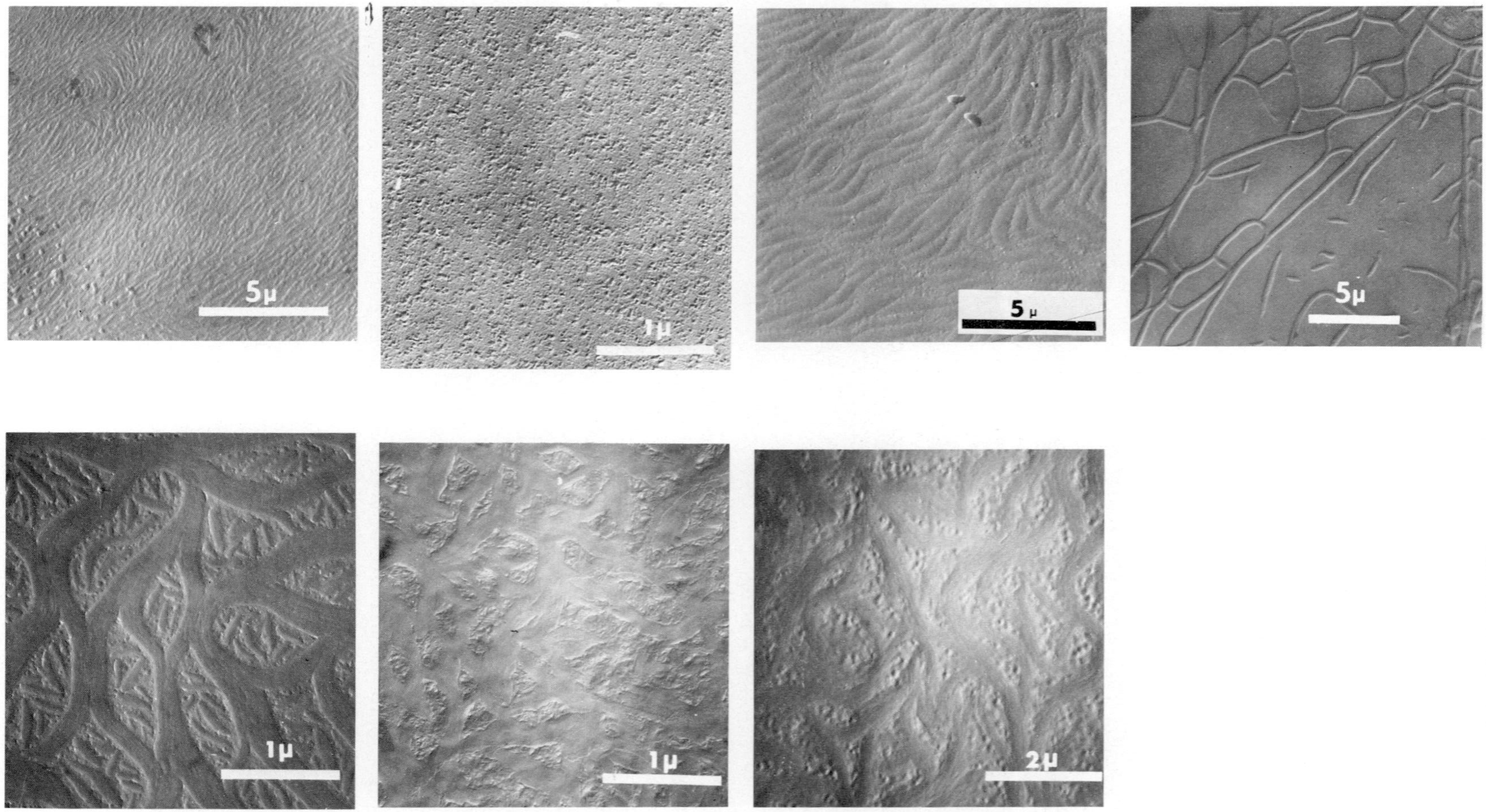

Figure 22. Electron micrographs of Pt-C shadowed dioxane cast films of, a) $G_{30}L_{60}G_{30}$, b) $L_{30}G_{60}L_{30}$(cast from a benzene-dioxane solvent), c) $G_{60}L_{30}G_{60}$, d) $G_{60}L_{30}G_{60}$, e) $G_{120}L_{60}G_{120}$, f) $G_{30}V_{60}G_{30}$, g) $G_{60}V_{30}G_{60}$ [26].

peaks at ca. 50°C and 145°C, the former at about the same temperature as in PBLG and the latter slightly higher (PBLG is at 125°C). However, in the random GL copolymer, as shown in Figure 13, the latter peak increases with L content, up to 194°C at GL-11 and above with increasing L. On the basis of these results (i.e. x-ray, IR and dynamic mechanical) it is concluded that phase separation has occurred in these polymers, as in the GBG polymers. The effect of annealing and solvent also is consistent with this interpretation, being similar to that for homopolymer PBLG.

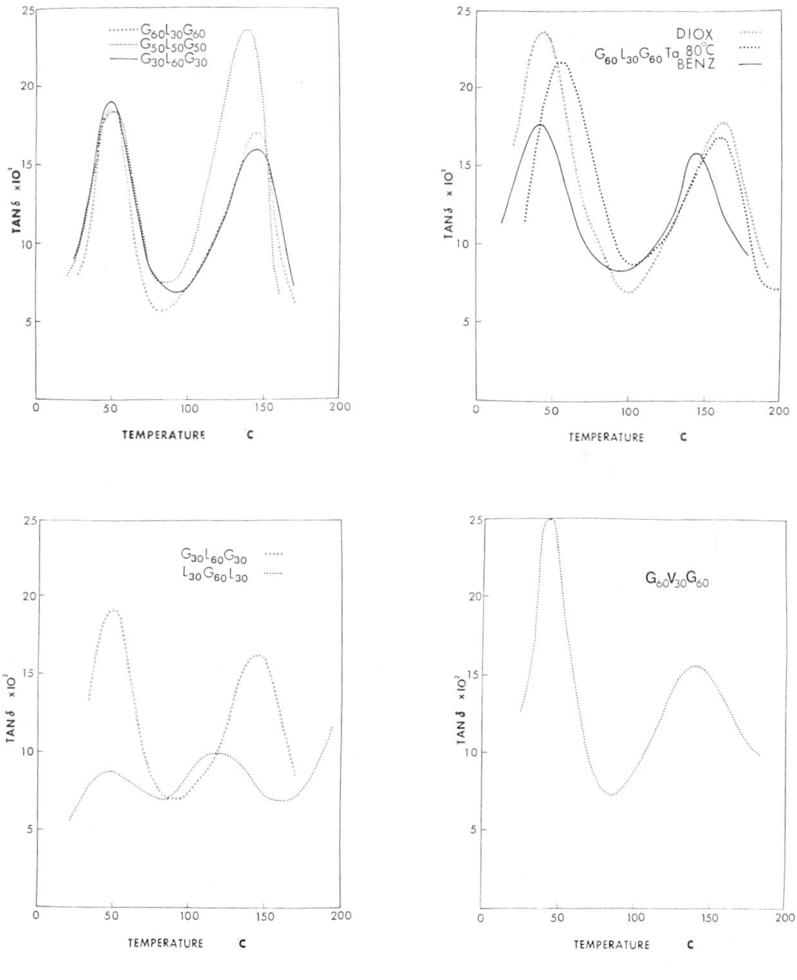

Figure 23. Dynamic mechanical spectra for G-L and G-V block copolymers cast from dioxane unless otherwise specified [26].

In the $L_{30}G_{60}L_{30}$ copolymer the 50°C peak is much lower in intensity than in the similar composition $G_{30}L_{60}G_{30}$ copolymer. This suggests that the L component forms the matrix in this copolymer, whereas the G component is the matrix in the other copolymers; likewise there is a substantially higher α peak, at a slightly lower temperature, in the $G_{50}L_{50}G_{50}$ polymer than in the other two shown on the same graph. This may be due to the differences in molecular packing, as seen by the smaller 100 spacing in Table V, but is in need of further examination.

As indicated above, a wide range of morphologies can be obtained in polypeptides, homopolymers, and block copolymers. The physical properties depend not only on the composition and the sequence, but also on the morphology; i.e. on the primary, secondary and tertiary structure. Characterization of these factors is only just beginning, but, based on experience gained with synthetic polymers, both the need for it and the techniques to accomplish it are well known and are following closely behind the synthesis of new materials. However, as pointed out above, considerable research is still needed to understand the basic morphological features of the fibrillar morphology observed in many polypeptides.

ACKNOWLEDGEMENT

Various portions of this work were supported by the National Science Foundation, Grant Number DMR-76-20089 and the National Institutes of Health, Grant Number HL 15195. Appreciation is expressed to Professors J. M. Anderson and A. Hiltner for advice and assistance.

References

1. P. H. Geil, "Polymer Single Crystals", Interscience, Wiley, New York, 1968.
2. B. Wunderlich, Macromolecular Physics, Vol. I & II, Academic Press, New York, 1973, 1976.
3. a) H. D. Keith, G. Giannoni and F. J. Padden, Biopolymers, 7 (1969) 775.
   b) F. J. Padden, H. D. Keith and G. Giannoni, Biopolymers, 7 (1969) 793.
   c) F. J. Padden and H. D. Keith, J. Applied Phys., 36 (1965) 2987.

4. J. Bungs and P. H. Geil, unpublished date
5. H. Hasagawa and P. H. Geil, to be published.
6. e.g. R. G. Quynn and H. Brody, J. Macromol. Sci., (Phys.) B5 (1971) 721.
7. G. Capaccio and I. M. Ward, Polymer, 15 (1974) 233.
8. P. Soni and P. H. Geil, to be published.
9. N. G. McCrum, B. E. Read and G. Williams, Anelastic and Dielectric Effects in Polymeric Solids, Wiley, New York, 1967.
10. J. J. P. B. Blais and P. H. Geil, J. Ultrastructure Res., 22, (1968) 303.
11. F. Rybnikar and P. H. Geil, Biopolymers, 11 (1972) 271.
12. A. McKinnon and A. Tobolsky, J. Phys. Chem., 70 (1966) 5.
13. A. McKinnon and A. Tobolsky, J. Phys. Chem., 72 (1968) 1157.
14. D. Parsons and K. Marlius, J. Mol. Biol., 10 (1964) 530.
15. C. Bamford, A. Elliott and W. Hanby, "Synthetic Polypeptides", Academic Press, New York, 1956.
16. E. Samulski and A. Tobolsky, Biopolymers, 10 (1971) 1013.
17. D. Parry and A. Elliott, J. Mol. Biol., 25 (1967) 1.
18. T. Watanabe, Y. Tsujita and J. Uematsu, Polym. J., 7 (1975) 181.
19. Y. Mitsui, Y. Iitaka, M. Tsuboi, J. Mol. Biol., 2Y (1967) 15.
20. T. Fukuzawa, I. Uematsu and Y. Uematsu, Polym. J., 6 (1974) 431
21. S. Barenberg, Ph.D. Thesis, Case Western Reserve University, 1976.
22. T. Hayashi, J. M. Anderson and P. A. Hiltner, to be published.
23. E. Fukada, T. Furukawa, E. Baer, P. A. Hiltner and J. M. Anderson, J. Macromol. Sci. (Phys), B8 (1973) 475.
24. G. Pezzin, G. Ceccorulli, M. Pizzoli and E. Peggion, Macromol., 8 (1975) 762.
25. A. Tsutsumi, K. Hikichi, T. Takahashi, Y. Yamashita, N. Matsushima, M. Kanke, and M. Kaneko, J. Macromol. Sci. (Phys), B8 (1973) 413.
26. W. M. Wong and P. H. Geil, unpublished data.
27. T. Hayashi and J. M. Anderson, to be published.
28. P. A. Hiltner, J. M. Anderson and E. Borkowski, Macromol., 5 (1972) 446.

29. a) "Block Polymers", ed. S. L. Aggarwal, Plenum, New York, 1970.
b) Copolymers, Polyblends and Composites, ed. N. A. J. Platzer, Am. Chem. Soc., Weshington, 1975.

30. H. Kawai, T. Soen, T. Inoue, T. One and T. Uchida, Mem. Fac. Eng., Kyoto Univ., 33 (1971) 383.

31. M. Matsuo, Japan Plastics, 2 (1968) 6.

32. B. Lotz, A. J. Kovacs, G. A. Bassett and A. Keller, Kolloid-Z., 209 (1966) 115.

33. R. D. B. Fraser, B. S. Harrap, R. Ledger, T. P. Macrae, F. H. C. Stewart and E. Suzuki, Biopolymers, 5 (1967) 797.

34. P. A. Hiltner, unpublished data.

35. K. Riew, private communication.

36. J. Beecher, L. Marker, N. Bradford and S. Aggarwal, J. Polymer Sci., C, 26 (1969) 117

37. Samples were supplied by J. M. Anderson and G. Sigler, see reference 27 for preparation technique.

38. J. L. Koenig and P. L. Sutton, Biopolymers 10 (1971) 89.

39. D. B. Fraser, B. S. Harrap, T. P. MacRae, F. H. C. Stewart and E. Suzuki, J. Mol. Biol., 12 (1965) 482.

STRUCTURAL ASPECTS OF EDIBLE MATERIALS

W. E. MARSHALL
General Foods Corporation, Tarrytown, N.Y. 10591

SUMMARY

A stepwise approach is presented to study the structural aspects of biopolymers in foods. The concept of primary, secondary, tertiary and quaternary functionality is presented and discussed. The primary functionality of a food biopolymer relates to its molecular functionality under simple conditions. Biopolymer-biopolymer interaction is termed secondary functionality. The functionality resulting from biopolymeric interactions with low molecular weight food constituents represents tertiary functionality. The quaternary functionality is that resulting from processing.

Mathematical clustering is employed as an aid towards understanding functionality. Examples of each type of functionality is presented as well as cluster analysis.

INTRODUCTION

Understanding the structural aspects of food materials is more important today than ever. In prior years of knowledge of how meats, fruits and vegetables were constructed and how this architecture changed in processing and transportation was important to insure optimum quality at the lowest cost. But today studying food structures may be an important factor in the survival of large populations.

Crop failures and the bidding-up of grain prices by the wealthier nations will lead to predicted food shortages among the Third World. The relationship between economic status and meat consumption provides a particular opportunity to the food industry through studying the structures of foods. As nations become wealthy consumers turn more to meat consumption. Animal foods - such as meat and dairy, rather than vegetable and grain foods provide some of the most desired meals worldwide. Unfortunately they are beyond the reach of an ever-growing number of impoverished people.

The objective in studying the structures of foods is to understand the relationship between these molecular structures and the eating qualities of a food. These qualities are organoleptic or sensory in nature, such as chewy, crunchy, moisture and oil release, and cohesiveness.

The purpose in attempting to understand the relationship between structure and sensory qualities is twofold: The first is to attempt to apply nature's technology to other foods. For example, although a carrot contains as much water as a melon, it is crisp rather than soft. This suggests that the water-holding structure of a carrot is more efficient than that of a melon. To understand this water-holding capacity of the carrot may allow for the development of what is termed "an intermediate moisture food". Intermediate moisture foods are bacteriologically stable and more palatable upon rehydration inasmuch as they have not been dried below a critical point. This critical point is that at which the water is bound tightly to the structural polymers of the food. If that bound water is removed by dehydration some of the structural integrity is lost and cannot be regained through hydration. The crispy character of a carrot would indicate that it contains an architecture and/or biopolymers capable of holding large amounts of water very strongly. Biopolymers differ in their ability to bind water and this competition for water dictates the bacteriostatic nature of an intermediate moisture food. If the water-binding capacity of a food is stronger than that of the bacteria, the microorganism cannot utilize the water for their own growth. The study of the natural water-binding systems - such as found in vegetables and fruits - can lead to technologies for other purposes, e.g. preservation. This has been the classical and first purpose for studying the structures of foods - to apply the knowledge to other food processing or distribution. We might call that the 20th Century purpose.

The second purpose served by studying the structural aspect of a food is to learn how to duplicate that food with its unique structure and its components by using other food raw materials or other processes. This might be called the 21st Century purpose. The need has been brought about by the acute and chronic shortages of certain foods predicted in the years ahead. These foods are generally those of animal and seafood origins. The animals that inefficiently convert a large amount of cereals, legumes and grains into palatable end products are those predicted to be in short supply. Also, the popularity of these foods can cause an increase in demand for agriculture crops as feed. These crops are needed as food by much of the world. The seafoods are increasing in cost and decreasing in availability due to energy, capital and labor costs as well as over-fishing in some waters.

These animals are also basic sources of proteins and fats used in a wide range of foods. Therefore their availability and cost indirectly effects the availability and cost of non-meat foods. For example, dairy and egg components are used extensively in many foods because of their unique characteristics.

The second purpose served by studying food structure is therefore to duplicate meat and seafood as well as to duplicate the characteristics of their biopolymers which are used in making other foods.

We have been studying the structure of foods for over a hundred years - trying to relate nature's technologies to new foods. Food science is familiar with many of the common cellular and fibril structures as well as the structural biopolymers. However, now these structures and biopolymers must be examined not for what technologies they can offer to other food systems but what are they intrinsically and how can that structure and biopolymer be duplicated using other food raw materials.

Let us consider one approach - an observation, evaluation and duplication procedure. Fortunately we can see the microstructure of a food by eye or through the microscope. Brightfield, darkfield and phasecontrast microscopy are commonly used to allow visualization of the heterogeneous nature of food. Some structural qualities such as hardness and elasticity can also be meaured and assist in the interpretation of visual data. Organoleptic values are quite essential, i.e. the sensory evaluation of mouthfeel and breakdown of the product. This is not an exacting science however. Many foods are too complex, too varied and too sensitive to be rigorously studied by laboratory methods.

For observation, a heavy reliance is placed on the electron scanning microscope. Its major advantage lies in its ability to present three-dimensional high resolution views of a material's surface topology. Through experience and histological staining we can form some hypotheses as to what type of structure is being viewed. Imagination and three-dimensional perspectives also help. Figure 1 shows the closely packed fibrils of lean meat. They appear to be hexagonal in cross section. Figure 2 shows the cellular nature of meat fat. The fat-containing cells appear to be quite large in relation to their membranes. The assignment then becomes how to duplicate the close-pack hexagonal fibrils with a certain quality of hardness and elasticity to provide the eating quality of meat. Or the question is how to duplicate a large cellular structure containing stabilized fat.

There is an assumption here that if a physical structure is observed and duplicated, the duplication will have similar properties to the

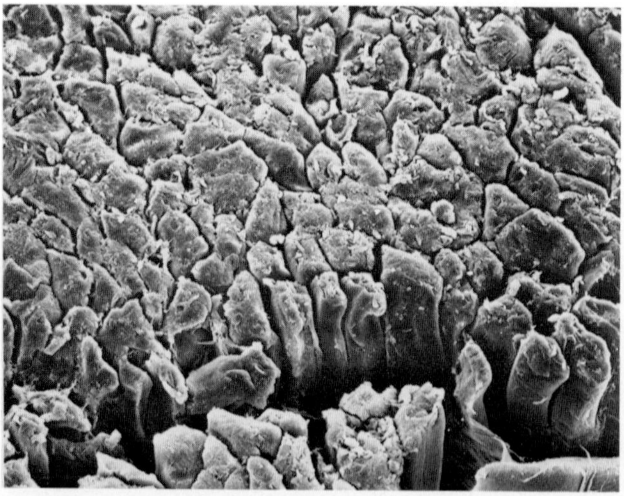

Fig. 1. Porcine Muscle Tissue (200x)

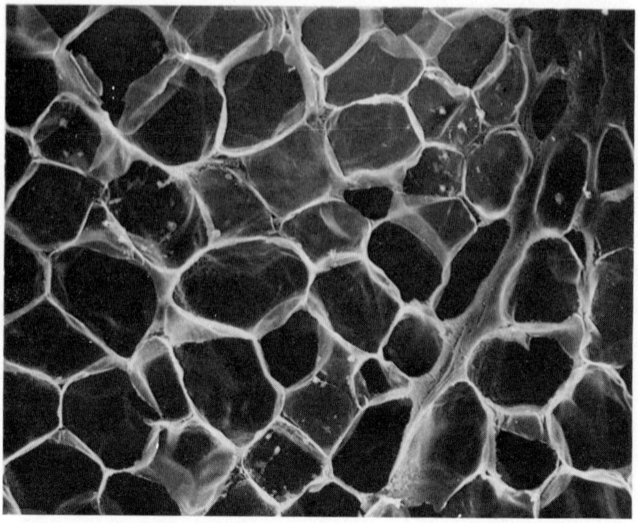

Fig. 2. Porcine Fat Tissue (200x)

original. This is only true if the replacement ingredients have functionalties similar to those of the ingredients being replaced. An analogy can be found in architecture - a simple photo of a bridge cannot reveal the nature of the steel or the composition of the concrete. If the nature of the steel and concrete are known and duplicated by another material the resultant bridge will possess the proper strength with the same design. We attempt to measure and catalog these biopolymers used in the structure by what are referred to as functionality tests.

There is also a related assumption, which is that the only way to obtain a certain organoleptic quality in a food is by duplicating the same unique biostructure found in nature. This is again only partly true. A similar organoleptic quality can be attained by using biopolymers with a functionality similar to those being copied but in a slightly different architecture.

PRIMARY FUNCTIONALITY

Functionality is a measurement of the biopolymer's special physicochemical properties of interest to the food researcher. Proteins - because of their diverse character - are usually the subject of functionality tests. They are conducted in simple model systems with standardized test conditions. Ten tests are commonly used:

Solubility in water as measured across the pH scale. It is a value slightly higher than that of true solubility inasmuch as colloidal biopolymer is included in the reading.

Heat gelation is a property to illustrate the change in consistency of a biopolymer solution during heating. If a gel is formed by heating, its texture and break point are measured.

Emulsification capacity is the ability of the biopolymer to convert an oil-in-water emulsion to a water-in-oil emulsion. A vegetable oil is employed as the test oil.

Foaming properties are measured at constant pH by bubbling nitrogen gas through a biopolymer solution for a given time and temperature and determining the volume of the resultant stable foam.

Viscosity characteristics are determined over a range of pH and concentration values.

Heat coagulability is the loss in solubility after heating at various pH levels at 100°C for 30 min.

Film formation is a subjective test performed by drying a known concentration of a biopolymer on a crystallizing dish.

Surface tension is determined by the tensiometer on a solution at a given pH and temperature.

Water-holding capacity is evaluated under conditions of pH and concentration, heating for 30 min. at 60°C followed by cooling and centrifugation. It is expressed as water held by the biopolymer.

Wettability is performed on insoluble biopolymers and graded subjectively.

This series of functionality testing has been conducted on several hundred food biopolymers, especially proteins. These characteristics are all important - to varying degrees - in most of the commonly eaten food products. The ranges of data are very wide. Differences of up to 500-fold between biopolymers are not uncommon. The data suffers in that it represents an ideal system much unlike the complex system of a food.

I would like to draw an analogy as to the usefulness of this data. It is similar to the usefulness of using the physicochemical data of individual amino acids to predict the characteristics of a protein. Only in relatively few cases can it be successfully extrapolated. Knowing the amino acid composition of collagen one can expect a rather unique protein. Likewise, looking at the functionality date of a protein one can envision a particularly unique role. The data for egg albumin for example dictates its use as an ingredient in a food desired to be set by heating. This heat setting characteristic usually overrides the characteristics of other biopolymers in a food.

Returning to the analogy to protein composition, the secondary structure is helpful. Knowing the sequence of amino acids one can predict turning points, disulfide bond formation and certain hydrophilic convolutions. The tertiary structure is even more helpful, indicating areas of specific chemical characteristics of binding or enzyme active sites. The quaternary structure is most helpful indicating exposed or buried reactive sites.

So it is with food protein functionality. Studies of protein - protein interactions, and protein-non-protein interactions are more helpful than individual protein functionality. It must be remembered that foods cover a wide range of pH, water compositions and salt concentrations. Foods also contain low molecular weight constituents such as emulsifiers, conditioners, oxidizers and reducers.

SECONDARY FUNCTIONALITY

The biopolymer-biopolymer interaction can be considered the secondary functionality. In foods, the biopolymers other than protein which are important are starches, gums, nucleic acids, and non-digestible carbohydrates. An individual protein can interact with itself, with other proteins, anylose, anylopectin, mannans, carrageenan and nucleic acids, to mention only a few.

Proteins and starches are commonly used in food technology in the building of a new product or in attempts to copy a natural product. Functionality tests are used to evaluate the many rheological characteristics of starches. The water binding capacity, viscosity and heat setting are used to classify their origin - tapioca, potato and corn for example. The interaction between soy protein and tapioca starch can be expected to be different than that between soy protein and corn starch. Specialty starches and modified starches can also be expected to provide different interaction data. In all some two dozen starches are commercially available. Investigating the interactions of 100 proteins with 25 starches in 10 functionality tests provide valuable but still only partial answers towards predicting its usefulness in building a food or as a replacement for a more expensive biopolymer.

The gums and non-digestible carbohydrates are another group of polymers often used in food development. With the new knowledge relating to the efficacy of dietary fibers even greater use of these substances will be made. To date the functionality testing of gums and pectins has been evaluated in typical food systems - especially puddings and desserts. Here the formulas are simple - both in the number of ingredients and the energy required in the form of heat and work. The conditions can be controlled, many samples can be prepared and the variables evaluated simply by mouthfeel. Extensive testing and analysis is replaced by kitchen methods.

A study of the functionality of nucleic acids has only recently been of importance - since the consideration of SCP's - single cell proteins. Yeasts have a relatively high level of nucleic acid which has to be recognized when developing a food using a yeast base. The protein behaves differently when the sample is free of nucleic acid. It is unclear whether this is a result of the processing to lower the nucleic acid content or to the nucleic acid - protein interaction. This interaction may be further complicated inasmuch as the nucleic acids are strong water binders. The nucleic acid fraction may remove water from around the protein thereby altering the functional performance of the protein.

TERTIARY FUNCTIONALITY

The biopolymeric components of a food total 50 to 75% of the food on a dry weight basis. The balance consists of a number of lower molecular weight entities which have a profound influence on the structural aspect of the biopolymers. In fact, these low molecular weight components can have a greater influence on structure than the biopolymers have on each other. The amount of free water, ionic strength and pH are strong influencing factors. Sugar, salt, and water are the most commonly used substances to regulate these influencing factors. Water is bound tightly by sugars and oligosaccharides. They compete successfully for the available water in a food process. Because of their solubility they hydrate immediately upon coming into contact with water whereas some proteins do not hydrate well until heat or work is introduced. The biopolymers must then remove water from the low molecular weight components which requires more energy than simple hydration.

To summarize to this point: the evaluation of the intrinsic functionality of an individual biopolymer is termed primary functionality; the evaluation of polymer-polymer interaction is the secondary functionality; the polymer-non-polymer functional interactions represent the tertiary functionality.

Let us consider the important compounds in tertiary functionality. Besides sugar, salt and water, fat is of considerable importance in regulating the degree of primary and secondary functional interactions. This regulation is achieved by the lipid interfering with hydrophilic and hydrophobic interactions between polymers. This interference is most pronounced in protein-protein interactions. The interfering lipid material can be native to the protein source or can be added to the formula in processing. Soy protein is rather hydrophobic in the natural state - the soya bean. Gentle extraction of the lipid - soy protein yields a protein with certain functional characteristics which are quite different if the lipid had not been first removed. Also well-known is the observation that the polarity of the solvent influences the

FUNCTIONALITY FACTORS

Primary Functionality

The molecular properties of the biopolymer

Secondary Functionality

Biopolymer-Biopolymer Interactions

Tertiary Functionality

Biopolymer-Solute Interactions

Quaternary Functionality

Structural changes in the conformation of the biopolymer due to food processing, e.g.

. heating

. freezing

. homogenization

. extrusion

. mixing

Fig. 3.

physicochemical characteristics of the resulting lipid-free protein.

Much of our oil seed protein functionality data reflects denaturation in the starting material as a result of commercial isolation. As the biopolymer folds and unfolds, reactive sites change. Depending on the processing conditions of protein extraction the type as well as degree of denaturation can differ. In the extraction of soy protein, its many buried sulhydryl groups can be exposed with subsequent polymerization by interacting with itself. Some of the buried sulhydryl groups are exposed during processing allowing structural changes to occur.

Fat is commonly used in most food formulas to provide a smooth texture. When used, it is important to control its distribution so as to not interfere with a desired polymer interaction. Frozen whipped toppings are an excellent example in which the polymeric materials must interact sufficiently to maintain the foam structure. However, the lipid content must be high to provide a smooth mouthfeel. A sensitive balance exists between the amount of polymer and oil. The best polymers for these types of products are the milk proteins, particularly caseinate. It possesses good emulsifying properties, good film formation and is not significantly effected by temperature.

The fat in meat products is quite different structurally than in whipped desserts. In meat the fat must maintain its own structure in a macro sense. Cubed sections up to one inch must be an integral part of the food. Its micro structure must allow for a portion of the fat to be rendered while the remaining fat must provide a smooth mouthfeel during eating. The role of certain polymers here is to bind the fat in a heat sensitive manner releasing it at about 40-50°C. Other polymers must bind the fat loosely but not in a heat sensitive manner. In this way it is perceived oily in the mouth at temperatures about 50°C.

The last class of compounds in the consideration of tertiary functionality are the minerals. Here the most profound effects are seen with calcium and sodium interfering with the functionality of caseinate. A proper ratio of sodium to calcium is often critical to the performance of the protein. Calcium can form bridges between anionic groups on gums and pectins as well and many alginates will form stable gels through the interaction of calcium.

Before discussing quaternary functionality I would like to introduce the term "primary functionality splitting". A biopolymer known to be an excellent emulsifier and heat-gelling agent has these two different primary functionalities. A particular food application may require its excellent emulsifying qualities but without the heat-gelling qualities. Can functionalities be altered independently? One would expect this to be possible but it has proven to be difficult. These different functionalities appear to be the sum of a number of molecular factors which are either common or inter-related. Many interfering agents must be screened before the functionalities can be split independently.

QUATERNARY FUNCTIONALITY

The last degree of functionality - the quaternary functionality - is that effected by processing. The intrinsic properties of a biopolymer are represented by the primary functionality. The polymer-polymer interactions are the second degree of functionality. The interactions between the polymers and the low molecular weight materials represent the tertiary functionality. The physical and chemical changes seen in the primary, secondary or tertiary functionalities as influenced or caused by energy exchanges represents the last consideration of quaternary functionality.

Food processing is not simply the application of heat energy to a system. Only certain food processes such as simple beverages require only heat for pastuerization. Typical food processes today are extrusion, freeze-drying, flaking, drying, homogenization, and shearing to name a few examples. Extrusion is one of the more complicated and gaining in usage. It has common application in the plastic industry by which components under high temperature and pressure are conveyed at variable times by a screw inside a barrel through a die at the end. The important physical factors of temperature and pressure are produced not only by steam pressure but also by the configuration of the screw-rod, its pitch and frequency. The die also influences back pressure - a smaller opening, a greater back pressure.

One might expect that the extremely high temperature of $140^{\circ}C$ and pressures of 500 psi could easily swamp the subtleties of free water and pH. However, this is not the case. The extruder is rather a reaction vessel to complete a particular sequence of polymeric reactions. Because of the extremely high energy input almost all potential reactions are realized. The experiment must therefore be set up to increase the chances of one functionality being maximally altered rather than only one reaction occurring. This approach is similar to an organic ring substitution in which the researcher attempts to maximize the yield by both increasing what is desireable and decreasing what is undesireable.

The sulhydryl and disulfide bonds in soy protein are good examples to illustrate how primary, secondary, tertiary and quaternary functionalities are important. Soy protein can interact with itself in the presence of a transient reducing agent during extrusion. Sodium bisulfite is used as the reducing agent. Under the high energy conditions of extrusion the bisulfite reduces the disulfide bonds and the protein is unfolded to a maximum extent. At this point in extrusion it can be smeared with itself or another polymer. The sulfite is driven off as a gas allowing new disulfide bonds to form. Or if desired, the sulhydryls can be blocked by other polymers or low molecular weight entities. Biopolymeric films can be formed in this manner.

A second example - that of homogenization energy - uses high pressure without excessive heat. In homogenization the oil droplets can be altered in size and a biopolymer "driven" into the droplet allowing for a stable oil in water emulsion. Here again the choice of polymer is more critical rather than the pressure, time or temperature of reaction. If the product is then to be whipped the polymer must have the functionality not only to emulsify but also to form a stable foam. Here we have an example of two key functionalities which are inter-reactive. To utilize two biopolymers for these two functionalities usually requires a component or process which blocks the interactions. Fat can often be used if the hydrophobic/ hydrophilic ratio is favorable. The successful use of the component is based most often on trial and error while controlling the other variables in the formula.

A process can be used to block an undesireable interaction. Altering the order of ingredients going into a process is used to insure that the proper secondary or tertiary functional influence will occur. Reactive hydrophobic groups can be blocked by utilizing a small amount of oil or surfactant before the polymer is used in the reaction.

Food systems ordinarily contain 10 to 20 ingredients. Most of these are capable of affecting the functionality of the biopolymers. Some are themselves polymers added for a particular functional and/or structural purpose. Their interactions must be known and controlled. As mentioned their synergy or antergy can be controlled by formula changes, order of ingredients and type of process.

Similarly, when duplicating a food the right biopolymers must be selected and manipulated. The primary functionality - that intrinsic to the biopolymer - overrides secondary, tertiary and quaternary functionalities. Controlling the structural aspect of food starts with understanding primary functionality. Primary functionality is necessary in order to make a good beginning towards an overall understanding. Continuing to use protein structure as an analogy one fits the secondary, tertiary and quaternary structural data with primary or amino acid compositions. The primary biopolymer functionality tests cannot be used to extrapolate to finished product functionality any more than amino acid composition can be used to predict enzyme activity. It is simply a necessary start.

The primary functionality tests do however have some immediate applications and I would like to discuss these now in greater detail. They have a particular value in identifying replacement biopolymers especially proteins. Protein costs are rising steadily and the entire industry is constantly searching for new inexpensive nutritionally equivalent proteins to replace the more expensive ones. A statistical technique of clustering is used to aid in this search.

MATHEMATICAL CLUSTERING

Clustering is the grouping of similarities. Linnaeus in the 18th Century clustered or classified the animal and plant kingdom. For example, man is clustered or classified as a mammal within the animal kingdom.

Classifying and clustering occur constantly in thought and speech. Objects that differ only in insignificant details are given the same name, can be treated the same, and can be expected to act the same. For example, a wife notices that the man coming in the door differs only in insignificant details from her husband that left in the morning, and so she expects him to answer to the same name.

Cluster analysis is an effective tool towards generating hypotheses about a category's commonality which prior misconception and ignorance might otherwise preclude. As a concept clustering is easily understood. However, it is not a term for a single integrated technique with well defined rules of utilization; but rather a collection of procedures and elements of statistics. The major steps of cluster analysis are choosing the proper variables, selecting the correct number of clusters and the interpretation of results. Let us consider each step with examples.

The choice of proper variables in primary functionality is not difficult regarding proteins. The functionality tests including gelation, emulsification and solubility are all strong discriminators to illustrate differences between proteins. Proteins differ markedly in their ability to form stable gels. In general, the albumins form excellent gels whereas many proteins - such as caseinate - do not gel at all under the test conditions. The oilseed proteins are intermediate in their gelation characteristics. Using the gelation test, proteins will cluster themselves into groupings rather than a continuous spectrum of values. However, test results of emulsification values distribute themselves more as a continuum from excellent to poor. As might be expected solubility data of the commercial protein materials is also distributed as a continuum from 100% solubility to less than 10% solubility. It is a straight-forward task to group proteins by gelation data alone but requires some judgment to cluster by all three variables.

One might expect a protein with an emulsifying capability similar to caseinate to be an excellent candidate for replacement of caseinate in a food system. However, in actual practice it may fail because of its gelation characteristics. And there are other functionalities which can contribute significantly to a protein's performance in a food system - heat-setting, foaming, water-binding, etc. The replacement of proteins based on one functionality characteristic is usually insufficient in a complex food system.

Based on the gelation variable alone, proteins could be divided into four clusters - excellent, good, fair and poor. When the variables of emulsification and solubility are included, the number of clusters increase to 10 to 15. Mathematical devices are available for assistance in this further clustering but judgment based on a knowledge of the significant ranges is critical. If 3 groups per variable are used - e.g. high, medium and low - 27 clusters based on the functionality data for gelation, emulsification and solubility would result. A knowledge of the critical variables and significant ranges must be used to reduce the total numbers of clusters to a manageable level. A staggering number is seen resulting from using 10 primary functionality tests rather than 3.

The statistical method for sacrificing information in this multi-variable setting is called principle component analysis. The basic concept is to transform the original data in a manner which identifies those measurement variables which contribute least to the total variability of the data. The original data is transformed into a new set to minimize the influence of these least informative values.

Many statistical procedures for clustering are available to the chemist. Using the Ward Procedure we have examined the characteristics of 125 proteins in the 10 primary functionality tests. In this Procedure each of the 125 proteins is mathematically considered a unique group. Clustering begins by considering the 2 proteins which have functionality data most similar and joining them into one group, leaving 124 groups. This procedure is continued sequentially until one final grouping is attained. In cases where one or more proteins appear to be dissimilar to other group members a new group consisting of these dissimilar proteins is created.

Using gelation data alone one may judgmentally see 4 distinct clusters. When the data from emulsification and solubility are added Principle Component Analysis and mathematical clustering is needed to arrive at a significant and manageable number of clusters.

By utilizing Principle Component Analysis and mathematical clustering the data from 125 proteins tested over three functionalities was grouped into 15 clusters. Figure 4 shows the profiles of nine clusters of proteins for the primary functionality tests of solubility, emulsification and gelation. Each group contains between 3 and 20 proteins. One is immediately struck by the range of data between clusters. Proteins have an unlimited capability in this regard within food systems.

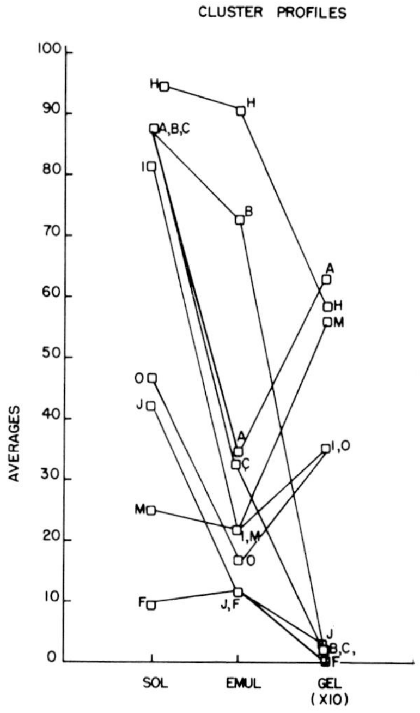

Fig. 4.

Group B contains a number of caseinate and caseinate-like proteins and shows high solubility, good emulsification and no gelation. The proteins within Group B can replace caseinate in most food systems. However, they are as expensive as caseinate or suffer in taste and odor. Therefore we must look to other groups for a replacer protein.

Comparing Groups B and H one sees similarity in solubility and in emulsification but a wide variance in gelation. One might anticipate that a protein of Group H could replace caseinate as an emulsifier in a food system. However, the high gelation characteristic of Group H interferes or fails to synergize with the emulsification property in many food systems.

However, the proteins in these two clusters could be interchanged in certain food products. But if a heat step were employed at a critical point in the process and proteins from Group B would be selected to avoid a gelling of the product. On the other hand, if gelation is desired a protein from Group H would be selected. This is an excellent example of the influence of quaternary functionality in predicting product performance.

An example of the influence of tertiary functionality can also be given in regards to caseinate in food systems. The presence of calcium - often found in food systems for nutrition - can alter the emulsification characteristics of caseinate.

Only nine of the 15 groups have been represented here for clarity. The data from all groups indicates almost all combinations can be obtained.

Since proteins are an important nutritional factor their substitution must be done with proteins of a similar or higher nutritional value. Therefore the Protein Equivalent Ratio, or PER should also be considered as a variable in cluster analysis.

Cluster analysis of the data from the remaining 10 primary functionality tests provides an informative but yet a partial picture. Foaming viscosity, heat coagulation, film formation, surface tension, water binding, and wettability are generally lower - order discriminators but critical in certain food systems. Of far greater importance is the consideration of secondary, tertiary and quaternary functionalities. The cluster profiles change when other biopolymers and low molecular weight materials are introduced into the test systems. The variables in the primary functionality data are not uniformly altered by interactions with other biopolymers or low molecular weight components. For example, proteins A and B may show synergy in gelation but no effect or antergy in emulsification, foaming or viscosity.

In order to consider the influence of processing, or quaternary functionality, all types of work must be considered. Mixing, extruding, auto-claving, freezing and homogenizing are a few of the many processes used. Their effects on biopolymer functionality are studied under ideal conditions and extrapolated to simulate more complex food systems.

Conversely, adjustments in a process can be made based on knowledge gained in evaluating the secondary and tertiary functionalities.

Food research today is concentrating on understanding the structural and functional aspects of biopolymers. The objective is to provide more food in a world threatened by shortages. One approach has been presented here requiring a knowledge base built upon the intrinsic characteristics of biopolymers, their interactions, and their alterations during processing. Some technical accomplishments have been seen but to have a lasting value they must be rewarding to the consumer. As mentioned earlier, the organoleptic values of chewiness, hardness and moisture release in the food are the ultimate tests. This research requires the involvement of many scientific disciplines - biochemistry, mathematics, engineering and psychology to name only the obvious. Irrespective of the approach the task is difficult but one that must be accomplished.

REFERENCES

1. F. Gejl - Hansen and J. M. Flink, Journal of Food Science., 41 (1976) 483.
2. J. E. Kinsella, Functional Properties of Proteins in Foods: A Survey, Critical Reviews in Food Science and Nutrition (1976).
3. E. Balmaceda, M. Kim, R. Franzen, B. Mardones and J. Lugay, Journal of Food Science, to be published.
4. M. R. Andesberg, Cluster Analysis for Applications, Academic Press (1973).

PROPERTIES OF MODIFIED COLLAGENOUS MATERIALS

Edward J. Kramer
Department of Materials Science and Engineering and the Materials Science Center, Cornell University, Ithaca, NY 14853

SUMMARY

The structure-property relationships of reconstituted collagen hollow fiber membranes have been investigated and compared with those of native collagen fibers. In the dry fiber the tropocollagen molecules aggregate to form large fibrils with the molecular axis predominantly along the fiber axis. On immersion in aqueous media however these break down laterally into much smaller fibrils which bend between crosslinks to accept water in a pore structure. Fibril diameters from small angle and wide angle x-ray scattering are about 30Å. The mechanical and transport properties reflect this structure. At low strains a small "toe" is observed in the stress-strain curve of the wet hollow fiber which can be accentuated by swelling the fiber further (by decreasing the pH of the solution). The toe results from the straightening of the bent fibrils at low applied stress and a mathematical elastica model of the straightening can reproduce the stress-strain curve. The permeation of solutes of various molecular size, as well as the pressure-driven flow of water, through the membrane has been investigated. Both quantities are reasonably well explained by models of transport through a forest of 30Å diameter fibrils. The fact that the hydraulic permeability can be decreased by about a factor of five by applying a tensile strain of only 10% provides dramatic confirmation for the structural model.

INTRODUCTION

The study of structure-property relationships of modified collagenous materials covers an extremely broad area, ranging from investigations on leather to cold cast gelatins. Rather than attempt to superficially cover this entire field this paper will review in depth what is known about the structure-property relationships for a particular type of modified collagen, enzyme solubilized collagen that has been reconstituted from acid solution. This discussion will highlight the experimental techniques that can be brought to bear on such a material and the results that can be expected from them. Whenever possible a comparison will be made between the structural features of the modified collagen and those of native collagen.

Most of the information reviewed here has been generated by a tripartite research group made up of the Japan Leather Company (principally Dr. Tereo Miyata) who have developed the processes for manufacturing the material, the New York Hospital/Cornell Medical College (principally Drs. A. L. Rubin and K. H. Stenzel) who have clinically evaluated the materials and the Cornell Engineering College (Drs. Stevenson, Ast, Viswanadham, Schwartz and Agrawal) who have investigated the structure-property relationships.

PREPARATION METHODS

Collagen solutions are prepared from calfskin by enzyme treatment with proctase. This treatment removes the telopeptides [1,2], short nonhelical appendages at the end of the triple-helical tropocollagen molecule. The resulting tropocollagen molecules (∿2800Å long and 10-15Å in diameter) may be monomolecularly dissolved in acid solution. Removal of the telopeptides also eliminates immunological reaction with antibodies to native collagen [3]. The tropocollagen can be precipitated from acid solution by raising the ionic strength or pH and can be subsequently crosslinked by ultraviolet irradiation [4]. By such methods the enzyme treated collagen may be restructured into a variety of forms including films, hollow fibers and gels. Alternatively native collagen fibrils, produced by mechanically digesting calfskin, can be mixed in a slurry with the acid solution of enzyme-treated collagen so that the resulting material after precipitation and crosslinking is a polymeric composite of native fibrils and reconstituted collagen. In addition to a high volume use as edible sausage casing, these reconstituted collagen materials are candidates for various biomedical applications [5], especially as a membrane for artificial kidney dialysis.

For the most part our structure-property investigation has centered the most promising of the membranes, a hollow fiber approximately 400μm in outer diameter and 15 to 20μm thick, which is produced from 100% enzyme treated collagen. It is prepared by extruding a 2 to 3% solution (∿pH 3) of collagen molecules through an annular nozzle into a coagulation bath consisting of an aqueous solution saturated with NaCl. The hollow fiber is then irradiated with ultraviolet light which crosslinks it, neutralized with NaOH, washed and then dried.

STRUCTURE

Wide angle x-ray scattering (WAXS) from the dry hollow fiber reveals that the tropocollagen molecules are oriented with their long axes parallel to the hollow fiber axis (extrusion direction). The two most prominent reflections in the WAXS pattern shown in Fig. 1 are the 2.8Å meridional arc, which has its origin in the basic repeat distance along the axis of the triple helix, and the 11Å equatorial arc which represents the lateral

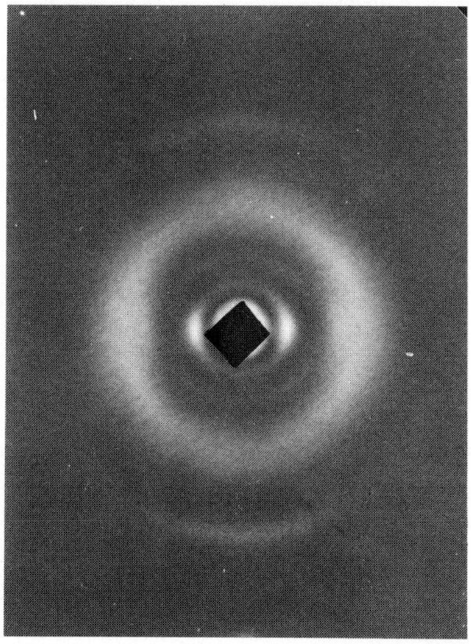

Fig. 1. Wide angle x-ray scattering pattern from dry hollow fiber. The fiber axis is vertical.

separation between nearest neighbor tropocollagen molecules [6]. These features, and indeed the whole pattern, are very similar to what one would obtain from an oriented collagen fiber from bovine tendon [1]. The equatorial reflection is sensitive to relative humidity with the spacing increasing from 11.6Å to 15.3Å [6] as the relative humidity is increased from 0 to ~100% whereas the 2.8Å helix repeat is almost unaffected, behavior which also directly parallels that of native collagen [7]. The water absorption isotherm of the hollow fiber (Fig. 2) is also almost identical to that obtained by other investigators for native tendon [8]. The hollow fiber exhibits a large anomalous increase in water content if it is transferred from a ~100% relative humidity vapor environment to an aqueous environment [6] with the water uptake increasing by almost a factor of 3.

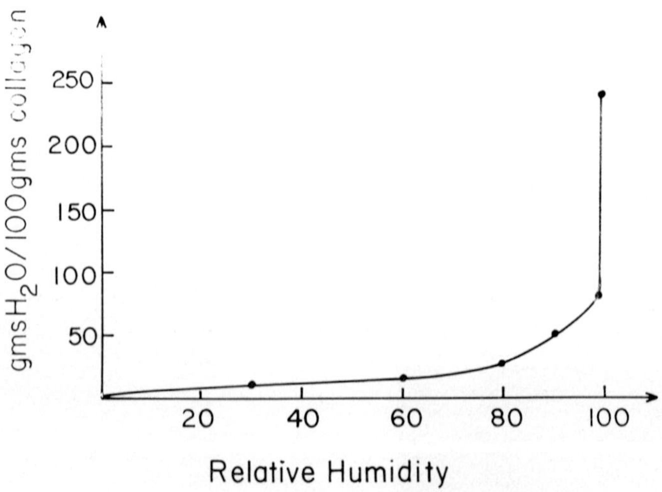

Fig. 2. Water absorption isotherm for the hollow fiber.

(This anomaly, known as Von Schroeder's paradox, is also well documented in gelatin, elastomer, and synthetic fiber systems [9,10]. It apparently is due to the very steep absorption isotherm near 100% relative humidity in these materials and to the fact that unless unusual steps are taken to control temperature, the temperature of the vapor is usually ∿.1 to .2°C

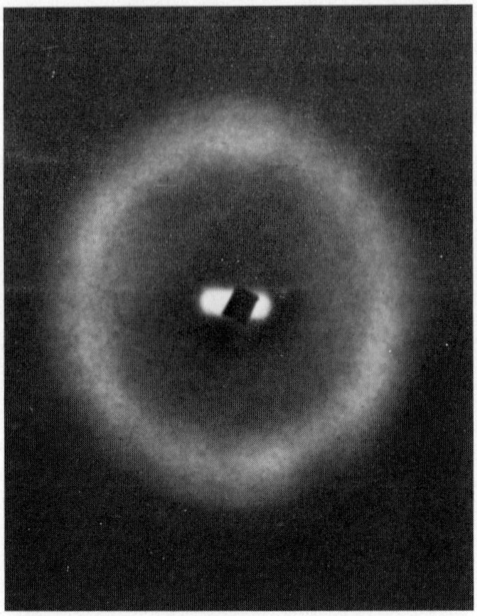

Fig. 3. Wide angle x-ray scattering pattern from wet hollow fiber. The fiber axis is vertical.

above that of the liquid, leading a small undersaturation of the vapor. The result is a large increase in water absorption on moving the sample from apparently saturated vapor to the liquid.) Corresponding to this large increase in water content there is an abrupt change in the WAXS pattern (Fig. 3). The 15Å equatorial reflection becomes so broadened it is difficult to distinguish from the background. A microdensitometer trace of the pattern shows the presence of the 2.8Å meridional reflection, indicating that the triple helix, or at least portions of it, is still intact. In marked contrast to this behavior the equatorial arcs and the overall features of the WAXS pattern of native tendon become sharper on soaking in water [7].

There are equally dramatic changes in the small angle x-ray scattering (SAXS) pattern of the hollow fibers on wetting. In the dry condition, Fig. 4a, there is a weak equatorial arc which corresponds to a Bragg spacing of about 48Å, approximately four times the WAXS 12Å Bragg spacing. The existence of this reflection, as well as the relatively small breadth of both equatorial reflections along the equatorial direction indicates that regions of relatively good collagen "crystal" (a relatively perfect periodic lateral arrangement of tropocollagen) larger than ∿400Å exist in the dry fibers. [The exact structure of this lateral arrangement is uncertain at this time, although since only two equatorial reflections are present it is clearly different from the lateral arrangement of tropocollagen molecules

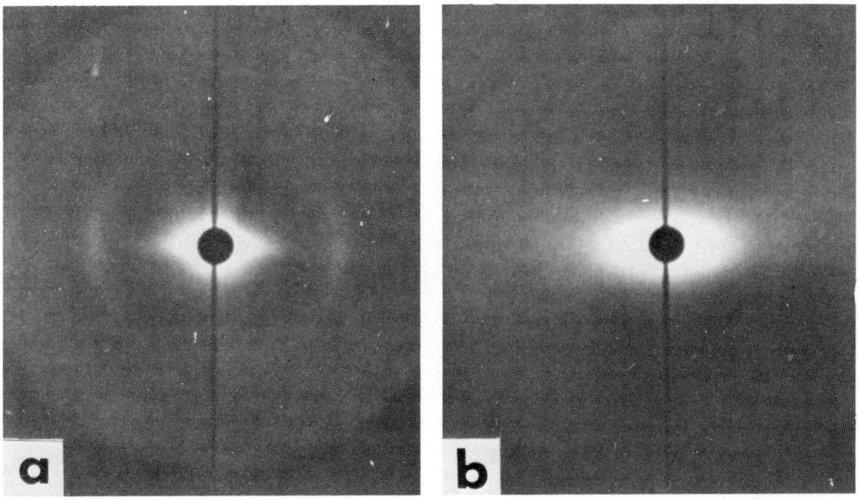

Fig. 4. a) Small angle x-ray scattering (SAXS) pattern from dry hollow fiber. b) SAXS from wet hollow fiber. The fiber axis is vertical in each case.

in rat-tail tendon [11,12]. Similarly models of lateral packing devised for native collagen such as the five molecule tubular protofibril of Smith [13], the cylindrical lattice of Sasisekharan and Ramachandran [14], and the hexagonal lattice of Burge [15] can not account for the equatorial WAXS and SAXS pattern of the reconstituted collagen.] On soaking the hollow fiber in water the 48Å equatorial reflection disappears completely, Fig. 4b. There is also no trace in the SAXS pattern of either the dry or wet hollow fiber of the meridional reflections that arise in native tendon due to the 640Å periodicity along the fibril axis [16]. Since these reflections arise primarily from the periodic array of holes in the quarter-staggered arrangement of tropocollagen molecules that make up native collagen fibrils [17], it is evident that this periodic hole arrangement is absent in the hollow fibers. This lack of a periodic hole structure is probably due to the precipitation under high supersaturations (saturated NaCl solution) rather than to the removal of the telopeptide chains by the enzyme solubilization since precipitation of enzyme solubilized collagen under lower supersaturation conditions results in fibrils with a periodic hole structure [18].

Although there is no periodic hole structure, the very strong continuous scattering around the origin (Fig. 4a) can be best interpreted in terms of elliptical microvoids, located at random in the fiber, but predominantly oriented with long axis of the ellipse along the fiber axis. The SAXS pattern of the wet fiber is different in two respects (Fig. 4b). The aspect ratio of the continuous scattering ellipse is smaller than in the dry fiber indicating either more nearly circular (water-filled) pores or a less oriented pore structure. In addition the intensity in the high angle tail of the continuous scattering for the wet fiber decreases much more slowly than that for the dry fiber. This increase in the extent of the high angle tail signifies a large increase in the specific surface area of the collagen phase in the fibril. A structural model of the fiber and its changes on wetting that accounts for both the changes in the WAXS and SAXS pattern is shown in Fig. 5. The dry fibril is made up of large diameter, well oriented fibrils (with the axis of the tropocollagen triple helix predominantly along the fibril axis) and occasional small, high aspect ratio voids between fibrils. Upon wetting these large fibrils break down into many smaller fibrils which bend to accept water in pores between them. This bending is necessitated by the UV-induced crosslinks, which connect one fibril to another at various points along its length. Since the volume fraction of collagen in the wet fiber is only ∿20% and since most of the

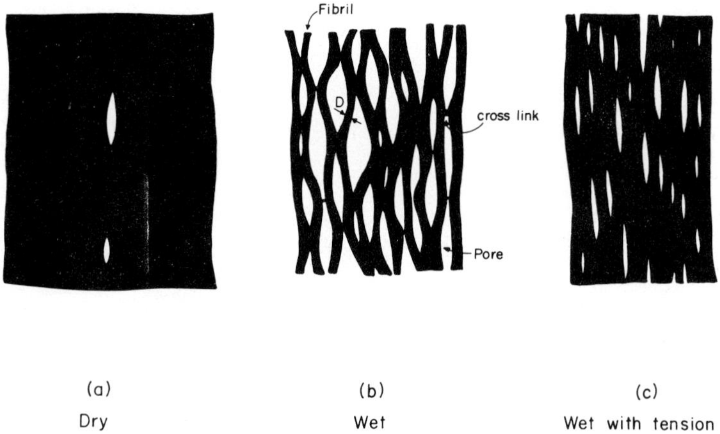

Fig. 5. Schematic drawing of the fibril and pore structure in the a) dry hollow fiber b) wet hollow fiber c) wet hollow fiber under tension.

water must reside in the pore phase [6], the fibril bending will be quite significant.

This qualitative picture of the structure is confirmed by a number of observations. The most important is the anisotropy of swelling by water. Although the volume of the hollow fiber increases by ∿400%, the fiber actually shrinks slightly along the fiber axis as a consequence of the bending of the fibrils. Even greater swelling of the collagen hollow fiber can be produced by soaking it in acidic solution [19] which positively charges the collagen molecules in the fibrils and should therefore produce a larger bending of the fibrils between crosslinks. The greatly increased lateral swelling of the fiber caused by decreasing the pH is accompanied by an increased shrinkage along the fiber axis. On the other hand increasing the ionic strength of a low pH solution by adding NaCl results in an increased screening of the charge on the fibrils and therefore decreases the lateral swelling and the bending of the fibrils. An increase in length of the hollow fiber accompanies this decrease in lateral swelling [19].

The fibril diameter can be determined from the SAXS pattern from an unoriented array of hollow fibers. For our two-phase collagen-water system we find that Porod's law [20] holds for the high angle tail of the continuous scattering [6], i.e.,

$$\lim_{s\to\infty}[I(s)s^4] = \text{constant} \tag{1}$$

where I(s) is the intensity at the scattering vector $\underline{s}$ ($|s| = 2\sin\theta/\lambda$ where λ is the x-ray wavelength and θ the scattering angle). One can then define a correlation distance ℓ_p as

$$\ell_p \equiv \frac{2}{\pi^2} \frac{\int_0^\infty I(s)s^2 ds}{\lim_{s\to\infty}[I(s)s^4]} \tag{2}$$

The integral in the numerator is the scattering invarient which is proportional to the volume of material irradiated while the Porod's law limit is proportional to the surface area between the two phases. The correlation distance is thus proportional to the inverse of the specific surface area S, viz.

$$\ell_p = 4\, v_f(1-v_f)/S \tag{3}$$

where v_f is the volume fraction of the fibril phase. The correlation distance is in turn related to the range of inhomogeneity in the two phases, ℓ_f and ℓ_w, as

$$\ell_p = \ell_f(1-v_f) = \ell_w v_f \tag{4}$$

The ℓ_f and ℓ_w are the average lengths of segments of straight lines drawn at random in the fibril and water phases respectively [20]. For an array of straight long cylinderical fibrils ℓ_f is exactly the fibril diameter D. [For a distribution of fibril diameters D_i, $\ell_f = \Sigma D_i^2/\Sigma D_i$.] The ℓ_w cannot be as directly interpreted but if we assume the fibrils are arranged on a rough triangular lattice with a center to center spacing L then

$$L = \ell_w \left(\frac{\pi}{2\sqrt{3}}\right)^{\frac{1}{2}} \frac{v_f^{\frac{1}{2}}}{1-v_f} \tag{5}$$

The fibril diameters have been determined in various environments [19] and for various crosslinking treatments [21]. As expected from the qualitative discussion above, the fibril diameter decreases from being more than 1000Å for the dry hollow fiber to ∿30Å in the wet neutral fiber. This decrease in lateral fibril dimension on wetting is consistent with the loss of the 48Å equatorial reflection and the extreme broadening of the 12Å equatorial reflection in the WAXS pattern. Charging the collagen molecules by decreasing the pH to 2.2 causes a further breakdown in apparent fibril lateral dimensions to about 20Å, not much larger than a single collagen molecule diameter. On the other hand increasing the ionic strength at pH 2.2 and thus screening the charge on collagen molecule results in an increase in fibril dimensions (to ∿50Å at pH 2.2, 0.8 M NaCl). At still higher values of ionic strength in acid solution the fibrils become so large that the 12Å equatorial reflection becomes sharp enough to be again a prominent feature in the WAXS pattern [22]. This result is still more support for the idea that the changes in the WAXS pattern on wetting are primarily due to decreases in the diameters of the coherently diffracting fibrils. As one also might expect for decreasing crosslinking times there are modest decreases observed in the fibril diameters in neutral water. The decreases however are small enough as to be at the limits of detection with the SAXS technique so that further analysis of them is probably not warranted.

MECHANICAL PROPERTIES

Our interest in the mechanical properties originates in the very poor wet strength of the reconstituted collagen product which confines its use as a membrane to geometries, such as the small hollow fiber, which give rise to only small stresses in the membrane. We soon found that this

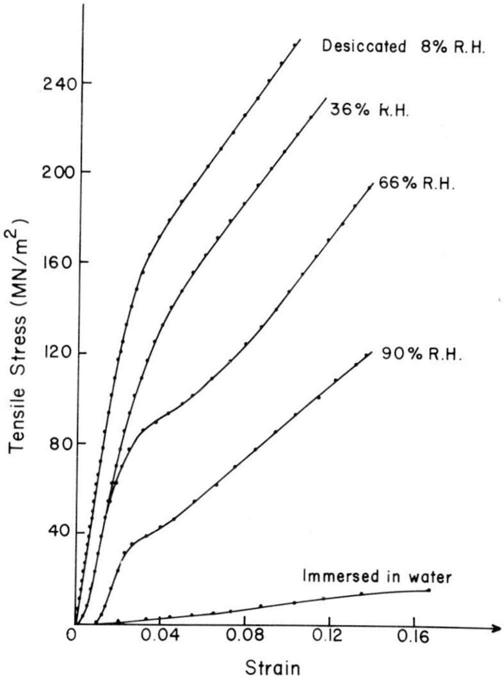

Fig. 6. Tensile stress-strain curves of the hollow fiber at various relative humidities.

material has large changes in its mechanical properties with relative humidity, pH, ionic strength and crosslinking that parallel the changes in structure just discussed. The fibers were tested in tension along the fiber axis and the stress in the fiber was computed based on the dry cross-sectional area of the fiber. [This procedure is appropriate since in the wet fibers the water in the pores between fibrils is incapable of bearing any load.] The striking changes that occur in the stress-strain curve as a function of relative humidity are shown in Fig. 6. At relative humidities of more than ∿30% a small "toe", a region of positive curvature at low strains, develops and increases with increases in humidity. There is a distinct increase in the extent of this toe on taking the fiber from an apparently 100% relative humidity atmosphere and immersing it in water, just as there is a large increase in swelling (Fig. 2) as a result of this treatment. The extent of the toe correlates well with increases and decreases in swelling produced by other treatments. Fig. 7 shows that decreasing the pH of the aqueous environment, a treatment which greatly increases the swelling, also dramatically increases the extent of the toe in the stress-strain curve. On the other hand screening the charge on the fibrils by adding NaCl to an acidic environment decreases the swelling and also decreases the extent of the toe in the stress-strain curve as shown in Fig. 8.

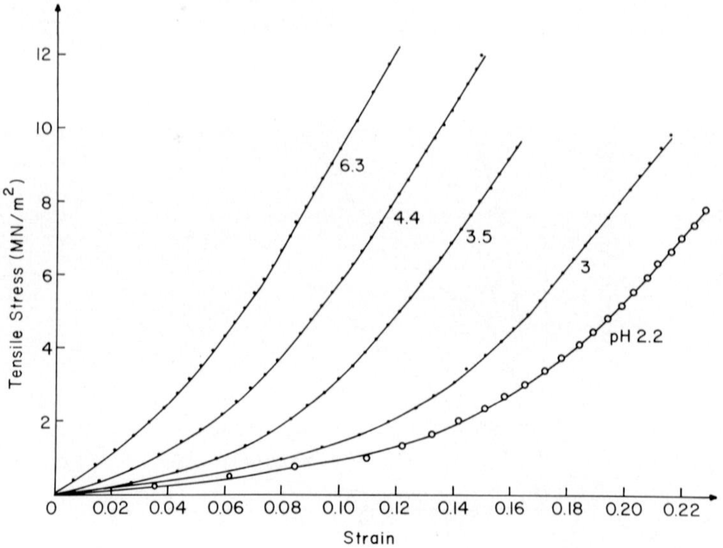

Fig. 7. Tensile stress-strain curves of the hollow fiber aqueous solutions for various pH values.

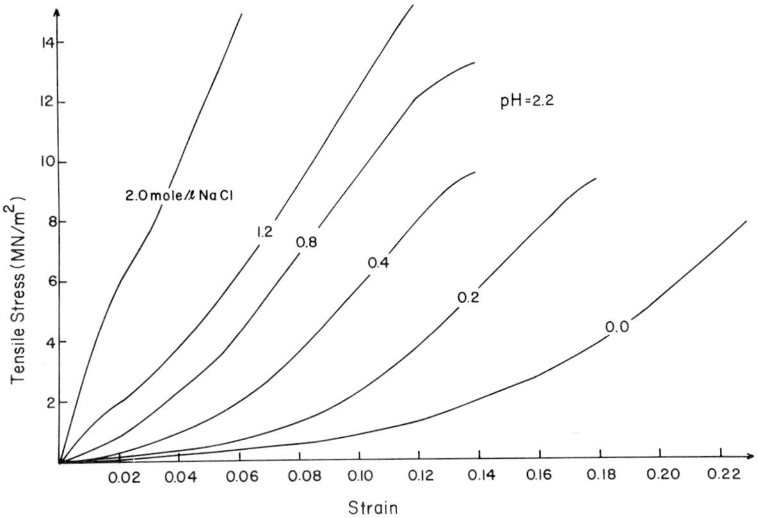

Fig. 8. Tensile stress-strain curves of the hollow fiber in aqueous solution (pH 2.2) for increasing ionic strength.

Recalling the structural model in Fig. 5 the most straightforward explanation of the toe in the stress-strain curve is that it arises due to the straightening of the fibrils bent to accommodate water between them. To see if this idea works quantitatively we assume that the linear region of the stress-strain curve past the toe represents the true elastic stretching of fibrils and thus the slope of this linear region represents the true elastic (Young's) modulus E of the fibrils. The extrapolation of this line to zero stress yields a strain ε'_∞ which should represent the strain necessary to completely straighten the fibrils. The first test of this idea is to compare the longitudinal shrinkage strain ε_s of the fiber on swelling with ε'_∞. There exists a good qualitative correspondence between these two quantities although ε'_∞ tends to be systematically somewhat higher than ε_s (e.g., at pH 2.2 ε'_∞ = 14.4% whereas ε_s = 12%.) Apparently environments which cause lateral swelling also cause a slight lengthening of the fibrils themselves, partially offsetting the decrease in fiber length due to fibril bending. Additional evidence for the proposed change in the fibril/pore structure of the membrane comes from the change in the SAXS pattern of a wet fiber with increasing tensile strain [6]. The continuous scattering ellipse becomes noticeably flattened along the tensile axis (along the meridian of Fig. 4b). This change in aspect ratio of the SAXS is exactly what one would expect to observe by straightening a dilute system of bent or microcrimped fibrils.

We can also attempt to mathematically represent the toe region of the stress-strain curve as the extension of a curved elastica. One such model is to consider each fibril as being initially an idealized zig-zag with each segment of the zig-zag being uniformly of length ℓ and diameter D and lying an angle of either + or - θ_o to the fiber (tensile) axis. This model, illustrated in Fig. 9a, was first worked out by Diamant, Keller, Baer, Litt and Arridge [23] in their now classic paper on the extensibility of macroscopically crimped rat tail tendon collagen fibers. It predicts that an equation of the form

$$\varepsilon - \frac{\sigma}{E} = \varepsilon_\infty - \lambda\sqrt{E/\sigma} \qquad (6)$$

will hold over a reasonable range of strain. Here $\lambda = (D/\ell)[1-\cos(\theta_o/2]/\cos\theta_o$ and $\varepsilon_\infty = (\sec\theta_o - 1)$. Empirically Eq. 6 produces a good fit to our

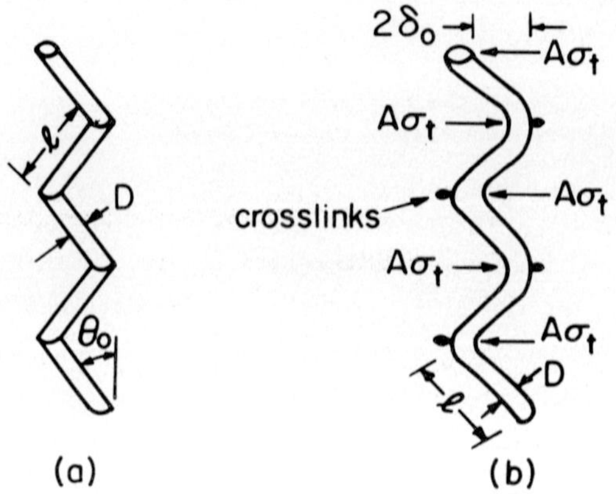

Fig. 9. a) Schematic drawing of the zig-zag crimp model of Diamant et al. [23]. Initial zig-zag is unstressed. b) Schematic drawing of a single fibril bent by lateral forces, $A\sigma_t$, acting between crosslinks.

σ-ε data except at very low strains where because of the mathematical approximations made in its derivation it cannot be expected to hold [19,24]. An example of the fit is shown in Fig. 10 which is a plot of data at pH 2.2 and various ionic strengths.

On further reflection however this fit is disquieting since in the analysis of Diamant et al. the zig-zag crimp is initially unstressed whereas the bending of the fibrils between crosslinks in the reconstituted collagen must arise due to a lateral repulsion between fibrils. The straightening of a fibril crimped by lateral forces is not governed by the same equations as the straightening of the stress-free crimp. Even though the analysis of Diamant et al. produces a good empirical fit to the data,

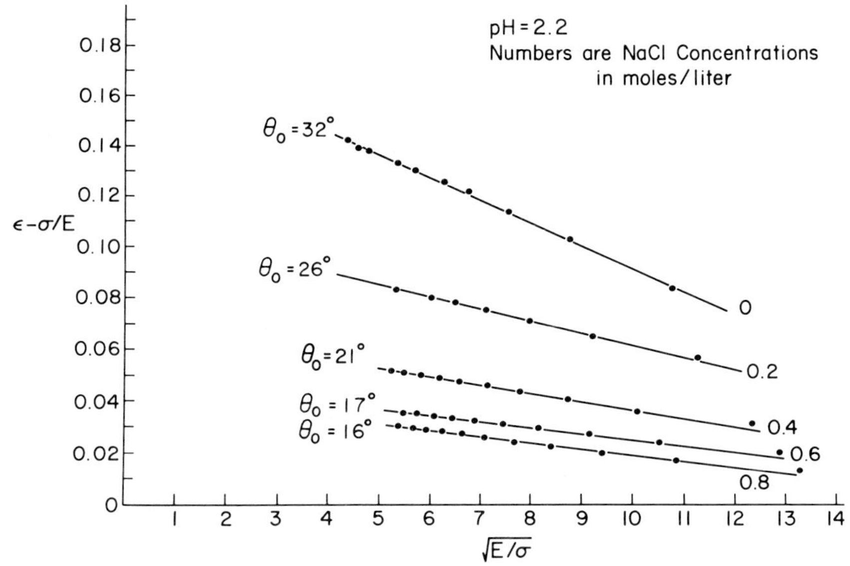

Fig. 10. Plot of the parameters of model of Diamant et al. [Eq. 6] for the data at pH 2.2 and various ionic strengths.

it is not clear that the crimp parameters derived from this analysis have any basis in reality. Accordingly numerical solutions for the tensile deformation of a single fibril crimped by lateral forces (Fig. 9b) have been obtained [19]. The results are of the form

$$Q_1(\ell/D,\ \sigma_t,\ \varepsilon_\infty) = 0 \tag{7}$$

$$Q_2(\sigma, \varepsilon, \ell/D,\ \sigma_t,\ \varepsilon_\infty) = 0 \tag{8}$$

where $A\sigma_t$ is one of the lateral forces responsible for bending, A is the fibril cross-sectional area and Q_1 and Q_2 are functions of the indicated variables involving elliptic integrals of the first and second kind [19]. Providing ε_∞ can be estimated (e.g., by extrapolation of the linear portion of the stress-strain curve) a single σ, ε point suffices to determine the ℓ/D ratio of the fiber and σ_t. The entire stress-strain curve then can be computed from these data assuming σ_t and ℓ/D constant. The fit to the experimental stress-strain curve at neutral pH is excellent as seen on Fig. 11.

When the fibrils are charged (at low pH), one does not expect to be able to fit the entire stress-strain curve with a single value of σ_t since the lateral repulsive forces between charged fibrils will increase with decreasing lateral separation. The data in fact confirm this expectation.

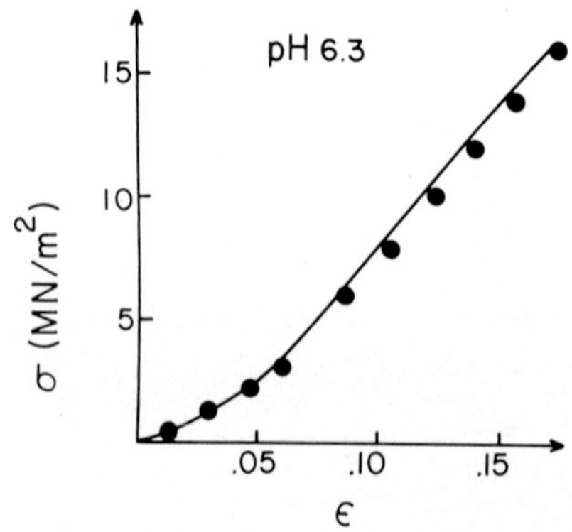

Fig. 11. Experimental stress-strain curve for pH 6.3 (solid line) and the stress-strain curve computed from the lateral force model (solid circles).

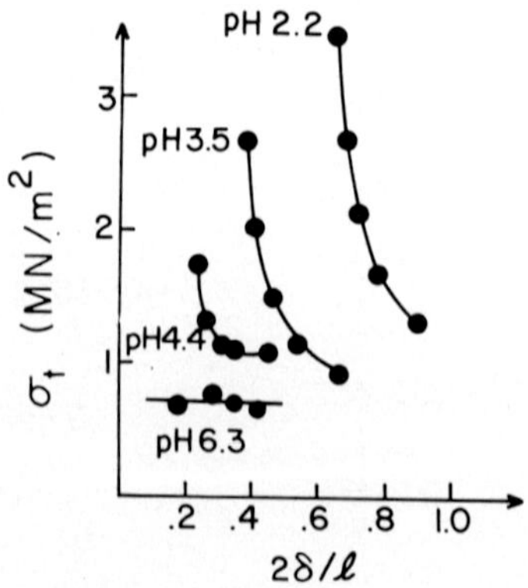

Fig. 12. Lateral stress σ_t needed to match experimental stress-strain data versus the normalized average separation between fibril $2\delta/\ell$ for several pH values.

To demonstrate this departure from the simple model the σ_t's needed to make the theoretical curve fit the experimental points are plotted in Fig. 12 versus the normalized average separation between fibrils $2\delta/\ell$ which also can be computed at each value of strain from the model [23]. As implied by Fig. 11 σ_t is independent of $2\delta/\ell$ for pH 6.3 but increases strongly with decreasing $2\delta/\ell$ for the lower pH curves. This increase in σ_t is more rapid than the δ^{-2} dependence of a unscreened Coulomb repulsion, a result which is also expected since at smaller fibril separations the screening by counter-ions becomes less, increasing the effective charges seen by one fibril on its neighbors.

The magnitude of σ_t is also in reasonable agreement with calculated values assuming all lysine, histidine, arginine and hydroxylysine residues are all singly ionized at pH 2.2 and are screened by the mobile ions in solution [19].

One further important check on the mechanical deformation model can be made using the SAXS data. If we arrange the crosslinked bent fibrils in a unit-cell we can compute the specific surface S and the SAXS correlation distance ℓ_p. Figure 13 shows one such arrangement of fibrils which has 2 fibrils per unit cell. (We do not imply that the bent fibrils are actually arranged on a regular lattice; Fig. 13 simply allows us to compute the specific surface of a structure which has the same v_f, $2\delta_o$ and D as the more random arrangement achieved in practice.)

From the correlation distance, the fibril volume fraction v_f, and ε_∞ the fibril diameter D and the unit cell lattice parameter L can be computed to be

$$D = \frac{\ell_p}{1-v_f} \tag{9}$$

$$L = \frac{\ell_p}{(v_f)^{\frac{1}{2}}(1-v_f)} \left(\frac{\pi}{2}\right)^{\frac{1}{2}} (\varepsilon_\infty+1)^{3/2} \tag{10}$$

Further from consideration of the geometry in Fig. 13

$$2\delta_o = (L^2-D^2)^{\frac{1}{2}} - D \tag{11}$$

so that the lateral deflection of the fibrils can be predicted from the SAXS and ε_∞ data. On the other hand values of $2\delta_o/D$ can be computed from the stress-strain curve and ε_∞ using the mechanical model [19]. If this ratio is combined with the SAXS values of fibril diameter a second independent estimate of $2\delta_o$ is obtained. These two estimates for pH 6.3 and pH 2.2 together with the SAXS values of L and D and the values of $2\delta_o/D$ computed from the stress-strain curve [19] are shown in Table I. The

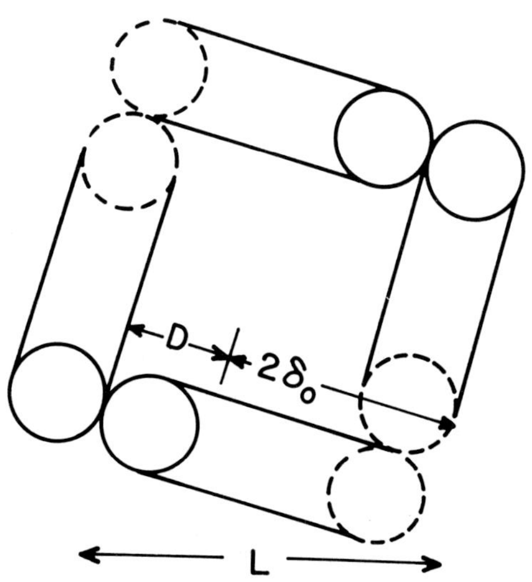

Fig. 13. A possible average "unit cell" of bent fibrils viewed along the hollow fiber axis.

Table I

pH	L	D	SAXS $2\delta_o$	Mechanical $2\delta_o$	Model $2\delta_o/D$
	Å	Å	Å	Å	
2.2	91	20	70	68	3.48
6.3	79	29	45	44	1.54

excellent agreement between the two estimates of $2\delta_o$ must be considered fortuitous in view of the simplified nature of the model and the possible errors in the SAXS data but nevertheless it is a further indication of the basic correctness of the model. Note also that when the increased bending of the fibrils (which decreases ℓ_p) is taken into account the average fibril diameter decreases only slightly as the pH is decreased from 6.3 to 2.2. The fibril length to diameter ratios derived from the stress-strain data using the model are also approximately independent of pH. Apparently the fibrils bow out between crosslinks without appreciably changing diameter as the pH is decreased.

It is interesting to compare the elastic behavior of the reconstituted hollow fiber with that of native tendon collagen. In the latter case it is well known that the fibrils are macroscopically crimped. A toe in the stress-strain curve of tendon is observed and has been attributed to the straightening of this macroscopic crimped structure [23]. On the other hand it is known that native collagen fibers in acidic or basic media shrink reversibly along the fiber (molecule) axis but swell laterally resulting in a net increase in the volume fraction of water in the fiber [25,26]. If held at constant length a considerable tensile force is developed by the fiber [26]. It is not unreasonable to expect such a fiber at low pH to develop a microcrimp similar to that developed in the reconstituted collagen hollow fiber. Furthermore such a native fibril would exhibit very similar changes in the stress-strain curve with pH to those demonstrated for the hollow fiber. Whether such microcrimp structure, actually exist and contribute to the observed stress-strain curve is not yet definitely established but in view of the interest in the mechanical properties of tendon it is a problem well worth examining in the future.

The origin of the large changes observed in the fibril elastic modulus E (defined as the slope of the stress-strain curve in the linear region past the toe) with relative humidity is more obscure. Figure 14 shows that a major decrease occurs in E as the relative humidity is increased with a very low value ($\sim$100MN/m^2) being reached when the fiber is soaked in water

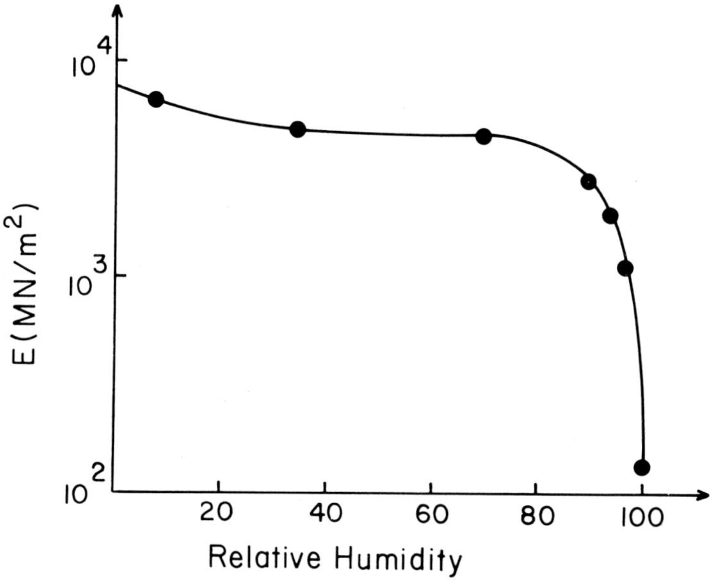

Fig. 14. Young's modulus of the fibrils (based on dry crossection of the hollow fiber) versus relative humidity.

[24]. This value is more nearly typical of a heavily crosslinked elastomer than a glassy or semicrystalline fiber. Relative minor decreases in wet modulus occur with decreases in crosslinking time or pH of the aqueous medium.

One possible explanation for the low modulus is that portions of the tropocollagen molecule unwind into constrained random coils in the presence of water and thus deform in a predominantly rubbery manner in this environment. It was previously suggested that the ultraviolet irradiation crosslinking treatment might lower the stability of the triple helix locally to allow such constrained random coils to form. Indicating such a possibility was evidence from the 221 nm circular dichroism absorption peak of UV irradiated dilute solutions of collagen which showed a small decrease in this peak (decrease in helical content of the molecule) at room temperature and a small shift in the helix coil transition) with UV irradiation. Recent differential scanning calorimetry results however demonstrate that the helix-coil transition (shrinkage transition) is at $\sim$45°C for the hollow fibers and more importantly that the enthalpy of the transition is approximately the same as that for native tropocollagen molecules in dilute solution [27]. This evidence appears to rule out UV irradiation-induced denaturation of significant portions of the tropocollagen triple helix at room temperature in the hollow fiber and thus UV denaturation cannot account for the large decrease in modulus on wetting. It might still be however that both the native and the reconstituted collagen fibrils have non-helical (or constrained random coil) regions along their lengths which could give rise to a low modulus. However comparison of the apparent modulus of the wet hollow fiber (calculated on the basis of the total wet cross sectional area rather than the dry cross sectional area as all previous stresses and moduli in this paper) with that calculated from measurements on rat-tail tendon [23] show that the tendon modulus is about a factor of ten higher than the hollow fiber modulus. This comparison (which of course is not a true comparison of fibril moduli) does not support the hypothesis that the low fibril modulus in the hollow fiber is due to regions of constrained random coil or elastomeric regions along the molecule. The fact that the modulus decreases with increasing temperature, indicating that there are important enthalpic contributions to the modulus, also is not in accord with an elastomeric explanation although one can rationalize the temperature dependence by supposing the T_g of these regions to be slightly below room temperature.

If the elastomeric hypothesis of the low modulus is difficult to defend, it is also difficult to support the alternatives. One possibility that is difficult to assess quantitatively is that the fibrils all cannot be straightened so as to be fully loadbearing at the same time. To account for the ten-fold decrease in apparent modulus between the native rat-tail tendon and the hollow fiber would require that the load is borne only by $\sim$1/10 of the total fibrils with 9/10 of the fibrils having enough slack that when the load bearing fraction is pulled taut they do not contribute. In the dry state these small fibrils would be laterally bonded to each other over most of their length promoting the same kind of load transfer between fibrils that allows high extensional moduli to be attained in composites made of discontinuous filaments. As this lateral bonding and load transfer weakens with increased relative humidity the modulus would drop. This explanation implies that it might be possible to increase the modulus of the wet hollow fiber by plastically deforming the fiber (and perhaps recrosslinking by UV irradiation afterwards) in the dry state. This treatment might plastically extend the tightly stretched fibrils and redistribute the load over the remainder. At low relative humidities considerable plastic deformation of the fiber is possible, taking place at stresses above σ_y on Fig. 6. It will also be worthwhile to examine the mechanical properties of native collagen fibrils more closely so as to clearly define the improvement possible in reconstituted collagen fibers.

The very low modulus of the wet hollow fiber also probably has a important effect on its fracture strength. Since the critical stress intensity factor for crack propagation in the membrane can be approximated by

$$K_{I_c} = \sqrt{G_{I_c} E} \tag{11}$$

where G_{I_c} is the critical strain energy release rate (or the total work necessary to advance a crack over a unit area) the factor of >500 decrease in apparent E on wetting will produce a factor of >20 decrease in K_{I_c} if the G_{I_c} is constant. This decrease is of the same order of magnitude as the actual fracture strength decrease on wetting. It is likely that significant increases wet strength of the membrane can be produced if a method for increasing the apparent modulus of the wet membrane can be found.

TRANSPORT PROPERTIES

The diffusivities of solutes through the membrane are of great importance for clinical applications. Measurements of the ratio of the effective diffusivity D_{eff} of a solute in the membrane to the diffusivity D_l of the same solute in the solution have been made using an transient diffusion experiment employing radioactive solutes [28]. Figure 15 shows the results for solutes of three different molecular weights, urea, sucrose and polyethlylene glycol (MW = 4000). For comparison data for diffusion of the same solutes through conventional cellulose acetate hollow fiber dialysis membranes are also shown [28]. The collagen hollow fiber has significantly higher permeability than cellulose acetate although this apparent advantage is somewhat offset in practice since the thickness of the current collagen hollow fiber wall is greater than that of the cellulose acetate hollow fiber.

One expects that D_{eff}/D_l will begin to decrease rapidly when the diameter of the solute molecule becomes comparable to the average distance between collagen fibrils in the membrane. On the basis of the mechanical model one might represent this average distance by $2\delta_o$ (46Å at pH 6.3) and compare it with the diameter of the polyethylene glycol molecule (∿33Å) since at this point D_{eff}/D_l is only ∿1/5 its value for small solutes. Again it would appear that the dimensions of the fibril structure determined from SAXS and the mechanical model are in reasonable agreement with the properties measured. Preliminary data for higher molecular weight solutes indicate much lower D_{eff}/D_l values ($\frac{D_{eff}}{D_l} < 10^{-2}$ for bovine serum albumen,

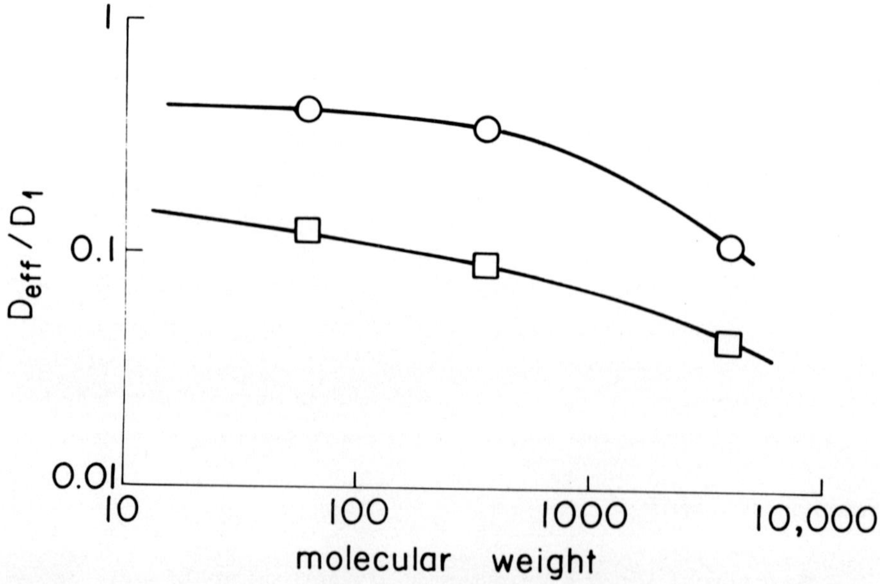

Fig. 15. The ratio of the effective diffusivity D_{eff} of the solute through the membrane to its diffusivity D_l in the solution as a function of molecular weight of the solute. The circles represent data on the collagen hollow fiber membrane; the squares represent data for cellulose acetate hollow fibers [28].

MW $\simeq$ 70,000). For the very low molecular weight solutes one expects the D_{eff}/D_l to be primarily a function of the volume fraction of the continuous water phase. Models of the diffusivity based on analogies to the problem of the electrical conductivity of a system of non-conducting cylinders in a conducting medium [29] appear to predict the magnitude of D_{eff}/D_l in the limit of zero solute size reasonably well [30].

Another clinically important, and more structure-sensitive transport property is the hydraulic permeability κ of the membrane which is given as

$$\kappa = \mu Q \ln(r_o/r_i)/2\pi L \Delta P \tag{12}$$

where Q is the volume flow rate of water through, and ΔP is the pressure difference across the wall of the hollow fiber; r_i, r_o and L are its inner and outer radius, and length respectively. If the flow rate of water is measured for a given imposed pressure drop, κ can be computed. It is important in a membrane which has as low an elastic modulus as the present hollow fiber to correct for the change in r_i and r_o with pressure. If this correction is made κ is essentially independent of pressure.

Figure 16a, b show the changes in κ that take place with increased

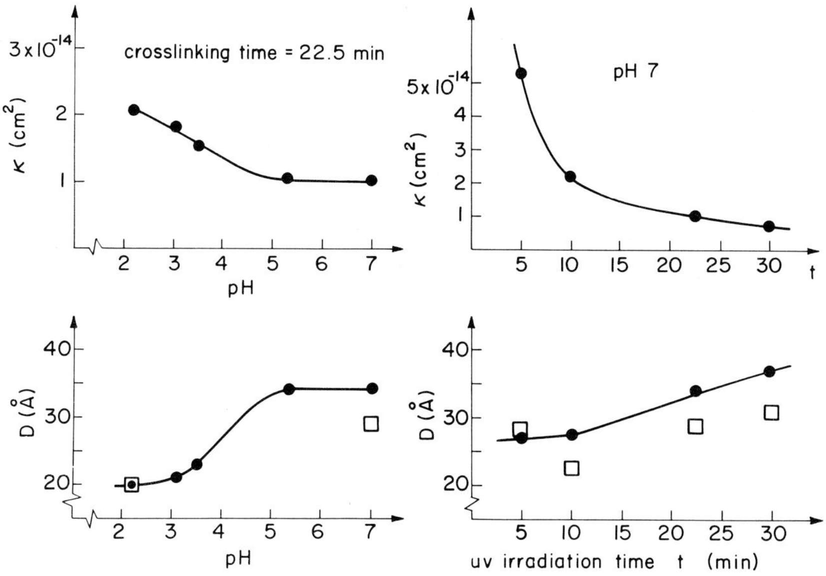

Fig. 16 a,b Hydraulic permeability κ and fibril diameters (filled circles) computed from κ via Eq. 11. Square data points are SAXS values of fibril diameter assuming straight cylinderical fibrils.

crosslinking and decreased pH. The hydraulic permeability decreases rapidly at first with increased crosslinking but then apparently saturates at somewhat less than $1 \times 10^{-14} cm^2$. Decreasing pH increases κ but not as much as one might anticipate from the large decrease in v_f (from 0.215 at pH 6.3 to 0.0875 at pH 2.2).

If the mechanical model and the schematic drawing shown in Fig. 5 are correct, it should be possible to decrease κ substantially by applying a tensile strain. The results of measurements at various tensile strains show a very marked decrease (Fig. 17) with κ being reduced by approximately a factor of five by a tensile strain of only 10%. These changes in κ are reversible, in that if the strain is removed κ recovers its original value. The large decrease in κ with tensile strain is a striking confirmation of the bent fibril model of the membrane.

It is of interest to see if the changes in κ from one membrane to another (e.g., Fig. 16) can be accounted for quantitatively in terms of structural differences in the membrane, i.e., differences in v_f and D. It seems reasonable based on Fig. 5 to model the flow as the flow of water through a forest of straight, long, solid cylinders [21]. An approximate

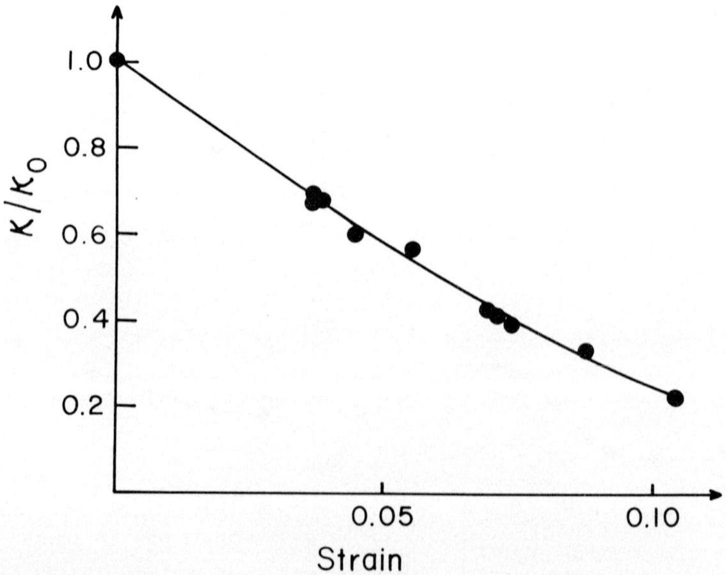

Fig. 17. The ratio of hydraulic permeability under strain to hydraulic permeability under zero strain versus applied strain along the hollow fiber axis.

solution given by Happel [31] yields

$$\kappa = \frac{D^2}{32v_f}\{\ell n(1/v_f) - \frac{1-v_f^2}{1+v_f^2}\} \quad (13)$$

This model can be tested by using the experimental values of κ and v_f in Eq. 13 to estimate values for D which can then be compared with the SAXS values. This comparison is made in Fig. 16a, b as a function of pH and UV irradiation time. The fibril diameters directly determined from SAXS (assuming straight cylinderical fibrils) are shown as open squares whereas the values determined from κ and v_f data are shown as the filled circles. The very good agreement, as well as the dramatic decrease in κ with tensile strain, is further evidence for the structural model pictured in Fig. 5.

The sensitivity of the κ of the membrane to structure has both its good and bad features. On one hand the sensitivity magnifies quality control problems. One particularly vexing problem is that the membrane apparently undergoes additional crosslinking when aged at low humidities which leads to a decrease in κ with time. On the other hand it is often desireable to be able to decrease κ relative to the solute permeability of the membrane, which is much less structure sensitive, at least for small solutes. If κ is too high relative to D_{eff}, patients may become dehydrated during dialysis. The possibility of decreasing κ by up to five-fold by applying a small tensile strain parallel to an array of the fibers, should be useful in adjusting κ to an optimum value.

ACKNOWLEDGEMENTS

The research reviewed here and the writing of this review, was supported by the National Science Foundation (NSF) through grant DMR75-02897. The research also benefitted from the use of the facilities of the Cornell Materials Science Center which is also funded by NSF. The cooperation of Dr. Tereo Miyata of the Japan Leather Co., who furnished numerous specially processed hollow fibers as well as very cogent advice, is gratefully acknowledged. I especially thank Prof. James F. Stevenson for many discussions on the interpertation of the transport properties and Prof. Dieter Ast for his views and data on the thermodynamic properties of the hollow fiber. Most of the experimental data reviewed here was the work of two very energetic postdoctoral associates, Dr. R. K. Viswanadham and Dr. D. C. Agrawal. Finally I would like to acknowledge useful conversations with Prof. J. Jenkins, Dr. A. L. Rubin, M.D., Dr. K. H. Stenzel, M.D. and Prof. H. H. Johnson.

REFERENCES

1 G.N. Ramachandran, Treatise on Collagen, ed. G.N. Ramachandran (Academic Press, New York, 1967).

2 A.L. Rubin, M.P. Drake, P.F. Davidson, D. Pfahl, P.T. Speakmam and F.O. Schmitt, Biochemistry, 4 (1965) 181.

3 F.O. Schmitt, L. Levine, M.P. Drake, A.L. Rubin, D. Pfahl and P.F. Davidson, Proc. Nat. Acad. Sci., 46 (1960) 197.

4 K.H. Stenzel, A.L. Rubin, W. Yamayoshi, T. Miyata, T. Suzuki, T. Shode and M. Nishizawa, Trans. Amer. Soc. Artif. Internal. Organs., 17 (1971) 293.

5 A.L. Rubin and K.H. Stenzel, in Biomaterials, ed. by L. Stark and G.C. Agrawal, Proc. of Univ. of Illinois Workshop (Plenum Press, New York, 1969).

6 R.K. Viswanadham and E.J. Kramer, J. of Materials Science, 10 (1975) 1472.

7 M.A. Rougive and R.S. Bear, J. Leather Chem. Assoc., 48 (1953) 735.

8 H.R. Elden, Int. Revs. Connect. Tissue Res., 4 (1968) 283.

9 P. VonSchroeder, Z. Phys. Chem., 45 (1903) 75.

10 J.W.G. Musty, J. Appl. Chem., 16 (1966) 221.

11 A.C.T. North, P.M. Cowan and J.T. Randall, Nature (London) 174 (1954) 1142.

12 A. Miller and T.S. Wray, Nature (London) 230 (1971) 437.

13 J.W. Smith, Nature (London) 219 (1968) 157.

14 V. Sasisekharan and G.N. Ramachandran, Proc. Indian Acad. Sci., Sect. A 45 (1957) 363.

15 R.E. Burge, J. Mol. Biology, 7 (1963) 213.

16 R.S. Bear, O.E.A. Bolduan and T.P. Salo, J. Leather Chem. Assoc., 46 (1951) 107

17 A.J. Hodge, in Treatise on Collagen, ed. G.N. Ramachandran (Academic Press, New York, 1967).

18 T. Miyata, private communication.

19 R.K. Viswanadham, D.G. Agrawal and E.J. Kramer, J. Polymer Sci. 14, Nov. 1976 (Materials Science Center Report #2628).

20 A. Guinier and G. Fournet, Small Angle Scattering of X-rays (Wiley, New York, 1969).

21 R.K. Vishwanadham, D.G. Agrawal, J.F. Stevenson and E.J. Kramer, "Water Transport Through Collagen Hollow Fiber Membranes" Materials Science Center Report # 2736, submitted for publication.

22 A. Marciente, unpublished research at Cornell University.

23 J. Diamant, A. Keller, E. Baer, M. Litt, and R.F. Arridge, Proc. Roy. Soc. London B, 180 (1972) 293.

24 R.K. Viswanadham and E.J. Kramer, J. Materials Sci., 11 (1976) 1254.

25 A. Ciferri, L.V. Rajagh and D. Puett, Biopolymers, 3 (1965) 461.

26 I.V. Yannas and A.J. Grodzinsky, J. Mechanochem. Cell Motility, 2 (1973) 113.

27 A. Schwartz, G. Rosenblatt and D.G. Ast, private communication.

28 J.F. Stevenson, M.A. von Deak, M. Weinberg and R.W. Schuette, AIChE. J., 21 (1975) 1192.

29 H.B. Keller and D. Sachs, J. Appl. Phys., 35 (1974) 537.

30 J.F. Stevenson, unpublished results.

31 J. Happel, AIChE. Journal, 5 (1959) 174.

First Cleveland Symposium on Macromolecules

MECHANICAL PROPERTIES OF NATIVE FIBROUS MATERIALS

LUDWIG REBENFELD

Textile Research Institute, Princeton, New Jersey

SUMMARY

The utilization of native cellulosic and protein fibers in natural products as structural elements, and their extensive use as commercially important industrial materials, is due mainly to their favorable combination of mechanical properties. The mechanical properties of these fibers, in turn, reflect their morphology and fine structure. Both cellulosic and protein fibers are structurally complex. The molecular chains are aggregated into various structural subunits, such as crystallites, micelles, fibrils, and lamellar layers, which are then organized in a precise geometric arrangement in the ultimate fiber.

In the case of native cellulosic fibers (cotton), the deformation behavior appears to be strongly influenced by inter-fibrillar hydrogen bonds. At the same time, recent studies have shown that the mechanical properties of these fibers are strongly dependent on chemical environment. In the case of protein (keratin) fibers, such as wool and human hair, the mechanical properties similarly reflect their complex structure. Special attention, however, has recently been given to surface properties. These fibers have a unique scalar surface structure which is responsible for many of their interesting inter-fiber characteristics. New measurements of frictional properties and of surface energies of these fibers are described and interpreted in terms of their morphology and fine structure.

INTRODUCTION

Native cellulosic and protein fibers are frequently found in natural products as structural elements providing strength and mechanical integrity. In many cases, naturally occurring fibers, principally cotton, wool, and paper fibers, have also become commercially important industrial materials. The extensive utilization of these fibers in the textile, paper, and allied industries is due mainly to their favorable combination of mechanical properties. These fibers fulfill industrial requirements of strength, extensibility, toughness, resilience, and flexibility. The mechanical properties of these fibers, in turn, reflect their morphology and fine structure. Both cellulosic and protein fibers are structurally complex. The molecular chains are aggregated into various structural subunits, such as crystallites, micelles, fibrils, and lamellar layers, which are then organized in a precise geometric arrangement in the ultimate fiber. The morphological foundations of fiber properties were reviewed in 1965 by Rebenfeld [1] based upon our understanding of fiber structure at that time. Since then numerous features of fine structure have been revealed and clarified. It is these aspects of structure and particularly their relationships to mechanical behavior that is emphasized in this paper.

Probably one of the most important developments that has taken place in the field of fiber science during the past decade is the recognition that a special characteristic of fibrous materials is their large surface area per unit mass. Compared to other solids, such as bulk substances and films, the overall characteristics of fibrous materials are dominated by surface properties. Accordingly, the interactions of fibrous materials with their liquid and vapor environments are of extreme importance and the subject of intensive studies. Interactions with liquids in particular are of special interest, and measurements of fiber wettability properties have been accorded much attention. The purpose of such studies is two-fold. First, information is gained about the manner in which a fiber will respond when it is exposed to a liquid environment. This is of importance in biological systems, such as when natural fibrous tissue interacts with various body fluids or when fibrous prosthetic devices are implanted in the body. This is also of importance in industrial processing of fibrous materials, such as in the paper and textile industries. Secondly, measurements of fiber wettability properties are undertaken to provide fundamental structural information about the surfaces of fibers. As in all solids, the intermolecular organization at the surface differs from that in the bulk of the material. This underlying fact takes on special importance in the case of fibers where the surface area per unit mass is so high. Accordingly, special attention in this paper will be given to the surface properties of keratin fibers, particularly to new measurements and concepts of surface free energies.

COTTON FIBER GROWTH AND STRUCTURE

The morphology of cotton fibers and its attendant fine structure have been the subject of scientific studies for many years. In summarizing the essential features of cotton fiber growth and structure, one must bear in mind that a cotton fiber is a single biological cell that grows as a seed hair on the plant belonging to the genus Gossypium of the tribe Hibisceae. There are several cultivated species grown in the warmer climates of the temperate zone around the globe. The growth of the fiber proceeds in several distinct stages. The initial stage involves the lengthening of the epidermal cell to its full length of approximately 20 to 40 mm. The hollow tube consisting at this state only of a thin fibrillar membrane (0.2 μm), referred to as the primary wall, encloses the living protoplasm. During the second stage of fiber growth or development, the secondary wall of the fiber, comprising the bulk of the fiber, is deposited epitaxially in fibrillar form on the inner surface of the primary wall. The fibrils have been variously described as macrofibrils, microfibrils, and elementary fibrils depending on their apparent diameter [2]. The fibrils are essentially completely crystalline units of cellulose, poly (1,4-β-D-anhydroglucopyranose), having the Cellulose I lattice.

The deposition of cellulose fibrils in the development of the secondary wall takes place in the form of concentric layers of fibrils, known as growth layers, which are associated with the diurnal cycles during the growth period. Studies by Berriman [4] of cotton fiber growth and development under controlled laboratory conditions have shown that temperature variation is a major factor in the production of growth layers in a cotton fiber. This observation suggests that it is the temperature variation during a diurnal cycle that is the cause of growth layers in a cotton fiber. The importance of the growth layers in terms of mechanical properties will be discussed later.

The secondary wall normally does not completely fill the cell leaving a center canal in the fiber which is referred to as the lumen. When the cotton boll opens, thereby exposing the seed hairs to the atmosphere, the highly hydrated cotton fiber dries out and collapses to produce the characteristic flat ribbon-like cross-section of a cotton fiber. The essential features of cotton fiber morphology are schematically shown in Figure 1.

An important feature of cotton fiber development during the deposition of the secondary wall is the fact that the fibrils are epitaxially deposited on the inner surface of the primary wall in a helical configuration, with the sense of the helix reversing periodically and rather frequently

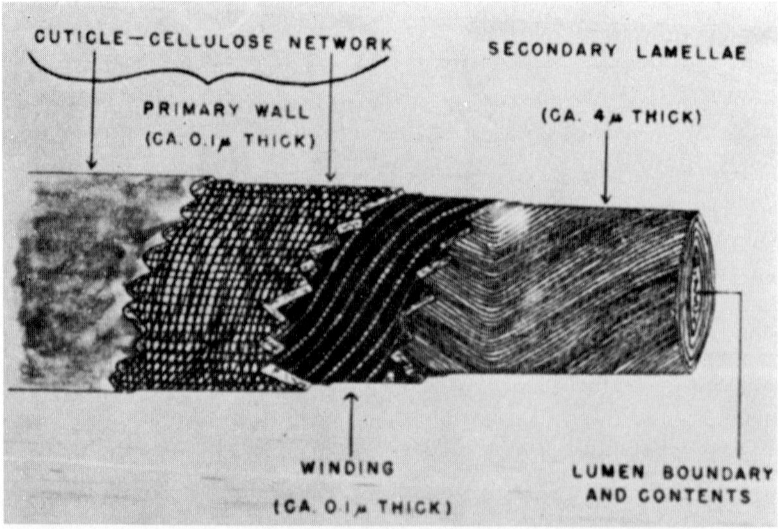

Fig. 1. Schematic diagram of layered components of a cotton fiber [3]. Photomicrograph courtesy of U. S. Department of Agriculture, Southern Regional Research Center.

along the length of the fiber. The points of reversal are uniform throughout the fiber cross-section and occur with a frequency ranging from 1 to 3 times per mm of the fiber. The origin of the helical reversals has been the subject of much study and conjecture. The most recent and extensive studies have been those of Raes and coworkers [5,6] who conclude that reversals are not growth accidents or the result of the dessication or collapsing process, but that they are in fact genetically determined. The helical reversals either contribute to or are in fact the cause of the highly convoluted form of a cotton fiber which manifests itself after boll opening. During the drying out process, the collapsed fiber twists upon itself producing many convolutions along the fiber which are presumably associated with the helical reversals.

The angle of the helix formed by the cellulosic fibrils appears to be constant throughout the fiber cross-section and also along the length of the fiber. For many years it was believed that the helix angle was a varietal characteristic varying over a fairly wide range, possibly from 20 to 45°. Recent studies by Morosoff and Ingram [7] and by Herbert [8] have shown that the helical angle is in fact constant for all cotton varieties having a value of approximately 22 degrees. The frequently measured x-ray angle of cottons, describing the azimuthal distribution of intensities of the 002 cellulosic diffraction arc, does vary over a wide range (20° to 45°). However, the x-ray angle as measured on fully grown fibers is the result of superposition of the helical and convolution angles. The x-ray angle is essentially characteristic of genetic variety, but it is somewhat influenced by growth conditions and environmental effects. Since the helical angle of the deposited fibrils has been shown to be constant for different varieties and strains of cotton, Peterlin and Ingram [2] conclude that the convolution angle, which is inversely related to the convolution frequency, is the varietal characteristic. This conclusion places even greater importance on the helical reversals which presumably determine the convolution frequency. As indicated, Raes and coworkers [5,6] have also concluded that reversals are genetically determined, and thus it would appear that there is rather consistent agreement on this particular aspect of cotton fiber structure.

In 1969, Kassenbeck [9,10] introduced a provocative and interesting concept concerning the structural uniformity of a cotton fiber. Using special techniques of ultramicrotomy and histochemical staining, Kassenbeck has shown variations in the density of fibrillar associations in the secondary wall which give rise to a bilateral structure. Kassenbeck has identified three separate zones in cross-sections of mature and fully grown cotton fibers as shown schematically in Figure 2. These zones differ in their fine structure and may be distinguished by their accessibility to chemical reagents. Zone A shows the highest packing density of fibrils, Zone B reveals a well defined system of concentric layers and is somewhat more accessible than Zone A, and Zone C constitutes the collapsed part of the structure and is the most highly accessible region of the cross-section. The proposed bi-lateral structure presumably arises from differential compressive stress distributions during the initial drying and collapse of the fiber after boll opening.

For obvious reasons most studies of cotton fiber structure and properties have been performed on fully grown fibers after initial drying and collapse of the fiber following boll opening. Pioneering studies of the structure and properties of undried fibers which were removed from unopened bolls were performed by Berkley and Kerr in 1946 [11]. Using available techniques at the time, these workers concluded that undried fibers contained essentially no regularly organized structure. The hydrated cellulose structure was considered to be amorphous, and this

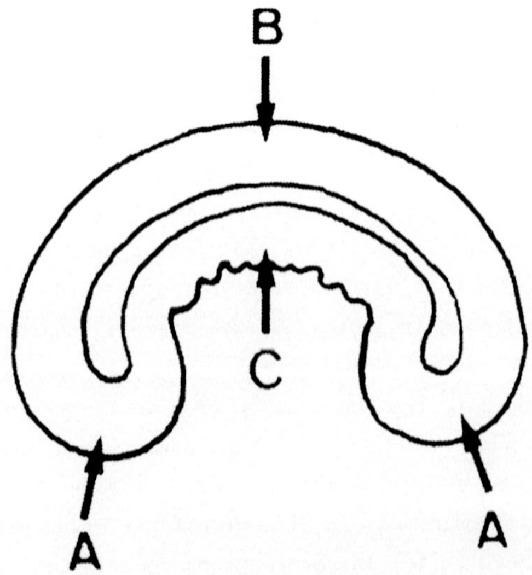

Fig. 2. Schematic diagram of a cotton fiber cross-section indicating bilateral asymmetry [9,10].

lack of crystalline structure was reflected in the high degree of plasticity of these fibers when subjected to tensile deformation [12,13]. More recent studies have suggested that the hydrated cellulose structure is not completely amorphous but in fact may be characterized as having a relatively high degree of crystallinity [14].

A series of intensive studies of never-dried cotton fibers were recently undertaken by Ingram and his coworkers [15-18]. From a structural point of view, these workers conclude that never-dried fibers are highly crystalline. Since the fibrils have been shown to be fully crystalline units even in the never-dried fibers, Ingram et. al. postulate that the key process during the drying is hydrogen bond formation between surfaces of these fibrils. These workers have also shown the high degree of plasticity of never-dried fibers revealing inter-fibrillar slippage during deformation. The fracture tips of never-dried fibers exhibit an extreme degree of tailing, characteristic of ductile failure, in contrast to the macroscopically brittle rupture of dried fibers and even of dried and rewet fibers. Similar conclusions regarding the initial drying process were recently drawn also for the case of jute fibers [19].

MECHANICAL PROPERTIES OF COTTON FIBERS

The mechanical properties of individual cotton fibers have been studied by many workers, and attempts have been made to relate these properties to various features of fiber structure. Mechanical properties are normally determined under uniaxial tension in an apparatus designed to provide a constant rate of strain. The recorded load-extension or stress-strain curve is characterized in terms of such quantities as the stress and strain at rupture, the tensile elastic modulus or stiffness, the energy or work to rupture, and yielding. In uniaxial tension, the stress-strain curve of a dry cotton fiber is essentially Hookean over the entire strain range, i.e., the stress is directly proportional to strain with no evidence of yielding. Since the fiber is composed of crystalline fibrillar units which are held together by strong interfibrillar hydrogen bonds, the stress-strain curve is that of a typical crystalline polymer solid and fiber failure is in the brittle mode of fracture.

When a cotton fiber is subjected to a tensile load, the various structural elements respond and deform in characteristic manner. The growth rings are apparently of some importance in the deformation response. Cotton fibers grown under controlled laboratory conditions with no growth rings are more extensible and less stiff than fibers of the same variety grown under normal field conditions which produce growth rings, as shown by the data in Table 1 for native and for mercerized fibers. The effects of growth rings on fiber strength appear to be minor.

Table 1

Effects of growth rings on cotton fiber mechanical properties [4]

	Field Grown		No Growth Rings	
	Native	Mercer.	Native	Mercer.
Fiber Strength (gpd)	3.82	3.85	3.32	3.50
Fiber Stiffness (gpd)	29.3	38.3	21.1	21.7
Extensibility (%)	12.3	13.4	15.7	16.8

The mechanical properties of cotton fibers are related to the x-ray angle which describes the orientation of the fibrils with respect to the effective fiber axis. This angle is not to be confused with the helical angle of the fibrils, since it is a composite orientation factor arising from the helical angle and the convolution angle. Nevertheless, the x-ray angle varies over a wide range for cottons of different varieties and strains and correlations between this structural characteristic and mechanical properties have been established by many workers [1,20,21]. Fibers with more perfect fibrillar alignment are stronger, stiffer, and less extensible than those with less perfect fibrillar orientation. It is important to remember, however, that the variations in fibrillar alignment or orientation do not indicate variations in the angle of the helix formed by the fibrillar deposition, but rather reflect variations in convolution angle which in turn is determined by reversal frequency and distribution.

The role of structural reversals in the mechanical deformation of cotton fibers has been the subject of much study and controversy. The point at issue has always been whether the reversal points represent points of weakness with failure occurring preferentially at points of reversal rather than between reversals. Arguing from a statistical viewpoint, Wakeham and Spicer [22] have concluded that failure occurs preferentially at reversal points. This conclusion was reached on the basis of microscopic examination in elliptically polarized light of broken fiber ends. Other workers have suggested that breaks do not occur preferentially at reversal points but near such points [23]. It has in fact been suggested that the reversal point is a point of enhanced strength since the fibrils at reversals are aligned perfectly with respect to the fiber axis and thus in a position to bear the applied load most efficiently [2,24]. These workers suggest that breakage occurs not at a reversal point, but at some position near the reversal which may be misinterpreted as a reversal breakage. On the other hand, Raes et. al. [5,6] have presented data, which is summarized in Table 2, to indicate that fiber strength between visible reversals is about 20% higher than the strength at reversals. Under any circumstance it is clear that reversals are points of structural irregularity. This irregularity in turn results in stress concentrations when an external load is applied either in the immediate vicinity or at some point close to a reversal resulting in fiber failure.

The mechanical properties of never-dried fibers are of some considerable interest, since they reveal the important effect of the interfibrillar hydrogen bonds that become operative after the drying process during fiber growth. In Figure 3 are shown load elongation curves of fibers in water that had never been dried and compared with those that had been dried once.

Table 2

The effects of reversals on cotton fiber strength [6]

Variety	At Reversals (mg)	Between Reversals (mg)	Strength Loss (%)
M 11	71.0	101.1	27.1
DPL 14	73.2	100.3	27.0
Smooth Leaf	91.5	114.5	20.1
Auburn 56	87.7	102.7	14.6
Stoneville 76	93.1	118.1	21.2

The never-dried fibers are much more extensible and less stiff than once-dried fibers, although it is noteworthy that fiber strength is approximately the same for these fibers. It is also of interest that the inherent mobility of never-dried fibers can be drastically reduced by stretching. Ingram and coworkers [15] conclude that this behavior reflects stress-induced interfibrillar hydrogen bonding which produces a structure which is similar to that produced by the normal drying process.

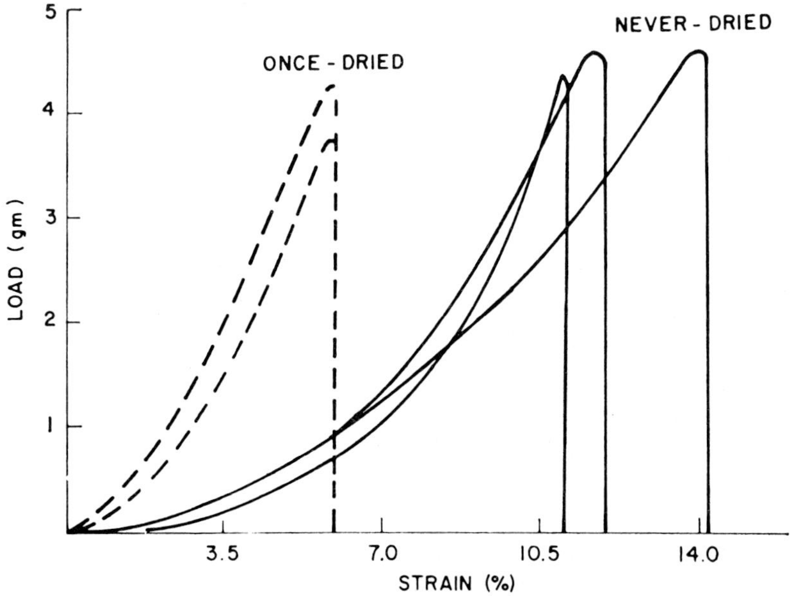

Fig. 3. Load-elongation curves of never-dried and once-dried cotton fibers measured in the wet state [15].

The importance of hydrogen bonding in cellulosic fibers and the effect of such intermolecular cohesive forces on mechanical properties is well known. Depending upon the nature and the source of the fibers, cellulose will absorb anywhere from 7 to 15% by weight water vapor from an ambient atmosphere at 65% RH and 21°C. The extent of sorption, of course, is dependent on temperature and on the partial pressure of water vapor in the atmosphere. The effect of the absorbed water is to decrease some of the intermolecular cohesive forces between cellulose chains because the hydrogen bonding capacity of the hydroxyl groups is satisfied by the absorbed water. The sorption process is accompanied by fiber swelling. In Table 3 are shown some data on the effects of moisture on cotton fiber mechanical properties. Stiffness decreases and extensibility increases with increasing moisture content; however, the effect of moisture on fiber strength is not clear-cut. This reflects the fact that fiber strength is indicative primarily of structurally uniformity and regularity. Fiber strength must be considered as an extreme-value characteristic and as a localized physical property.

Table 3
Effects of moisture on cotton fiber mechanical properties [25]

	35% RH	65% RH	100% RH
Strength (gpd)	2.22	3.21	2.90
Stiffness (gpd)	42.5	38.5	17.4
Extensibility (%)	6.00	8.60	10.2

Another manifestation of the effect of moisture on the mechanical properties of cellulosic fibers is evident upon examination of the stress-strain curves shown in Figure 4 for rayon yarns which may be taken as a model for native cellulosic fibers [26,29]. Clearly the effect of water on mechanical response indicates the important role of intermolecular cohesive forces. It should be noted that not all cellulosic fibers respond to water in the same manner. Variations in fiber structure and growth morphology influence the response of cellulosic fibers to water sorption, but in all cases the critical effects of water on intermolecular cohesive forces are evident.

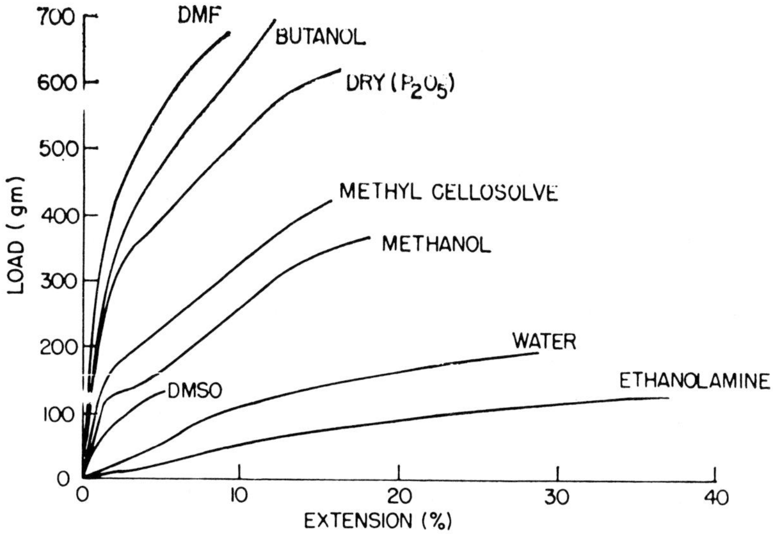

Fig. 4. Load-extension curves of rayon yarns in some polar solvents [29].

In view of the hydrophilic character of cellulose, nonaqueous organic solvents are generally poor swelling agents. A number of studies have been undertaken over the past several years to determine the factors that control nonaqueous swelling of cellulosic fibers [27,28]. In general only the highly polar solvents, such as dimethyl sulfoxide, dimethylformamide, tertiary and quarternary amines, and low molecular weight alcohols and acids, are capable of producing any significant levels of swelling. Weigmann and coworkers [29] have recently undertaken studies to systematize the interactions between organic solvents and a number of natural and synthetic textile fibers. These studies have relied principally on characterizing the interactions in terms of mechanical property effects. In the case of cellulosic fibers, these studies have been particularly productive and have revealed hitherto unknown effects and interactions.

The stress-strain or load-extension curves of cellulosic fibers (rayon yarns) in polar, water-miscible organic solvents, some of which are shown in Figure 4, can be readily explained in terms of the known hydrogen bonded structure of cellulose. These organic solvents are able to penetrate the cellulose structure and disrupt the intermolecular hydrogen bonding thereby plasticizing the structure, resulting in decreases in the elastic modulus and ultimate strength and increases in fiber extensibility.

In the examples shown in Figure 4, the effects of butanol and DMF are not consistent with the interactions described above, but they are still explicable in terms of a strongly hydrogen bonded structure of cellulose. It is believed that these solvents have a "drying effect", i.e., they have the ability to remove chemisorbed water which is known to remain entrapped in cellulose after the usual drying operations.

Of singular importance, however, have been the observations by Weigmann and coworkers [29] that a number of organic solvents which would be expected to be completely inert to the cellulose-water system are in fact not inert. These solvents were of the nonpolar, nonwater-miscible type and included such hydrocarbons as heptane, methylcyclohexane, and hexane. The effects of these solvents on the load-extension curves of rayon yarns are shown in Figure 5. Briefly, the effect of these solvents is to decrease the elastic modulus (slope of the load-extension curve), decrease strength, and decrease the extensibility. Obviously, this is not a simple plasticizing effect

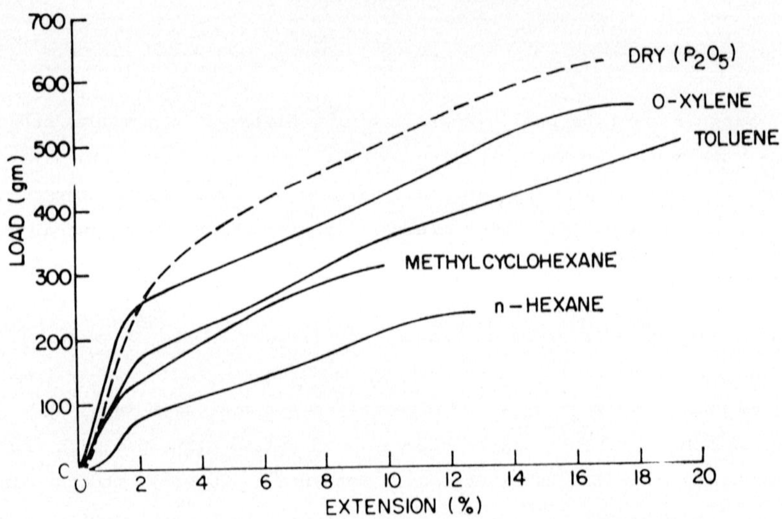

Fig. 5. Load-extension curves of rayon yarns in various hydrocarbon solvents [29].

which, similar to the effect of water and some other polar solvents, increases extensibility. Other observations about the interactions between these nonpolar solvents and cellulosic fibers are equally puzzling. For example, the interactions are apparently not accompanied by fiber swelling (increase in volume) as are the interactions between cellulose and water and other polar solvents. Furthermore, the effects of the nonpolar hydrocarbons are completely reversible. Upon removal of the solvent and re-establishment of an equilibrium with atmospheric conditions (relative humidity), mechanical properties and general character of the load-extension diagram are fully recovered. Weigmann [29] has suggested that these effects may reflect the existence of hydrophobic interactions within sheets of strongly hydrogen bonded cellulose chains that were postulated previously by Warwicker [30].

While no definitive conclusions regarding the mechanical behavior of cellulose fibers can be drawn at this time, it is clear that the behavior is complex, that it reflects the special morphology and fine structure of the fiber, and that it is strongly responsive to the chemical environment.

STRUCTURE OF KERATIN FIBERS

The keratins are a class of cellular proteins that have evolved as a protective outer barrier for the higher vertebrates. Keratin fibers such as wool, human hair, and a long list of specialty animal fibers have become important industrial materials because of their desirable combination of chemical and physical properties. Keratin fibers are structurally most complex both from a purely chemical point of view as well as from the point of view of their growth morphology and fine structure.

Attempts to establish the primary chemical structure of keratins are complicated by the fact that keratins are heterogeneous and composed of several different proteins. Investigators have isolated various components (high-sulfur and low-sulfur-containing fractions) from the native substance, and through amino acid analyses have been able to isolate at least 18 amino acids [31,32,33]. Great progress has also been made in efforts to establish the amino acid sequence in the various protein fractions. Nevertheless, the complexity of the primary structure of keratin fibers is so great that an exact description is not yet available.

The secondary and tertiary structure of keratin fibers is equally complex, and again great strides have been made in the elucidation of the unique intermolecular organization and architecture. The fiber consists of an assembly of three different cell types. These are the medulla, the cortex, and the cuticle cells which are connected with and among each other by the intercellular membrane complex. The cells are essentially concentrically arranged with the medulla as a center canal, surrounded by the cortex, and the cuticle cells forming an outer envelope of the fiber.

The cortex accounts for approximately 80% of the fiber, and each cortical cell is approximately 100 μm in length and 5 μm in diameter. A unique structural feature is the fact that the individual cortical cells in certain keratin fibers can be classified into two distinct categories on the basis of their response to chemical treatments. The portion of the cortex that is more receptive to dyestuffs, and more readily attacked by enzymes and other chemical reagents, is referred to as the orthocortex; the less dye-receptive and less reactive portion is called the paracortex. The paracortex is a higher sulfur-containing segment than the orthocortex, and since the sulfur arises principally from cystine, it can be assumed that the paracortex is more highly crosslinked.

The cortical cells are composed of closely packed microfibrils embedded in a matricular substance which is associated with the high sulfur content protein fraction [34,35]. The microfibrils are postulated to be essentially crystalline units composed of the α-helices which are probably associated in a multistrand rope and possibly in a protofibrillar structure. The existence of protofibrils and their geometric arrangement in the microfibrils has been the subject of intensive investigation and considerable controversy. It is generally accepted that the microfibril does contain a fine structure presumably of a skin-core type, but definitive evidence for the existence of discrete protofibrils has not been achieved.

KERATIN FIBER DEFORMATION

When a keratin fiber is deformed, the overall behavior reflects the response of the various structural elements to the deforming stresses. The stress-strain curve of a keratin fiber reveals three distinctly different regions. In the initial stages of fiber extension the behavior is essentially Hookean and only bond-angle deformations are involved. At a well-defined strain, between 1 and 2%, the elastic limit is reached, and the stress-strain curve enters the yield region. This region of the curve reflects the unfolding of the α-helices to the extended, sheet-like β-configuration. The process takes place against very little resistance, and the yield region of the stress-strain curve is therefore essentially flat. At approximately 30% extension the stress-strain curve reaches the post-yield region, with an attendant increase in the stress with increasing strain, resulting in fiber rupture at approximately 50% strain, before completion of the α-β transformation. The post-yield region represents the involvement of highly cross-linked domains in the fiber, which is superimposed of the α-β transformation taking place under the applied strain. Molecular interpretations of the stress-strain curves of keratin fibers have been the subject of numerous investigations and reviews [36,37,38].

The mechanical behavior as reflected in a stress-strain curve is strongly dependent on the chemical environment. For example, the effects of water and various levels of relative humidities are shown in Figure 6 from a paper by Hearle et. al. [36] based on data originally obtained by Speakman [39]. The hydrogen bonding network, presumably in the matrix, is broken or interrupted by water, resulting in a lower yield point a more clearly defined yield region, and also a clearer and more extensible post-yield region. The hydrogen bonding network obviously rigidifies the structure significantly. Since the water accessibility is restricted to the amorphous matrix, the importance of the matrix to mechanical properties is clearly evident.

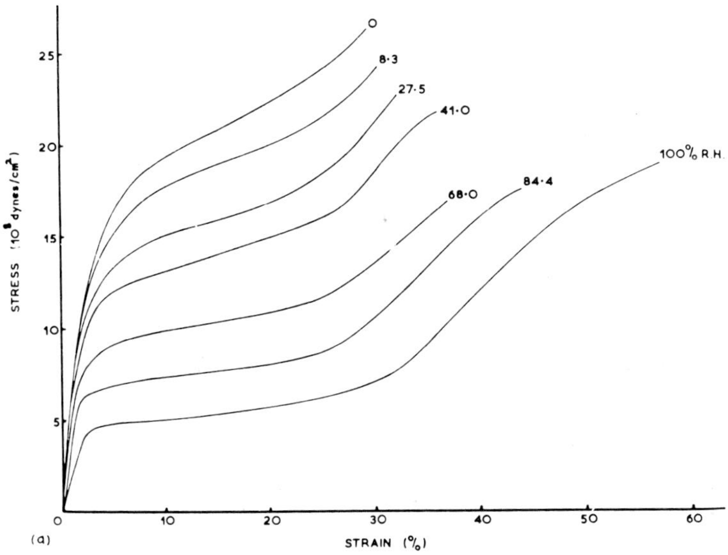

Fig. 6. Stress-strain curves of wool fibers at various humidities [39].

Weigmann and Dansizer [40] have studied the stress relaxation behavior of keratin fibers to reveal other aspects of the dependence of mechanical properties on the complex structure of the fiber. The stress relaxation phenomenon and its subsequent interpretation may be best accomplished by considering the relaxation modulus E_r which may be defined as the ratio between the stress and strain at any given time. In Figure 7 are shown the relaxation modulus values for keratin fibers which had been extended to a 20% strain level as a function of time for various temperatures ranging from 21° to 96°C. Zero time is taken as the time when the fibers are fully extended. At the lower temperatures, the relaxation modulus is essentially constant as a function of time and only a slight linear decrease is observed. With increasing temperatures, the departure from linearity becomes more pronounced and at 96°C the relaxation modulus shows no linear region along the time scale and decreases rapidly to values which would be expected for an amorphous, highly crosslinked polymer rather than for a semicrystalline polymer. If the relaxation process is allowed to take place in the presence of sulfhydryl-group blocking agent as in Figure 8, then the relaxation modulus shows only a small linear decrease with time, which is temperature-independent.

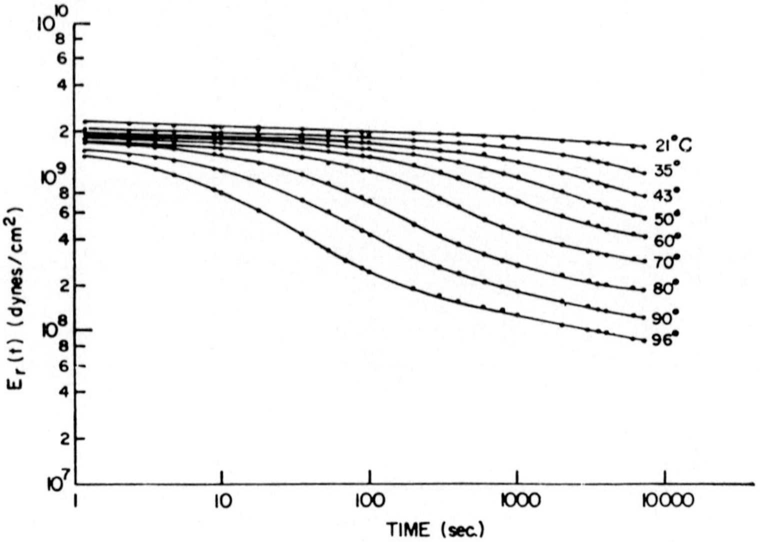

Fig. 7. Relaxation modulus as a function of time and temperature of wool fibers from 20% strain [40].

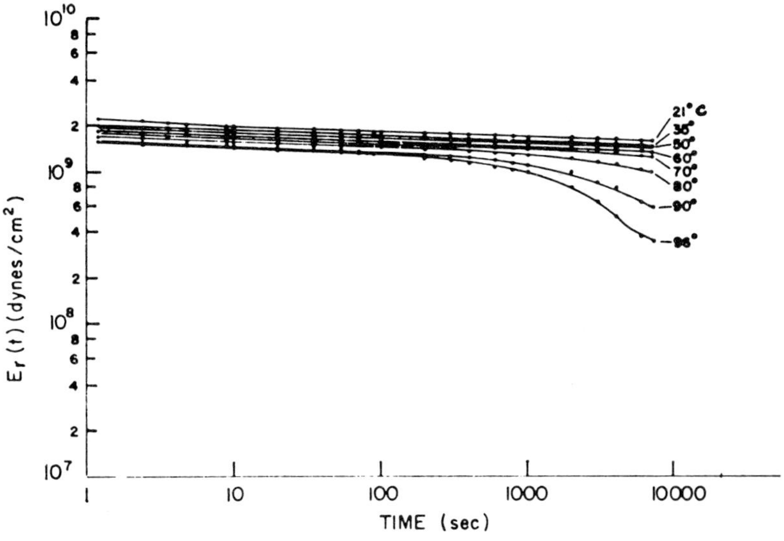

Fig. 8. Relaxation modulus as a function of time and temperature of wool fibers from 20% strain in the presence of a sulfhydryl group blocking agent, N-ethylmaleimide [40].

In this case the modulus values remain in the region characteristic of semicrystalline polymers over most of the time and temperature scale. These data show the importance of sulfhydryl-disulfide interchange reactions in the mechanical behavior of keratin fiber. The interchange reaction is also thought to be a controlling factor in the deformation process as reflected in the post-yield region of the stress-strain curve, and also in the supercontraction of keratin fibers [38].

SURFACE STRUCTURE AND FRICTIONAL PROPERTIES OF KERATIN FIBERS

Keratin fibers have a characteristic and unique scalar surface structure. The cuticle cells which account for nearly 20% of the fiber have a layered structure consisting of endocuticle and exocuticle, and an epicuticle surrounding each individual cell. The epicuticle, constituting approximately 1% of the total fiber mass, serves as an important inert protective surface layer for the entire fiber.

Keratin fibers exhibit unusual frictional behavior because of the overlapping scale structure. The magnitude of the coefficient of friction (μ) is normally higher when the fiber is moved in the against-scale (tip to root) than in the with-scale direction (root to tip). Measurements of frictional coefficients are strongly dependent on a number of system variables as forecfully pointed out in a recent authoritative review of wool fiber surface properties by Makinson [41]. Of major importance are such factors as sliding speed, nature of the rubbing surface, fiber surface contamination, moisture content, and chemical pretreatment of the surface. In the case of frictional measurements of wet fibers, the nature of the liquid medium in terms of temperature, pH ionic strength, and ionic character has a major effect also. Some typical values for against-scale (μ_a) and with-scale (μ_w) frictional coefficients are shown in Table 4. The difference in the two frictional coefficient values is responsible for some of the important technological properties of keratin fibers, for example, the felting properties of wool and the entanglements encountered under certain conditions with human hair. The existence of a differential friction effect ($\mu_a - \mu_w$) causes the fibers in an assembly of fibers to migrate preferentially in one direction, thereby causing felting and entanglement.

Table 4

Typical values of the coefficients of friction of wet and dry wool fibers [42].

Medium	μ_a	μ_w	$\mu_a - \mu_w$
Soap solution	0.62	0.23	0.39
Dry (65% RH)	0.30	0.19	0.11

The chemical modification of keratin fiber surfaces, either by direct chemical reaction or by the deposition of various polymers, can have an important influence in controlling felting and entanglement. This influence is at least in part due to the reduction of the differential friction effect, although it is known that other mechanisms are also operative. For example, Kamath et. al. [43] have studied the effects of various synthetic polymers on the frictional properties of wool fibers and correlated these effects with the dimensional stabilization of wool fabrics. These workers have found that polymers which form hydrogels on the fiber surface reduce friction, while the flexible elastomers increase friction. These results indicate that the mechanical properties of the polymer play an important role in the modification of fiber surface properties, and it is difficult to generalize about these complex and interrelated phenomena.

WETTABILITY PROPERTIES OF KERATIN FIBERS

An important aspect of fiber surface properties concerns their interaction with liquid environments. When a liquid is brought into contact with a solid, the interaction can lead either to a wetting or a nonwetting condition depending on the surface free energy of the solid and the surface tension of the liquid. Numerous quantities, some of which are essentially empirical while others are based on thermodynamic relationships, have been used to characterize the wettability properties of fiber surfaces. Miller [44] recently reviewed the concepts underlying fiber wettability measurements, and provided an analysis of the interrelationships among such quantities as surface free energy, critical surface tension, work of adhesion, contact angle, and specific wettability. The most extensively used technique to measure the surface interaction between solids and liquids (interfacial tension) is by means of a contact angle measurement. This geometric measurement is generally useful and appropriate for the case of planar, geometrically well-defined solids, but its application to fibers is questionable because of the dimensions and because of the irregular nonplanar surfaces involved. Miller [44] argues that contact angle measurements are wholly unsuitable for fibers, and recommends the use of wetting force measurements based on the Wilhelmy balance principle [45].

According to this principle, the pull (F_W) exerted on a vertical rod (i.e., filament or fiber) inserted into a liquid is expressed by

$$F_W = P\,\gamma_{LV} \cos\theta,$$

where P is the perimeter of the solid along the three-phase boundary line, γ_{LV} is the surface tension of the liquid, and θ is the contact angle. This pull can be downward (when the liquid rises up the solid) or it can be a "push" upward (when the meniscus is depressed at the interphase boundary). Normalizing for the length of contact between filament and liquid, we can describe a characteristic specific wetting force (w), as

$$w = \frac{\text{Force per filament}}{\text{Perimeter of filament}} = \frac{F_w}{P} = \gamma_{LV} \cos\theta$$

Thus, the determination of w, which has been called adhesion tension or specific wettability, will give a value for $\gamma_{LV} \cos\theta$ without any actual measurement of θ. Combined with an independently obtained value for γ_{LV}, this can be used to calculate θ or the work of adhesion (W_a) by the following relationship:

$$W_a = \gamma_{LV} (1 + \cos\theta)$$

Miller [44] claims that the work of adhesion is the basic energetic parameter that describes the interaction between a liquid and a solid phase. Accordingly, it is a quantity that should be particularly effective in revealing differences in the surface structure and properties of fibers.

The determination of the wetting force, F_w, can be carried out by arranging to measure the change in weight of a vertical fiber (e.g., with a microbalance) that occurs when it is placed in contact with a liquid. Miller and Young [46] recently described an experimental procedure to measure the advancing wetting force as a function of the depth of immersion of the fiber in the liquid medium. Increasing the depth of filament immersion in small increments ($\sim$ 1 mm) is a means of eliminating the effects of buoyancy which acts in opposition to the recorded wetting force. By measuring wetting force at increasing depths of immersion, it is possible to extrapolate to a zero immersion depth, thereby obtaining a true wetting force unperturbed by buoyancy effects. By reversing the experimental procedure, data can be obtained for the equilibrium between a liquid and a prewet fiber surface. This receding wetting force is required to establish the magnitude of hysteresis effects which are of importance in understanding solid-liquid interactions.

Typical data for the wetting by water of several fibers and filaments are presented in Table 5. If one uses the advancing work of adhesion as a measure of wettability, it is of some interest that the two textile fibers which we normally characterize as hydrophilic occupy the two extreme positions on the wettability scale. The extremely low wettability of wool fibers reflects the hydrophobic fiber surface despite the fact that the bulk fiber is indeed even more hydrophilic than cotton on the basis of moisture sorption behavior. The thin outer layer that forms a sheath around the cuticle cells known as the epicuticle is extremely hydrophobic

Table 5

Water wetting of single fibers and filaments [44]

	Work of Adhesion mN/m		
	Advancing	Receding	Hysteresis Ratio
Cotton (scoured)	148.3	161.2	1.09
Glass	138.5	138.5	1.00
Carbon	128.0	148.2	1.16
Nylon	106.3	131.4	1.24
Polycarbonate	94.1	165.0	1.75
Polyester	84.9	108.0	1.28
Polystyrene	81.6	106.6	1.31
Polypropylene	77.4	115.4	1.49
Teflon	56.1	128.5	2.29
Wool	45.6	111.3	2.44

presumably because of a relatively high lipid content [47]. It should be noted that the high wettability value for cotton is for a cleaned and scoured fiber. A native cotton fiber prior to scouring, which removes a waxy surface layer, would also exhibit very low advancing wettability or work of adhesion values.

Table 5 also shows hysteresis data for water with a variety of materials in terms of the ratio of advancing to receding work of adhesion. The data indicate that there is a general inverse correlation between the advancing work of adhesion and the magnitude of hysteresis. In other words, if the original surface is not very attractive to a liquid, forced wetting (that is, imposed immersion) will leave it in a state that is much more attractive to the adjacent liquid. This effect is particularly operative in the case of wool fibers. On the other hand, prewetting a glass fiber or a scoured cotton fiber does not change its attraction to water, which is considerable to begin with.

Kamath, Dansizer, and Weigmann [48] recently reported an interesting observation regarding the wettability of keratin fiber surfaces. These workers observed a time-dependent increase in the wettability (work of adhesion) of keratin fibers when these fibers have been exposed to water. This increase in wettability is attributed at least in part to an interaction between the surface molecules of the fiber and the liquid (water). One might visualize, for example, a relaxation of surface molecules allowing an orientation of polar groups toward the liquid and the nonpolar groups toward the interior of the fiber. It is of some interest in this regard that a similar although less pronounced time-dependent increase in wettability was observed for nylon fibers but not for polypropylene fibers [49]. Nylon may be viewed as an analog of keratin with the potential of reorienting polar groups after relaxation, while the completely hydrophobic polypropylene would not be expected to exhibit this behavior.

Kamath et. al. [48] also determined and reported critical surface tensions (γ_c) of keratin fibers using butanol-water mixtures based on the method developed by Feldtman and McPhee [50]. Critical surface tension values are shown in Table 6 for the case of initial wetting and for the case of final or equilibrium wetting. The time-dependent increase in wettability is clearly evident from these critical surface tension values.

Table 6

Critical surface tension of hair fibers

Scale Direction	Initial γ_c (mN/m)	Final γ_c (mN/m)
Against	20.9	26.0
With	20.0	24.9

Kamath et. al. [48] also were able to determine the dispersion and polar contributions to the critical surface tension using the method of El-Shimi and Goddard [51]. The values of the dispersion contribution to the critical surface tension (γ_c^d) and the polar contribution to the critical surface tension (γ_c^p) are shown in Table 7. The polar contribution for

Table 7

Dispersion and polar contributions to the critical surface tension of hair fibers [48]

	Initial		Final	
Scale Direction	γ_c^d (mN/m)	γ_c^p (mN/m)	γ_c^d (mN/m)	γ_c^p (mN/m)
Against	24.8	2.6	19.5	11.5
With	23.9	2.5	19.5	10.0

the initial wetting is small compared to the dispersion contribution. This indicates the strongly hydrophobic surface of the keratin fiber. In the case of the final wetting condition, after the fiber has been exposed to water for approximately 15 minutes, the polar contribution increases considerably and the dispersion contribution decreases slightly leading to a net increase in the total critical surface tension, as also shown in Table 7. As previously indicated, Kamath et. al. [48] attribute these time-dependent effects to an interaction between the keratin fiber surface and water leading to hydrogen bond breaking and consequent reorientation of the surface molecules.

CONCLUSION

The fine structure and morphology of native fibrous materials, such as the cellulosic and keratin fibers, are indeed complex. This complexity in turn is reflected in the mechanical properties of these fibers. It has become possible to relate certain properties with special features of structure. Recently, the surface properties of these fibers have been recognized as particularly important, and substantive research on these characteristics is in the beginning stages.

REFERENCES

1. L. Rebenfeld, J. Polymer Sci., Part C, No. 9 (1965) 91-112.
2. A. Peterlin and P. Ingram, Textile Res. J., 40 (1970) 345-354.
3. M. L. Rollins and V. W. Tripp, Textile Res. J., 24 (1954) 345-357.
4. L. P. Berriman, Textile Res. J., 36 (1966) 272-279.
5. G. Raes, T. Fransen, and L. Verschraege, Laboratorium de Meulemeester voor Technologie der Textielstoffen, University of Ghent, Final Report of PL 480 project No. UR-E40 (20)-1 (1964-1969).
6. G. Raes, T. Fransen, and L. Verschraege, Textile Res. J., 38 (1968) 182-195.
7. M. Morosoff and P. Ingram, Textile Res. J., 40 (1970) 250-255.
8. J. J. Hebert, Textile Res. J., 37 (1967) 57-58.
9. P. Kassenbeck, Symposium International Recherche Textile Cotonniere (SIRTEC), Institut Textile de France, Paris, April 1969, pp. 455-571.
10. P. Kassenbeck, Textile Res. J., 40 (1970), 330-334.
11. E. E. Berkley and T. Kerr, Ind. Eng. Chem., 38 (1946) 304-309.

12. J. R. Colvin, CRC Critical Reviews, Macromol. Sci., 1 (1972) 47.
13. A.N.J. Heyn, J. Polymer Sci., A-3, (1965) 1251-1265.
14. M. L. Nelson and T. Mares, Textile Res. J., 35 (1965) 592-603.
15. P. Ingram, D. K. Woods, A. Peterlin, and J. L. Williams, Textile Res. J., 44 (1974) 96-106.
16. J. L. Williams, P. Ingram, A. Peterlin, and D. K. Woods, Textile Res. J., 44 (1974) 370-377.
17. N. Morosoff, J. Appl. Polymer Sci., 18 (1974) 1837-1854.
18. J. L. Williams, H. Sugg, and P. Ingram, Textile Res. J. 46 (1976) 434-435.
19. P. K. Ray, A. C. Chakravarty, and S. B. Bandyopadhyay, J. Appl. Polymer Sci., 20 (1976) 1765-1767.
20. R. Meredith, J. Textile Inst. 42 (1951) T291.
21. L. Rebenfeld and W. P. Virgin, Textile Res. J., 27 (1957) 286.
22. H. Wakeham and N. Spicer, Textile Res. J., 21 (1951) 187-194.
23. J. O. Warwicker, S. C. Simmens, and P. Hallam, Textile Res. J., 40 (1970) 1051-1052.
24. H. Wakeham, T. Radhakrishnan, and G. S. Viswanathan, Textile Res. J., 29 (1959) 450.
25. L. Rebenfeld, Final Report to U.S. Department of Agriculture, Research and Marketing Contract No. 12-14-100-984 (31), Textile Research Institute, Princeton, N.J., November 12, 1959.
26. L. Rebenfeld, C. V. Oster, and L. F. Klurfeld, Textile Res. J., 41 (1971) 139-146.
27. L. Rebenfeld, Systematic Investigation of Nonaqueous Swelling Agents for Cotton Cellulose, U.S. Department of Agriculture, Agricultural Research Service, ARS 72-33, January 1965.
28. E. A. Colombo and E. H. Immergut, J. Polymer Sci., Part C, No. 31, (1970) pp. 137-156.
29. H.-D. Weigmann and A. S. Ribnick, Textile Res., J., 44 (1974) 165-173.
30. J. O. Warwicker and A. C. Wright, J. Appl. Polymer Sci., 11 (1967) 659-671.
31. W. G. Crewther and L. M. Dowling, Appl. Polymer Symposia, 18 (1971) 1-20.
32. H. Lindley, A. Broad, A. P. Damoglow, R. L. Darskus, T. E. Elleman, J. M. Gillespie, and C. H. Moore, Appl. Polymer Symposia, 18 (1971) 21-44.

33. A. Parisot and J. Derminot, Appl. Polymer Symposia, 18 (1971) 45-53.

34. R.D.B. Fraser, T. P. MacRae, G. R. Millward, D.A.D. Parry, E. Suzuki, and P. A. Tulloch, Appl. Polymer Symposia, 18 (1971) 65-83.

35. R.D.B. Fraser, T. P. MacRae, and G. E. Rogers, "Keratins - Their Composition, Structure, and Biosynthesis," Chas. C. Thomas, Springfield, Ill. (1972).

36. J.W.S. Hearle, B. M. Chapman, and G. S. Senior, Appl. Polymer Symposia, 18 (1971) 775-794.

37. M. Feughelman, Appl. Polymer Symposia, 18 (1971) 757-774.

38. H.-D. Weigmann and L. Rebenfeld in "The Chemistry of Sulfides," A. V. Tobolsky, Ed., John Wiley, New York, 1968.

39. J. B. Speakman, J. Textile Inst., 18 (1927) T446.

40. H.-D. Weigmann, L. Rebenfeld and C. J. Dansizer, Textile Res. J., 36 (1966) 535-542.

41. K. R. Makinson, "Surface Properties of Wool Fibers" in Surface Characteristics of Fibers and Textiles, M. J. Schick, Ed., Marcel Dekker, Inc., New York, 1975.

42. K. R. Makinson, Textile Res. J., 37 (1967) 763-771.

43. Y. K. Kamath, C. J. Dansizer, and H.-D. Weigmann, Proceedings, Fifth International Wool Textile Research Conference, Aachen, Germany, September 1975 (to be published).

44. B. Miller, "The Wetting of Fibers" in Surface Characteristics of Fibers and Textiles, M. J. Schick, Ed., Marcel Dekker, Inc., New York 1975.

45. J. Wilhelmy, Ann. Physik, 119 (1863) 177-217.

46. B. Miller and R. A. Young, Textile Res. J., 45 (1975) 359-365.

47. N.L.R. King and J. H. Bradbury, Aust. J. Biol. Sci., 21 (1968) 375.

48. Y. K. Kamath, C. J. Dansizer, and H.-D. Weigmann, Proceedings 9th IFSCC Congress, Boston, Mass., June 1976.

49. H.-D. Weigmann and Y. K. Kamath, Private communication, to be published.

50. H. D. Feldtman and J. R. McPhee, Textile Res. J., 34 (1964) 634-642.

51. A. El-Shimi and E. D. Goddard, J. Coll. Interface Sci., 48 (1974) 242-248.

MATERIALS FROM POLYSACCHARIDES

STEPHEN H. CARR
Department of Materials Science and Engineering
Northwestern University, Evanston, Illinois 60201

SUMMARY

Useful engineering materials have been derived from polysaccharides in their native and in reconstituted states. Although much information regarding their physical and chemical behavior is known, there are a number of areas related to solid state structures and properties which remain to be understood. Opportunities for creating materials having properties suited for new applications are afforded, in light of recent advances in macromolecular science, by many common polysaccharides.

INTRODUCTION

Polysaccharides generally are represented by high polymers created from pyranose and furanose carbohydrate molecules, such as D-xylose or D-glucose. These polymers occur in nature and are found in a very wide variety of forms. These forms include variations in the architecture of the polymer chains, variations in the sequence of carbohydrate moieties, and variations in the distribution of substituents found along the polysaccharide chain. This paper will focus on a very small portion of this family of high polymers, specifically cellulose and amylose, which are derived from D-glucose. The reasons for this restriction are simply that these polysaccharides are abundantly available, can easily be isolated and worked with, and make solids which are candidate engineering materials.

It is well known that both kinds of homopolysaccharides can be obtained in enormous quantities from agricultural sources, and much effort has been expended during the past century in learning how to manipulate the physical and organic chemistry of these polymers. It almost goes without saying that work designed to exploit the maximum properties of wood and other natural forms of cellulose, such as natural fibers, has been performed exhaustively. Nevertheless, there has been for a considerable period of time work aimed at making improved preparations of these polymers. Much of the following sections of this paper will be addressed to what has been learned about these macromolecules themselves and about the ways in which solids of a variety of characteristics can be prepared from them. It is the intention here to show possible ways in which future improvements can be made with polysaccharides in terms of superior engineering materials.

THE MACROMOLECULES

Polysaccharide chains come in a variety of chain configurations. Some are linear, such as is the case with cellulose, while others are branched, as is the case with amylose and glycogen. The nature of this particular chemical microstructure is determined by the biosynthetic route that corresponds to each one. In addition, those polysaccharides used as energy storage media have their architecture determined in part by the advantage it presents to catabolic processes. Other differences in the chemical microstructure are derived from the glucosidic linkage, from the point of substitution, and the nature of the monosaccharide units from which the chain is created. Generally speaking these chains are relatively stiff, due to their abundance of hydroxyl groups on each repeating unit and the physical bulkiness of these repeating units. The result is that these chains tend to be expanded when in a solution state. That this is so can be concluded from the relatively high Mark-Houwink exponents [1] on the order of 0.8 for cellulose and of 0.7 for amylose. Solvent media which tend to ionize hydroxyl groups increase the expansion, and hence the effective stiffness, of these chains.

The polysaccharides that we are dealing with here may exhibit a very regular conformation in either the solution or solid states. This is primarily a result of the fact that these polysaccharides are stereoregular, and amylose is particularly conspicuous in its ability to exhibit a specific conformation due to this stereoregularity. This material will in solution coil into very regular helical conformations, especially with the aid of small molecules which form complexes in the core of these chains. Studies on the complexation of amylose are extensively reported in the literature (see review by Foster, Ref. 2). It is instructive to note that when amylose macromolecules are complexed with iodine or n-butanol, they undergo retrogradation (precipitation due to interchain associations that displace water from their immediate environment) much more slowly [3]. This phenomenon is also a reflection of how stabilizing forces such as arise from polymer/solvent interactions influence solution conformations.

It seems correct to presume that adoption of a helical conformation by amylose in solution actually permits the chain to behave more in a flexible manner than it otherwise would (see Ref. 2, p. 374). Criteria governing the particular conformations encountered in cellulose are discussed by Sundararajan and Marchessault [4].

Cellulose and amylose dissolve in aqueous media essentially as hydrophobic species. Dobbins [5] has published a set of principles characterizing the dissolved state of such polysaccharides. It is observed for cellulose that if certain other chemical species such as copper ammonium salts are present, dissolution of this polysaccharide will occur. Amylose, on the other hand, will interact much more favorably with water, to the point that certain media effectively dissolve this polysaccharide [6]. It is understood that this solubilization of amylose arises,at least in part, from the presence of counter-ions in the aqueous phase. These ions can serve to dissolve amylose either by changing the solvent/polymer interaction or by actually ionizing part of the amylose itself. These ionic effects are not due to simple Coulombic processes, but rather they are the net result of the hydrophobic/hydrophilic qualities of the various constituents in the solution. Ions which interact very favorably with the macromolecule itself tend to stabilize the chain in an extended conformation. This means that the macromolecules would go into an unfavorably low entropy situation [7], and the result would be that these chains phase separate from the solution [8]. Ions which interact relatively mildly with the polymer will permit it to go through a normal array of conformations and thereby stay in the solution under a more wide range of conditions.

CRYSTAL STRUCTURES

Resulting also from the fact that these polysaccharides — amylose and cellulose — are stereoregular is their ability to undergo crystalline ordering in the solid state. Interchain hydrogen-bonding dominates much of the segmental packing scheme, and it also plays a role in the polymorphism [9] exhibited by polysaccharides. In addition, sorption of water can lead to complexation and/or disruption of the hydrogen-bonding pattern; often irreversible changes result from a hydration/dehydration cycle. A prime example of this latter phenomenon is mercerization of cotton [10], but complexation with other small molecules has also been reported [10].

Polymorphism in cellulose solids has been extensively investigated [11], and their interconvertability has been shown [11] to result from subtle changes in the format of their interchain packing. It is clear from X-ray diffractometry that the sharpness of scattering maxima varies [10, 12] between cellulose preparations. Such is thought to be due in part to relative amounts of ordered and disordered chains and to distortion of chains already packed into lattices. An extreme kind of preparation of cellulose has been reported [13] to scatter X-rays as amorphous material. A systematic study of the fraction of ordered material as a function of diffraction strength has been published [14], and it is seen from this work that cellulose in the native (never-dried) state is apparently the most highly "crystalline". Studies on crystallite size [15, 16] and lattice distortions [17] have been reported for work on cellulosic materials, and consequently one can now make successful interpretation of diffraction peak breadth changes found with these various polymorphic forms using either the above cited work or paracrystalline theory [18]. Infrared absorption spectroscopy also has capability [19] of distinguishing among several of the cellulose polymorphs. It is possible with this method to identify cellulose I with ease, but relative amounts of amorphous and form II are very difficult to resolve. However, recent improvements in understanding of details regarding cellulose in form II [20] make it a straight-forward task to characterize by X-ray diffraction methods.

Amylose is the polysaccharide that also undergoes considerable complexing as it forms different kinds of crystal structures [21]. Here one notes immediately that it is the complexing molecules and the chain conformations they induce that controls which crystal polymorph exists. This fact implies that interchain hydrogen bonding is not so important in the case of amylose (cf. the case of cellulose); rather intrachain hydrogen bonding is an important part of the stabilization of amylose in these crystalline forms. For a brief review of the four important crystalline forms of amylose see references 22 and 23. Very highly oriented fibers and films can be prepared from the V forms [24], which are complexed with such small molecules as dimethyl sulfoxide [25].

SOLID STATE MORPHOLOGIES

Organization of polysaccharide chains into solids appears to involve formation of microscopic ordered regions which are either fibrillar or lamellar in shape. Cellulose, of course, exists in its native forms as microfibrils [26]. These entities appear as the elementary configuration of cellulose chains as they result from their biosynthetic origin [27]. Basically, cellulose microfibrils are ⩾1000 Å in length and 140 - 180 Å in width [28, 29], but fragmentation of these fibrils will produce [30] what may be sub-units on the order of 35 Å in lateral size. There is considerable discussion about the exact molecular organization which such microfibrils might have, with a closely chain-folded scheme advanced by Manley [31] on the one hand and an essentially extended chain habit [26] on the other extreme. It does seem likely that such microfibrillar structures do possess an organization of some kind intermediate between these extremes.

During regeneration or precipitation of solubilized cellulose, the morphology which results is highly dependent on what were the prevailing conditions. Formation of fibers or films from viscose is seen to involve initially a sol → gel transformation [32] followed by creation of solid state morphology. Depending on stress supported by the gel, there will be a range of crystalline entities which appear. At low stresses it is probable that extensive chain folding occurs [33], but as stress increases, the microstructure forming from gelling viscose becomes increasingly that of oriented microfibrils [34-37]. In technical processes, there is a gradient in this morphology [38], from that on the surface of microfibrils highly oriented with the machine direction to that in the interior of foreshortened crystals probably more like common chain-folded lamellae. A "selective uniplanar" crystal texture is often observed in films of viscose cellophane [9, 39]. It is clear from experiments measuring hydrogen-deuterium exchange kinetics [10] and X-ray diffractometry [14] that there are ordered regions and disordered regions in regenerated cellulose; the as yet unanswered question relates to the possible existence of disordered regions within crystallites in these cellulosic solids. The ability of cellulose to form solids ranging from fibrillar [41] to lamellar has been studied by Chanzy and coworkers [42, 43]. In their work, progressive deacetylation of cellulose triacetate solution-grown single crystals was found to result (provided crystal structure of the precursor was properly selected) in well-defined chain-folded lamellae. This work emphasizes the ability of cellulose to make crystals of the habit encountered in common synthetic polymer solids, and it underscores the plausibility of the notion that chain-folded crystallites do exist in many preparations (low stress levels during coagulation) of regenerated cellulose.

Amylose will also form solids whose morphology ranges between that composed of fibrillar crystals to that containing exclusively chain folded lamellae. The excellent work of Marchessault, Sarko, and coworkers (see Refs. 44 and 45) has characterized much of the ability of these macromolecules to crystallize in various ways. Many solid-state preparations of amylose involve "rod-like" crystals, although it is not at all clear that their molecular organization is the same as that in cellulose microfibrils. It is true, however, that fibrillar precipitate particles can be obtained [46] from dilute solutions in which amylose had been complexed, but it was inferred from their work that some chain-folding exists in spite of the fibrillar (uniform 40 Å dia.) growth habit. It is important to note that amylose solids have low percent crystallinities relative to the case of cellulose (possibly owing to the less well-developed interchain hydrogen bonding in solid amylose), but additional crystallization in amorphous regions can be promoted either by heating or by exposure to moisture. In fact, fully developed spherulites, which are presumably composed of chain-folded lamellar crystals, can be made to grow during formation of amylose films. Infrared studies characterizing this chain folding have been reported [47, 48] and confirm the morphological behavior mentioned above.

PHYSICAL PROPERTIES

Mechanical properties of cellulosic materials has been investigated extensively (see Ref. 49). Load-extension curves reflect a wide variety of parameters related to the structure and properties in cellulosic materials [32, 50]. Molecular weight has been shown to have considerable effect, with the smallest chains being unable to form useful solids [51]. Temperature dependence, moisture dependence, deformation rate dependence, and cellulose pre-treatment all are readily related to changes in these mechanical stress-strain properties [51]. Creep and stress relaxation experiments have been reported [51] and reveal the dependence of material parameters on these bulk properties. Of considerable note is the fact that regeneration of cellulose under conditions that lead to very, very high orientations of chains can result in solids whose elastic modulus is exceptionally high [52]. In fact, for many years, highly oriented cellulosic fibers were regarded as representing some of the upper limits of performance of polymeric fibers. Films of regenerated cellulose exhibit less extreme mechanical properties, although the performance that they do have is obviously attractive. Fatigue properties of regenerated cellulose have also been reported [51].

Amylosic solids are characterized by mechanical properties comparable to cellophane, provided one takes into account the orientation differences in the crystal textures possessed by each. Amylose films obtained by simple drying from aqueous media usually have poor mechanical properties [53]. However, there are techniques by which water can be removed from amylosic systems that can produce very, very attractive materials (see,for example, Ref. 54). Better ways of producing amylose films involve casting from solutions with dimethylsulfoxide [55] as the solvent, with a complexing agent such as n-propanol usually being present. It is desirable to have amylosic systems containing little or no amylopectin, as the tensile strength is known [56] to be proportional to amylose content. The presence of the branched chain amylopectin is not well understood, but the deleterious effects probably are related to the way amylopectin makes defective both the crystals and the amorphous material between them. An alternative means of making films of amylose involves dissolving the material in an alkaline solution and then causing this system to undergo what is effectively a gelation reaction with the result that the water can be dried from an already established molecular structure. Gelation can be achieved by a variety of means, one of which is the shock application of high ionic strength media [57]. Films of amylose can also be made extrusion of aqueous pastes. It is important in the gelation of amylose to have the polymer concentration exceed 2% by weight. Careful control over fabrication conditions can vary the water resistance of amylose films from very good [58] to non-existent [59, 60].

Polysaccharide films are also known for their barrier properties. Amylosic films have unusually high resistance to the diffusion of oxygen, although cellulosic films seem to pass such gases very readily [61-63]. Cellulosic films such as cellophane can however be rendered moisture proofed by addition of overlayers of synthetic polymers, such as polyethylene [64]. It has been shown that derivitization of amylose will retard the oxygen barrier properties significantly for low levels of substitution [61]. Physical blends of amylose with viscose regenerated cellulose can result [65] in attractive films having mechanical, as well as vapor barrier properties intermediate between those of either constituent by itself. One suspects that because both amylose and cellulose exist as semicrystalline solids, the difference in oxygen (etc.) permeability is largely a reflection of differences in penetrant sorption coefficient [5].

Dynamic mechanical properties of polysaccharides [10] such as cellulose (cellophane), and amylose have also been studied extensively [66]. Recent work done in this laboratory [67] has reported a fairly comprehensive picture of these elastic constants in such polysaccharide films. A number of low temperature relaxation processes are reported to occur in these materials one related to the rotation of methylol groups and another related to motions involving water molecules complexed to methylol hydroxyls. Above room temperature it is apparent [68] that relaxation processes also occur, involving segments of backbones themselves. At the very highest temperatures, where chemical stability is still maintained, it is possible that motions of several sequences of glucose units in these chains can begin to move. This does not mean, however, that a glass transition or a melting of crystals can be observed in these materials. Water effects on these mechanically active loss processes can also be measured [10, 69]. Similarities in the mechanical coupling between structural units in cellulosic solids and amylose solids are obvious, as they relate to the semicrystalline nature of these polymers. One can tell that high degrees of orientation permit crystals to form which are very long in relationship to the amorphous regions that lie between them. This permits the transfer of loads most effectively onto the crystalline phases (as in the case of Fortisan fibers), unlike the situation in the randomly textured polysaccharide solids (close to the case of cellophane).

OPPORTUNITIES FOR THE FUTURE

The prospects for future development of polysaccharide solids seems exciting. At the present time volume usage of polysaccharides as engineering materials is stable or declining, but new understandings about the solid state of polymers in general may reveal clues to ways [70] to obtain superior properties from such polymers. Obviously chemical modifications of polysaccharides always represent possible ways to improve these materials. This paper has not been focused on all that is known in relationship to derivatised polysaccharides, but it is obvious that this is an enormously wide avenue of future development. What can be done with grafting, blending, or partial substitution in combination or in sequence is still to be found out.

What has recently been learned along the lines of processing of stiff chain molecules is also indicative of a possible avenue for future developments [71]. Specifically it is now recognized that liquid crystalline fluids [7, 8] can be created from chains that are intrinsically stiff. It is obvious from what has been reported above that cellulose and amylose molecules can be made to behave as stiff chains given the right solution properties of ionic strength, etc., so one can imagine that the application of flow to such liquid crystalline solutions of polysaccharides may improve the alignment of these polymer chains. Then it is simply a matter of causing the system to gel or otherwise phase separate from the water and remove the resulting product. What may be obtained might then be ultrahigh modulus polymer fibers or films, and if the material was in a conformation or in a hydrogen bonding scheme which prevented close interaction with water molecules, then one might even expect water-insoluble solids to be obtained in this manner. From considerations discussed in previous sections of this review it is clear that polysaccharides can be handled in the following states: solutions, sols, gels, dispersions, suspensions and possibly even melts. The kinds of material that can be obtained from all of these phases simply have not yet been explored.

One can envision some exciting applications that polysaccharides seem uniquely able to satisfy. This includes piezoelectric materials in which protons represent charged species which can be stabilized in just one part of the film. One imagines that protons can be stabilized on hydroxyl groups, or one can imagine that protons can be removed from hydroxyl groups. The resulting oxygen ions would be sites of immobilized charges which could be exploited in trapping charged species. Other performance properties of polysaccharides would complement this piezoelectric behavior. Another application is in high levels of water absorbancy, these are known already to be obtainable from very lightly crosslinked amylose gels or films. Finally, limited lifetime polymers for either biomedical or agricultural applications can be envisioned. In certain biomedical applications [72, 73] one can envision a limited lifetime agent of structure which serves either as a means for providing structure internally, or as a way for transporting drugs inside the body. The advantage in both cases is that the solid which is implanted does not need to be retrieved surgically, rather it undergoes hydrolysis (either by enzymes or by natural hydrolytic mechanisms). Because amylose can complex so readily with a wide variety of small molecules, it is an additional candidate for such implantable applications. In agriculture, there is a continuing need for crop covering films and soil mulches. Materials which disappear as the result of microbial degradation seem unusually well suited to such applications because of their susceptibility to certain microorganisms. An advantage of using microbe-degradable polymers is that they leave no residual polymeric material in the soil.

ACKNOWLEDGEMENT

This work was supported in part by the U.S. National Science Foundation with grant No. ENG 74-18131.

REFERENCES

1. J. Brandrup and E. H. Immergut, eds. Polymer Handbook, 2nd, Wiley, N. Y. 1975, p. IV-31.
2. J. F. Foster, in Starch, Chemistry and Technology, Vol. 1, R. L. Whistler and E. F. Paschall, Academic Press, N. Y., 1965, Chapt. XV.
3. V. S. R. Rao and J. F. Foster, Biopolymers 1, 527 (1963).
4. P. R. Sundararajan and R. H. Marchessault, Canadian J. Chem. 50, 792 (1972).
5. R. J. Dobbins, in Industrial Gums, R. L. Whistler, ed., Academic Press, N. Y., 1973, p. 19.
6. A. Sarko, F. J. Germino, and B. R. Zeitlin, J. Appl. Polym. Sci. 8, 1343 (1964).
7. P. J. Flory, Proc. Royal Soc. (London), Ser. A, 234, 60 (1956).
8. P. J. Flory, Proc. Royal Soc. (London), Ser. A, 234, 73 (1956).
9. D. W. Jones, in Cellulose and Cellulose Derivatives, Part IV, N. M. Bikales and L. Segal, eds., Wiley-Interscience, N. Y., 1971, p. 141.
10. A. H. Nissan, G. K. Hunger, and S. S. Sternstein, in Encyclopedia of Polymer Science and Technology, Vol. 3, H. F. Mark, N. G. Gaylord, and N. M. Bikales, eds., Wiley-Interscience, N. Y. 1965, p. 131.
11. O. Ellefsen and B. Tonnensen, in Cellulose and Cellulose Derivatives, Part IV, N. M. Bikales and L. Segal, eds., Wiley-Interscience, N. Y., 1971, p. 151.
12. A. Venkateswaran, J. Appl. Polym. Sci. 13, 2459 (1969).
13. A. Jeziorny and S. Kepka, J. Polym. Sci., B. 10, 257 (1972).
14. P. J. Hermans and A. Werdinger, J. Polym. Sci. 4, 135 (1949).
15. R. St. J. Manley and S. Inoue, J. Polym. Sci., B, 3, 691 (1965).

16. A. M. Hindeleh and D. J. Johnson, Polymer 13, 423 (1972).

17. S. A. Bradley, P. Engler, and S. H. Carr, Appl. Polym. Symp., No. 22, 209 (1973).

18. J. Haase, R. Hosemann, and B. Renwanz, Kolloid-Z.u.Z. Polymere 251, 871 (1973).

19. M. L. Nelson and R. T. O'Connor, J. Appl. Polym. Sci. 8, 1311 (1964).

20. F. J. Kolpak and J. Blackwell, Macromolecules 9, 273 (1976).

21. A. Sarko, Colston Papers, Vol. 26, 335 (1975).

22. B. Zaslow, in Starch: Chemistry and Technology, Vol. 1, R. L. Whistler and E. F. Paschall, eds., Academic Press, N. Y., 1965, Chapt. XI.

23. J. Blackwell, A. Sarko, and R. H. Marchessault, J. Mol. Biol. 42, 379 (1969).

24. H. F. Zobel, A. D. French, and M. E. Hinkle, Biopolymers 5, 837 (1967).

25. W. Winter and A. Sarko, Biopolymers 11, 849 (1972).

26. V. E. Stöckmann, Biopolymers 11, 251 (1972).

27. G. G. Leppard, L. C. Sowden, and J. R. Colvin, Science 189, 1094 (1975).

28. A. Bourret, H. Chanzy, and R. Lazaro, Biopolymers 11, 893 (1972).

29. K. H. Gardner and J. Blackwell, J. Polymer Sci., Part C, No. 36, 327 (1971).

30. K. H. Gardner and J. Blackwell, J. Ultrastruct. Res. 36, 725 (1971).

31. R. St. J. Manley, Nature 204, 1155 (1964).

32. J. W. Schappel and G. C. Bockno, in Cellulose and Cellulose Derivatives, Part V, N. M. Bikales and L. Segal, eds., Wiley, N. Y., 1971, p. 1126.

33. H. Dolmetsch and H. Dolmetsch, Kolloid Z. 185, 106 (1962).

34. E. Beebe and R. H. Marchessault, J. Appl. Phys. 35, 3182 (1964).

35. R. H. Marchessault, in Structure and Properties of Polymer Films, R. W. Lenz and R. S. Stein, eds., Plenum Pub. Co., N. Y., 1972, p. 25.

36. R. L. Coalson, A. Day, and R. H. Marchessault, Holzforschung 22, 190 (1968).

37. R. St. J. Manley and S. Inoue, J. Polym. Sci., Part B, 3, 691 (1965).

38. H. J. Gensrich and A. Grobe, Faserforsch. Textiltech. 21, 163 (1970).

39. M. Matsuo, S. Nomura, and H. Kawai, J. Polym. Sci., Polym. Phys. Ed. 11, 2057 (1973).

40. J. Haase, R. Hosemann, and B. Renwanz, Colloid Polym. Sci. 254, 199 (1976).

41. H. Bittiger and E. Husemann, Polymer Letters 10, 549 (1972).

42. H. D. Chanzy, Colston Papers 26, 417 (1975).

43. H. D. Chanzy, E. J. Roche, and R. K. Vuong, J. Polym. Sci., Polym. Phys. Ed. 11, 1859 (1973).

44. J. Borch, R. Muggli, A. Sarko, and R. H. Marchessault, J. Appl. Phys. 42, 4570 (1971).

45. J. E. Unbehend and A. Sarko, J. Polym. Sci., Polym. Phys. Ed. 12, 545 (1974).

46. H. Bittiger, E. Husemann, and B. Pfannemueller, Staerke 23, 113 (1971).

47. J. L. Koenig and P. D. Vasko, J. Macromol. Sci., Phys. B4, 347 (1970).

48. P. D. Vasko and J. L. Koenig, J. Macromol. Sci., Phys. B6, 117 (1972).

49. G. Wlodarski and E. Balcerzyk, Cellulose Chem. and Technol. 4, 651 (1970).

50. N. S. Wooding, in Rayon and Acetate Fibers, J. W. S. Hearle and R. H. Peters, eds., Butterworths, London, 1963, p. 462 ff.

51. H. Wakeham, in Cellulose and Cellulose Derivatives, Part III, E. Ott, H. M. Spurlin, and M. W. Grafflin, eds., Interscience, N. Y., 1955, p. 1247 ff.

52. J. N. Grant, in Cellulose and Cellulose Derivatives, Part IV, N. M. Bikales and L. Segal, eds., Wiley-Interscience, N. Y., 1971, p. 603.

53. I. A. Wolff, H. A. Davis, J. E. Cluskey, L. J. Gundrum, and C. E. Rist, Ind. Eng. Chem. 43, 915 (1951).

54. Nebraska Dept. of Agriculture and Inspection, Brit. Amended No. 965, 349 (1961).

55. A. M. Mark, W. B. Roth, C. L. Mehltretter, and C. E. Rist, Cereal Chem. 41, 197 (1964).

56. N. E. Lloyd and L. C. Kirst, Cereal Chem. 40, 154 (1963).

57. F. E. Carevic, U. S. patent No. 3,336,429, (1968).

58. A. Asanuma and K. Iwasa, Japan. Kokai No. 75,105,767 (1975).

59. A. Asanuma and K. Iwasa, Japan. Kokai No. 75,105,766 (1975).

60. A. M. Mark and C. L. Mehltretter, U. S. patent No. 3,553,196 (1969).

61. D. P. Langlois and J. A. Wagoner, in Starch: Chemistry and Technology, Vol. II, R. L. Whistler and E. F. Paschall, eds., Academic Press, N. Y., 1967, Chapt. XX.

62. A. M. Mark, W. B. Roth, C. L. Mehltretter, and C. E. Rist, Food Technol. 20, 75 (1966).

63. J. C. Romkin, I. A. Wolff, H. A. Davis, and C. E. Rist, Ind. Eng. Chem. 3, 120 (1958).

64. V. C. Haskell, in Encyclopedia of Polymer Science and Technology,

Vol. 3, H. F. Mark, N. G. Gaylord, and N. M. Bikales, eds. Wiley-Interscience, N. Y., 1965, p. 60 ff.

65. W. B. Roth and C. L. Mehltretter, J. Appl. Polym. Sci. **14**, 1387 (1970).

66. N. Tokita, J. Polym. Sci. **20**, 515 (1956).

67. S. A. Bradley and S. H. Carr, J. Polym. Sci., Polym. Phys. Ed. **14**, 111 (1976).

68. S. Nakamura and A. V. Tobolsky, J. Appl. Polym. Sci., **11**, 1371 (1967).

69. G. A. Baum, J. Appl. Polym. Sci. **17**, 2855 (1973).

70. F. C. Frank, Proc. Roy. Soc. London, Ser. A. **319**, 127 (1970).

71. A. Cifferri, Polym. Eng. and Sci. **15**, 191 (1975).

72. J. W. Frieders, U. S. Patent 2, 974, 049 (1961).

73. American Cyanimide, Japan. Kokai 16,496 (1960).

POLYNUCLEOTIDES AS INDUCERS OF INTERFERON

E. DE CLERQ
Rega Institute for Medical Research, University of Leuven, B-3000 Leuven, Belgium

SUMMARY

Polynucleotides are by far the most potent inducers of interferon. The salient features their interferon inducing capacity depends on are (1) a sufficiently high molecular size, (2) a double helical structure with sufficiently high thermal stability (Tm) and (3) the presence of a free hydroxyl at C(2') of the ribose moiety of both strands. Double-stranded (ds) RNA complexes are effective interferon inducers while triple-stranded RNA complexes are not. To induce a full interferon response, dsRNA complexes should exceed a threshold Tm of circa 60° (in 0.15M Na^+). The critical dependence of the interferon induction system on double-strandedness and Tm of the inducer has served as a tool to demonstrate the existence of a number of polynucleotide interactions (triplex formation, displacement reactions) which may be considered as models for the nucleic acid interactions involved in e.g. the control of gene expression and the transcription of DNA to RNA.

Prototypes of the polynucleotide inducers of interferon are poly(I).poly(C) and poly(A).poly(U). Both dsRNAs are active *in vitro* (cell culture) and *in vivo* (animals). The interferon induction process is thought to occur through a derepression mechanism. After the dsRNA has reacted with its receptor (presumably located at the cell surface), a message is transmitted to the cellular genome; the interferon gene is derepressed, interferon mRNA is transcribed and the mRNA is translated to the interferon protein. Before being released by the cell the interferon protein may be glycosylated.

Due to their interferon inducing properties, dsRNAs such as poly(I).poly(C) protect animals against a wide variety of experimental virus infections. Yet, the promises of poly(I).poly(C) as a useful antiviral agent in man have hitherto not been fulfilled. The poor performance of poly(I).poly(C) in man may seem related to the relatively low levels of interferon induced.

However, dsRNAs not only induce interferon, they also stimulate other host defense reactions such as cellular and humoral immune responses and macrophage activity. They also exert a direct toxic effect on rapidly dividing cells such as bone marrow cells and neoplastic cells. The latter obviously points to the potentials of dsRNAs as antitumor agents.

INTRODUCTION

Interferon has originally been described as a protein (or group of proteins) which are produced by the cell after this cell has been invaded by a virus and which block the recruitment of new cells by the virus. In this sense interferon production can be regarded as a control device by which viruses limit their own multiplication cycle.

Interferon is an extremely active molecule, the specific activity of which is only equalled by the polypeptide hormones. On the basis of antiviral assays the specific activity of interferon should largely exceed 10^9 units per mg of protein. Despite considerable efforts involving both classical physicochemical techniques and affinity chromatography, no interferon whether human, mouse or rabbit, has been obtained in a pure state. The interferons which have been characterized most thoroughly are human fibroblast interferon, human leukocyte interferon and mouse fibroblast interferon (1-6). The molecular weight of interferon is generally situated around 20.000. Human and mouse fibroblast interferon are glycoproteins, human leukocyte interferon probably is not (3).

To explain the inhibitory effect of interferon on the viral replication cycle, two possible sites of action have been proposed : virus transcription and virus translation. Which one prevails is still a matter of conjecture and beyond the scope of this review. In some instances, e.g. the release of type C virus particles from chronically infected cells, interferon may even directly interfere with the assembly or release of the virus (7,8).

Several basic concepts concerning the nature of the interferon molecule and its mode of action have been revised during the past few years. It has become increasingly clear that besides viruses various other agents including synthetic polynucleotides, polycarboxylates and aromatic low-molecular-weight compounds are able to trigger the production of interferon. It has also been shown that interferon, in addition to its antiviral action, exerts a wide variety of effects on cells, some of which, e.g. inhibition of cell division, stimulation of macrophage activity and enhancement of the expression of cell surface antigens, may underlie the alleged antitumor activity of interferon.

The general belief that interferon is species specific in its antiviral action, thus only active on cells of the animal species from which it has been derived, has been shaken thoroughly, since human leukocyte interferon was found to exhibit a far greater activity on bovine, porcine and cat cells than on human cells (9,10). Cross-species activity regularly occurs when interferons of phylogenetically higher animals (e.g. primate interferon) are assayed in cells of lower animals. The reverse, which may have potential value in human medicine, does, unfortunately, not occur.

Very schematically, interferon inducers could be divided into four distinct groups (11) (Table 1). Group I (viruses) and group II (polynucleotides) inducers share the ability to induce interferon in many different types of cells. Groups III and IV comprise a number of heterogeneous agents whose interferon inducing activity appears to be restricted to lymphocytes or other lymphoreticular cells (macrophages).

Table 1. Major groups of interferon inducers

Group	Agents	Types of cells which respond to interferon induction
I	Viruses (and other micro-organisms : e.g. chlamydiae, protozoa ...)	Wide variety of somatic cells
II	Polyribonucleotides (dsRNAs of viral or synthetic origin)	Wide variety of somatic cells
III	Agents stimulating T or B lymphocytes (e.g. phytohemagglutinin)	Lymphocytes
IV	Other agents : endotoxins, polycarboxylates, low-molecular-weight compounds (e.g. tilorone)	Lymphoreticular cells (macrophages, ...)

By far the most potent interferon inducers are double-stranded (ds) RNAs of either synthetic or viral origin. Only few viruses contain dsRNA as their genome : e.g. reoviridae (reo-, orbi- and rotavirus), mycophages and some insect and plant viruses. To explain the induction of interferon by viruses which do not contain dsRNA as their genome, the formation of a replicative dsRNA intermediate has been held responsible for the triggering of the interferon response. There are, however, several conditions in which single-stranded RNA viruses do not replicate, do not initiate the synthesis of complementary viral RNA, yet stimulate the production of interferon. From studies with temperature-sensitive mutants of reovirus it was concluded that the induction of interferon depended on a very late event of the virus multiplication cycle (12). Studies with temperature-sensitive mutants of Sindbis virus, however, implicated an early virus function in the interferon induction process (13). Hence, the role of virus-specified dsRNA in the induction of interferon by viruses remains equivocal.

The finding that synthetic dsRNAs such as poly(I).poly(C)* (Fig. 1) and

POLY (I)·POLY (C):

Double helix (A form)

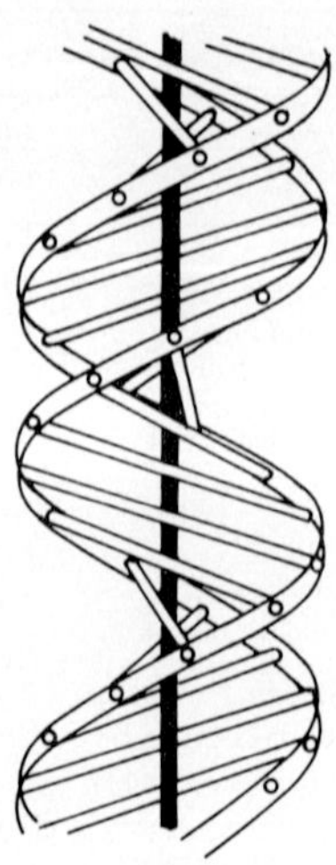

Fig. 1

---

* Abbreviations for synthetic polynucleotides conform to the recommendations of the IUPAC-IUB Commission (J. Mol. Biol. 55, 299, 1971. Less commonly used abbreviations are : poly(c^7A), poly(7-deazaadenylic acid); poly(c^7I), poly(7-deazainosinic acid); poly(L), polylaurusin or polyformycin B; poly(rT), poly(ribothymidylic acid); poly(br^5U), poly(5-bromouridylic acid); poly(br^5C), poly(5-bromocytidylic acid); poly(dUz), poly(2'-azido-2'-deoxyruridylic acid); poly(dUf), poly(2'-fluoro-2'-deoxyuridylic acid); poly(dCcl), poly(2'-chloro-2'-deoxycytidylic acid); poly(Um), poly(2'-O-methyluridylic acid); poly(ms^2I), poly(2-methylthioinosinic acid).

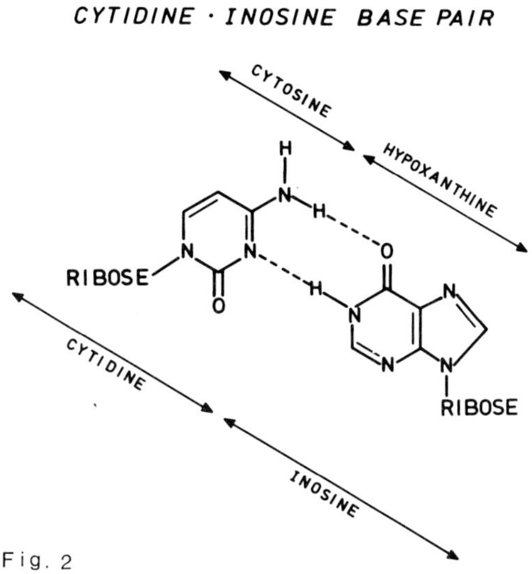

Fig. 2

poly(A).poly(U) are able to induce interferon (14,15) has offered the possibility to critically assess the structural features which govern the interferon induction process. Poly(I).poly(C) is the best known inducer of interferon. It occurs as a double-helical structure. The salient characteristics of this double helix are the ribose-phosphate backbone on the outside and the cytosine:hypoxanthine base pairs in the interior of the helix (Fig. 2). The structural criteria the interferon inducing capacity of poly(I).poly(C) depends on are summarized in Table 2. Most of these structural requirements have been dealt with in a previous review (16).

Table 2. Interferon induction by polynucleotides : requirements

1.	Sufficiently high molecular weight (strand continuity)
2.	Double-strandedness (base pairing) as opposed to single- and triple-strandedness
3.	Sufficiently high thermal stability
4.	Sufficiently high resistance to degradation by nucleases
5.	Presence of free 2'-OH groups in both strands of the double-stranded complex
6.	Intact purine ring (N-1, N-3, N-7, N-9)

First, the interferon inducing activity of poly(I).poly(C) is dependent on the molecular weight of the constituent homopolymers. If the molecular weight of either component, but more so of poly(I) than of poly(C), drops below a critical treshold (5 S or ∽ 100.000 daltons), the interferon inducing ability decreases commensurately (16).

As originally postulated by Field *et al*. (14,15), polynucleotide interferon inducers should be double-stranded. Examples of such interferon inducers are the homopolymer pairs poly(I).poly(C) and poly(A).poly(U), and the alternating copolymers poly(I-C).poly(I-C) and poly(A-U).poly(A-U). Single homopolyribonucleotides, whether single-stranded [e.g. poly(U)] or multi-stranded [e.g. poly(G)] induce little, if any, interferon (17).

Table 3. Triple-helical polynucleotide complexes* the existence of which has been suggested by interferon induction assays

Poly(U).poly(A).poly(I)
Poly(U).poly(A).poly(c^7I)
Poly(U).poly(A).poly(L)
Poly(rT).poly(A).poly(I)
Poly(rT).poly(A).poly(c^7I)
Poly(rT).poly(A).poly(L)
Poly(X).poly(A).poly(U)
Poly(U).poly(A).poly(U)
Poly(rT).poly(A).poly(rT)
Poly(rT).poly(A).poly(U)
Poly(br^5U).poly(A).poly(U)
Poly(br^5U).poly(A).poly(rT)
Poly(dUz).poly(A).poly(U)
Poly(rT).poly(A).poly(dUz)
Poly(dT).poly(A).poly(U)
Poly(U).poly(A).poly(dU)
Poly(Um).poly(A).poly(U)
Poly(dUf).poly(A).poly(U)

* The homopolymer written to the left of poly(A) is involved in Watson-Crick hydrogen bonding, whereas the homopolymer to the right of poly(A) is involved in Hoogsteen hydrogen bonding.

Triple-helical polyribonucleotide complexes such as poly(A).2 poly(U) are poor inducers of interferon (18). The poor interferon inducing performance of triple- as opposed to double-helical structures has led to demonstration of a number of triple-helical complexes (Table 3), the existence of which has not recognized before. These triple-stranded complexes may be composed of three fundamentally different heterocyclic bases [e.g. poly(U).poly(A).poly(I) (19); poly(X).poly(A).poly(U) (20)] or belong to the poly(A).2 poly(U) class [e.g. poly(dUz).poly(A).poly(U) (21); poly(dUf).poly(A).poly(U) (22)]. Triple-helical complexes can be formed through two mechanisms : (1°) simple addition of a third strand to a pre-existing double-stranded complex :

poly(Ux).poly(A) + poly(Uy) ⟶ poly(Ux).poly(A).poly(Uy)

if the Tm of poly(Ux).poly(A) is higher than the Tm of poly(Uy).poly(A) or (2°) displacement followed by addition :

poly(Ux).poly(A) + poly(Uy) ⟶ poly(Uy).poly(A) + poly(Ux);

poly(Uy).poly(A) + poly(Ux) ⟶ poly(Uy).poly(A).poly(Ux)

if the Tm of poly(Ux).poly(A) is lower than the Tm of poly(Uy).poly(A). As exemplified for poly(U).poly(A).poly(I) (Fig. 3), the homopolymer that is written to the left of poly(A) is involved in Watson-Crick hydrogen bonding, whereas the homopolymer to the right of poly(A) is involved in Hoogsteen hydrogen bonding.

BASE PAIRING SCHEME FOR
POLY (U) · POLY (A) · POLY (I)

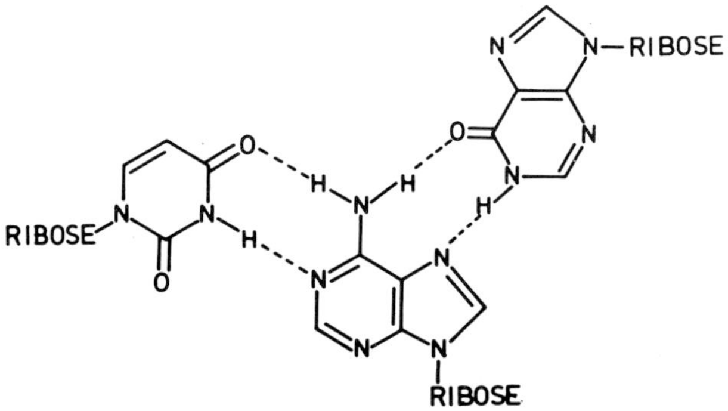

Fig. 3

The interferon inducing activity of dsRNA molecules largely depends on their Tm (melting temperature). A Tm value $\geqslant$60° (in 0.15M Na^+) appears necessary, but not sufficient, for a full expression of the interferon inducing activity (16,17). The notion that Tm is a critical property for interferon induction has been questioned occasionally (23,24) but reassured in various other reports (16). Within a series of partially thiolated poly(I).poly(C) analogs, thermal stability progressively decreased as the degree of thiolation of cytosine residues increased, and this decrease in Tm was accompanied by a loss of antiviral activity (25).

The most stable of all double-helical polyribonucleotides is poly(G). poly(C) (23). Poly(G).poly(C) has been reported to induce interferon and resistance to virus infection in some conditions (e.g. chick embryo cells, in the presence of DEAE-dextran) (23,26-28) but not in others (29). The latter failure was attributed to "improper purification of the polynucleotide preparations from protein and salt impurities" (26). However, repeated tests with properly characterized poly(G).poly(C) preparations in a variety of (human, rabbit and mouse) cell culture systems did not reveal any hint of interferon inducing activity (30). Thus, the claim that poly(G).poly(C) is as good an interferon inducer as poly(I).poly(C) (27) remains to be substantiated.

Interferon induction has been employed as a tool to monitor polynucleotide displacement reactions (Table 4). These displacement reactions were uniformly directed toward the formation of the double helix with the higher Tm (31). Since all displacement reactions were also accompanied by a rise in interferon production, these studies add further support to the importance of Tm in the interferon inducing capacity of polyribonucleotide duplexes. The helix-coil displacement reactions depicted in Table 4 were demonstrated at temperatures far beneath the Tm of the least stable helix, and, as evidenced for the reaction poly(A).2 poly(I) + 2 poly(C) $\longrightarrow$ poly(A) + 2 poly(I).poly(C), they readily occurred at the cellular level, presumably at the cell surface (31).

Another requirement for interferon induction that, to a certain extent, reflects the Tm requirement is adequate resistance to enzymatic degradation (16). The dsRNA should survive extra- and/or intracellular nucleases in order to be effective as interferon inducer. DsRNAs with low Tm either do not exist as such under physiological conditions or, due to extensive breathing of the double helix, are readily attacked by nucleases wherever they present single-stranded sequences.

Interferon induction is critically dependent on the nature of the ribose-phosphate backbone of the nucleic acid (Fig. 4). Substitution of thiophosphate for phosphate does not alter the interferon inducing ability

Table 4. Polynucleotide displacement reactions suggested by interferon induction assays

Poly(A).2 poly(I)	+ 2 Poly(C)	$\longrightarrow$	Poly(A)	+ 2 Poly(I).poly(C)
Poly(A).2 poly(I)	+ 2 Poly(br^5C)	$\longrightarrow$	Poly(A)	+ 2 Poly(I).poly(br^5C)
Poly(A).2 poly(c^7I)	+ 2 Poly(C)	$\longleftrightarrow$	Poly(A)	+ 2 Poly(c^7I).poly(C)
Poly(A).2 Poly(c^7I)	+ 2 Poly(br^5C)	$\longrightarrow$	Poly(A)	+ 2 Poly(c^7I).poly(br^5C)
Poly(L).poly(C)	+ Poly(I)	$\longrightarrow$	Poly(L)	+ Poly(I).poly(C)
Poly(L).poly(C)	+ Poly(c^7I)	$\rightleftharpoons$	Poly(L)	+ Poly(c^7I).poly(C)
Poly(L).poly(br^5C)	+ Poly(I)	$\longrightarrow$	Poly(L)	+ Poly(I).poly(br^5C)
Poly(L).poly(br^5C)	+ Poly(c^7I)	$\longrightarrow$	Poly(L)	+ Poly(c^7I).poly(br^5C)
Poly(I).poly(C)	+ Poly(br^5C)	$\longrightarrow$	Poly(C)	+ Poly(I).poly(br^5C)
Poly(c^7I).poly(C)	+ Poly(I)	$\longrightarrow$	Poly(c^7I)	+ Poly(I).poly(C)
Poly(c^7I).poly(C)	+ Poly(br^5C)	$\longrightarrow$	Poly(C)	+ Poly(c^7I).poly(br^5C)
Poly(c^7A).poly(U)	+ Poly(A)	$\longrightarrow$	Poly(c^7A)	+ Poly(A).poly(U)
Poly(c^7A).poly(rT)	+ Poly(A)	$\longrightarrow$	Poly(c^7A)	+ Poly(A).poly(rT)
Poly(c^7A).poly(I)	+ Poly(C)	$\longrightarrow$	Poly(c^7A)	+ Poly(I).poly(C)

of some dsRNAs [e.g. poly(I).poly(C)] (32,33), while increasing the interferon inducing ability of others [e.g. poly(A-U).poly(A-U)] (34,35). However, modification of the 2'-hydroxyl group leads to a dramatic decrease in activity. Thus, substitution of 2'-OH by either 2'-H (36), 2'-F (37), 2'-Cl (32), 2'-N_3 (38), 2'-O-CH_3 (39,40), 2'-O-CH_2-CH_3 (41) or 2'-O-CO-CH_3 (42) in either strand of poly(A).poly(U) or poly(I).poly(C) almost completely abolished the interferon inducing capacity of these complexes. Whether the 2'-OH group fulfils some specific role in the interaction of the polymer with the cellular receptor or merely generates a particular spatial configuration of the dsRNA remains an open question.

The interferon inducing activity of ds polyribonucleotides also depends on the nature of the heterocyclic bases in the interior of the double helix (Fig. 5). Double-helical polyribonucleotides based on the C-glycoside laurusin (formycin B) fail to induce interferon (43) although these complexes meet all previously established requirements for an effective inducer. Substitution of CH for N-7 of the purine ring has a differen-

CHEMICAL MODIFICATIONS OF RIBOSE-PHOSPHATE MOIETIES OF POLY (I) · POLY (C)

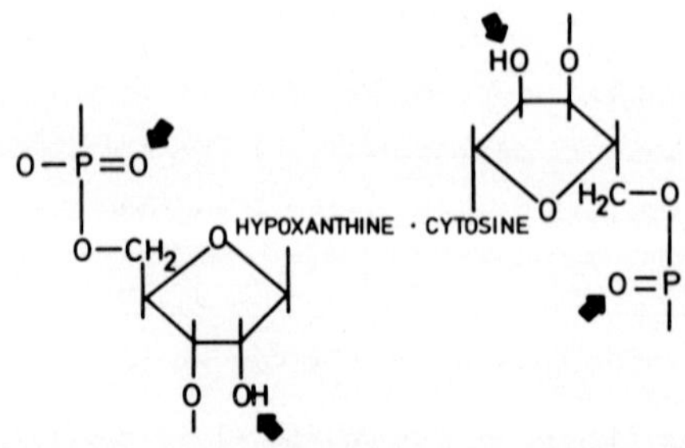

Ribose : C(2')–OH → –H, –F, –Cl, –N=N=N, –O–CH_3, –O–CH_2–CH_3, –O–CO–CH_3

Phosphate: –P=O → =S

Fig. 4

tial effect. Duplexes based on c^7A, e.g. poly(c^7A).poly(U) and poly(c^7A). poly(rT) are inactive as interferon inducers (18), whereas duplexes based on c^7I, e.g. poly(c^7I).poly(C) and poly(c^7I).poly(br^5C) are rather effective inducers of interferon (44,45). It now appears that the differential effects of c^7 substitution on the interferon inducing behavior of the poly(A).poly(U) and poly(I).poly(C) series can be accounted for by conformational differences. According to circular dichroism measurements (46), c^7 substitution should increase the base tilt of the poly(A).poly(U) series, whereas in the poly(I).poly(C) series c^7 substitution causes the base tilt to decrease. Base parameters such as "tilt" and "twist" have been defined on the basis of X-ray fiber diffraction studies (47). These parameters are illustrated in Fig. 6. The importance of the spatial organization of the dsRNA in the interferon induction process is further attested by poly(A). poly(X). This purine-purine duplex failed to induce interferon despite its high Tm ($\sim$ 85°, 0.10M Na^+) (48).

CHEMICAL MODIFICATIONS OF BASE MOIETIES OF POLY(I)·POLY(C)

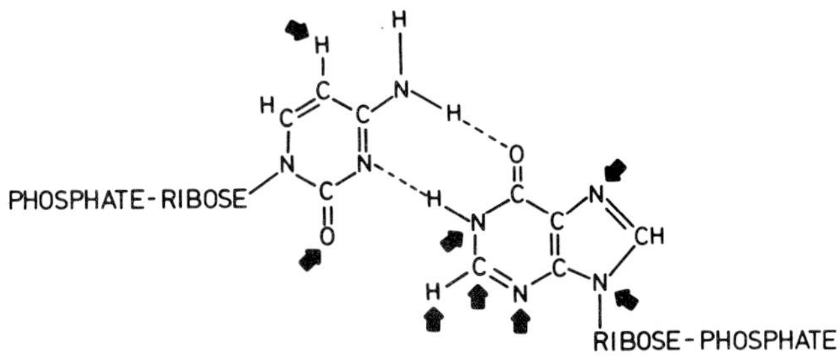

Cytosine:

C⑤-H → -I, -Br

C②-O → -S

Hypoxanthine:

N① → CH

CH② → N

N③ → CH

N⑦ → CH

C⑧-N⑨ → N②-C③ *(formycin B)*

C②-H → $-NH_2$, $-S-CH_3$

Fig. 5

Substitution of bromine at C-5 of cytosine (44,45) and substitution of sulfur at C-2 of cytosine (49,50) (Fig. 5) do not markedly affect the interferon inducing activity of poly(I).poly(C). Neither does 5-bromine substitution affect the base tilt of poly(A).poly(U) or poly(I).poly(C) (46). That poly(A).poly(br^5U), unlike poly(I).poly(br^5C), fails to induce interferon (18), could tentatively be assigned to rearrangement of poly(A). poly(br^5U) to poly(A).2 poly(br^5U) under physiological conditions.

Several other modifications have been carried out in the purine ring of poly(I) and/or poly(A) : implantation of a methylthio group at C-2 (51), substitution of CH for N-1 or N-3 (52) and replacement of C-2 by nitrogen (53). All these modifications result in a significant if not complete loss of interferon induction. To a large extent, this loss of activity may be accounted for by a perturbation of the base-pairing properties of the modified purine chain.

Polynucleotides are endowed with various biological activities other than interferon induction, e.g. anti-complement activity and anti-"reverse transcriptase" (RNA directed DNA polymerase) activity. The latter two activities are governed by similar, if not identical, structural requirements. However, these requirements are quite distinct from those which

BASE PARAMETERS : " TILT" and "TWIST"

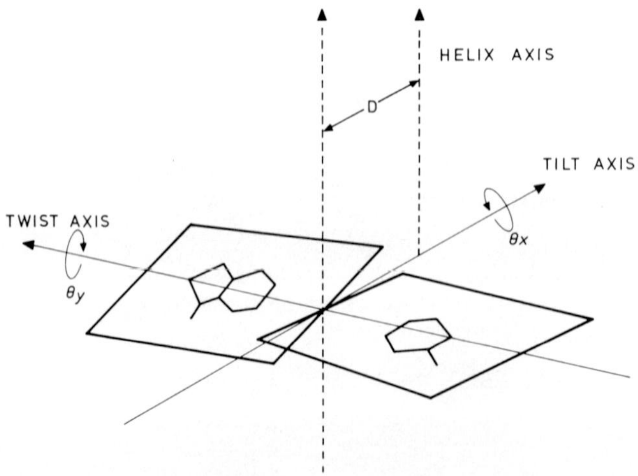

D : distance of helix axis from intersection of tilt axis and twist axis.

Fig. 6

determine interferon induction (Table 5). Single-stranded polynucleotides are invariably inactive as interferon inducers. Among these single-stranded polynucleotides some are quite effective inhibitors of complement and reverse transcriptase activity : e.g. (i) poly(I), poly(G) and poly(X) (54, 55) (these polymers tend to form self-aggregated structures); (ii) base-modified polynucleotides such as poly(ms^2I), poly(c^7I) and poly(n^2I) (52-54,56); and (iii) 2'-O-substituted polynucleotides such as poly(dUf), poly(dUz) and poly(dCcl) (54,57,58). Other single-stranded polynucleotides, e.g. poly(C) and polydeoxyribonucleotides are inactive as complement inhibitors (54) or reverse transcriptase inhibitors (57). Double-stranded polynucleotides can act as inducers of interferon, provided they comply with the requirements discussed above (Table 2). Hence, poly(I).poly(C), poly(A).poly(U), poly(A).poly(rT), poly(I).poly(br^5C) and poly(c^7I). poly(br^5C) are effective inducers, whereas poly(c^7A).poly(U), poly(c^7A). poly(rT), poly(L).poly(br^5C), poly(A).poly(dUz) and poly(I).poly(dCcl) are not. None of these double-stranded polynucleotides is capable of inhibiting complement or reverse transcriptase activity. Triple-stranded complexes do not induce interferon. Nor do they inhibit complement or reverse transcriptase activity.

Table 5. Biological activities of polynucleotides

	Interferon induction	Inhibition of reverse transcrip-tase activity	Anti-complement activity
1. Single-stranded polynucleotides			
- poly(I), poly(G) ...	-	+	+
- poly(U), poly(C) ...	-	-	-
- poly(ms^2I), poly(c^7I) ...	-	+	+
- poly(dU), poly(dT) ...	-	-	-
- poly(dUz), poly(dCcl) ...	-	+	+
2. Double-stranded polynucleotides			
- poly(I).poly(C), poly(A).poly(U) ...	+	-	-
- poly(c^7I).poly(C), poly(I).poly(br^5C)...	+	-	-
- poly(L).poly(C), poly(c^7A).poly(rT) ...	-	-	-
- poly(A).poly(dUz), poly(I).poly(dCcl)...	-	-	-
3. Triple-stranded polynucleotides			
- poly(U).poly(A).poly(U) ...	-	-	-
- poly(X).poly(A).poly(U) ...	-	-	-

In the process of interferon induction by poly(I).poly(C), poly(I) and poly(C) do not play an equal role. Several observations point to the relatively greater importance of poly(I). Within a range of well-defined molecular sizes, decrease of the size of poly(I) brings about a significantly greater loss of interferon induction than a similar decrease in the size of poly(C) (59-61). Introduction of mismatched bases in the poly(I) strand results in a significantly greater decrease of interferon production than similar substitutions in the poly(C) strand (60,62), and, when poly(I) and poly(C) are administered separately to the cells (63), a significantly greater interferon response is obtained if poly(I) is added first than if poly(C) is added first (64). Upon sequential addition to the cells, poly(I) and poly(C) reunite at the cell surface (63). The binding of poly(I). poly(C) to the cell is strongest if poly(I) is added first (64). In view of the central role of poly(I) in the interferon induction process of poly(I). poly(C) it is not surprising that poly(C) can be replaced by a distantly related analog [poly(1-vinyl-cytosine] without significant loss of activity

(65), and that poly(I) preparations have been obtained, which, even in the total absence of poly(C), prove capable of triggering an interferon response (66). Although the precise physicochemical basis for the interferon inducing activity of the latter has not been elucidated, the fact that apparently "pure" poly(I) preparations can function as interferon inducers rules out that a continuous high molecular weight base-paired structure is the conditio sine qua non for the induction of interferon by polyribonucleotides.

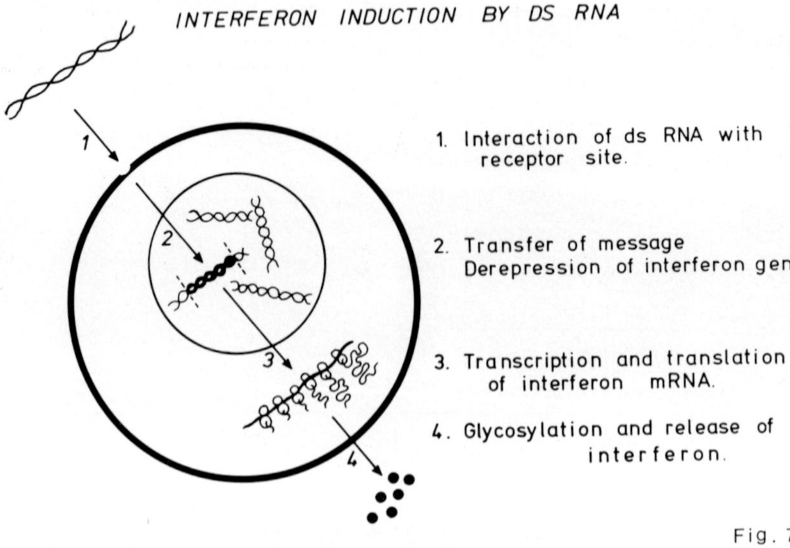

Fig. 7

The interferon induction process could be divided into 4 steps (Fig. 7), the first one being the interaction of the inducer with a specific cellular receptor site (23). This interaction would generate a message which, in turn, would lead to a derepression of the interferon gene, transcription and translation of the interferon mRNA and, eventually, glycosylation and secretion of the interferon molecule. There is circumstancial evidence for any of these steps except for the second one (transfer of the message from the receptor site to the cellular genome).

Since antibodies to dsRNA recognize many of the structural features that are responsible for the interferon inducing capacity of dsRNA, it has been postulated that the cellular receptor site for polynucleotide inducers of interferon is protein in nature (67). Studies with poly(I).poly(C) attached to cells [L cells (68), red blood cells (69)] or covalently coupled to insoluble supports (e.g. sepharose beads) (70-74) suggest, but do not prove, that the receptor for interferon induction is localized at the cell surface. Leakage of polynucleotide from the insoluble support during its incubation with the cells impedes an unequivocal answer as to whether penetration of the polynucleotide into the interior of the cell is indeed immaterial for the induction of interferon.

Table 6. Sequence of major events during interferon induction* by poly(I).poly(C)

Approximate time (hrs)	Event	Site
0 - 0.5	Triggering	Receptor on plasma membrane (?)
0.5 - 2 (or longer)	Transcription of interferon mRNA	Nucleus
1 - 3.5	Translation of interferon mRNA	Rough ER (or free polysomes ?)
1 - 3.5	Glycosylation of interferon	Rough or smooth ER (?), or Golgi apparatus (?)
1.5 - 4.0	Release of interferon	Plasma membrane (via secretory vesicles ?)
2 - ?	Transcription, translation and build-up of repressor	Nucleus and cytoplasm
3 - ?	Inactivation of interferon mRNA by repressor	Cytoplasm

* In human or rabbit cell cultures [according to Vilcek *et al.* (11)]

Once the receptor has been triggered by the interferon inducer, a signal is transmitted to the cellular genome and, as a result, the interferon gene is switched on. Evidence for this part of the interferon induction process is still lacking. Fact is that within 2 hours after exposure of the cells to poly(I).poly(C) most of the interferon mRNA has been transcribed (11) (Table 6). Indirect evidence for the need of interferon mRNA synthesis stems from studies with metabolic inhibitors (actinomycin D, cycloheximide) (75-78). Direct proof for the existence of interferon mRNA was provided by De Maeyer-Guignard *et al.* (79) who demonstrated that mRNA extracted from virus- or poly(I).poly(C)-induced mouse cells could be translated into interferon in heterologous cells. More recently, translation of both mouse and human interferon mRNA has been demonstrated in cell-free ribosomal systems and Xenopus laevis oöcytes (80-82).

Proteins made for export from the cell are, as a rule, synthesized on membrane-bound polysomes (rough endoplasmic reticulum). Much of the intracellular interferon is indeed membrane-associated (83). Yet the actual site of interferon synthesis has not been elucidated. Since 2-deoxy-D-glucose and D-glucosamine, two inhibitors of glycosylation, inhibit the production of biologically active interferon, it is reasonable to assume that interferon is glycosylated before its release (84). The release of interferon from the cells is an energy- and temperature-dependent process (85). It is inhibited by SH-blockers such as p-hydroxymercuribenzoate (86). Although the cellular organelles which contribute to the release of interferon have not been established, it is noteworthy that interferon secretion is inhibited by vinblastine (87), an alkaloid with high affinity for microtubules.

In human and rabbit cells interferon production peaks approximately 3 to 4 hours after exposure of the cells to poly(I).poly(C). By 6 to 8 hours the production of interferon is shut off (11). However, when metabolic inhibitors (e.g. actinomycin D, cycloheximide or 5,6-dichloro-1-β-D-ribofuranosylbenzimidazole) are added to the cells at the appropriate times and dosages, interferon production is not shut off and the interferon yields are increased (75-78,88-90). This phenomenon is generally referred to as "superinduction" (Fig. 8). The shut-off of interferon production has

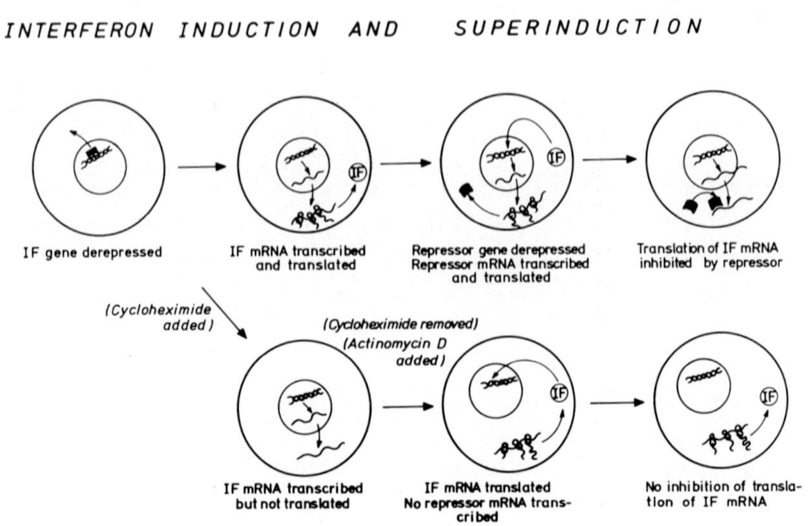

Fig. 8

been attributed to a repressor which inhibits the translation of interferon mRNA (91). The mode of action of this putative repressor molecule can only be speculated upon. Its action is thought to be largely irreversible (90).

Poly(I).poly(C) has been shown to protect animals against a wide variety of experimental virus infections (Table 7), encompassing nearly all major virus families (papova-, adeno-, herpeto-, pox-, picorna-, toga-, orthomyxo-, paramyxo-, retro- and rhabdoviridae) (92-98). In all instances poly(I).poly(C) was most effective when given prophylactically, that is prior to virus challenge. Little, if any effect was observed if treatment was initiated at the time symptoms of the disease had appeared (92-94). The whole antiviral protection imparted by a single dose of poly(I).poly(C) administered prior to virus challenge appears to be interferon-mediated (97,98). Due to hyporeactivity of the host, interferon production declines upon repeated administration of the interferon inducer. As a consequence, the antiviral efficacy of multiple doses of poly(I).poly(C) is rather limited (94,96).

Poly(I).poly(C) has also been shown to inhibit the growth of a large variety of experimental tumors (Table 8), whether these tumors were induced by viruses, chemicals or X-ray irradiation, had been transplanted or arose spontaneously (92,99-104). Although the mechanism of the antitumor action of poly(I).poly(C) has not been assessed unequivocally, it is certainly not or not only mediated by interferon production. In athymic nude mice poly(I).poly(C) caused a dramatic inhibition of L-929 tumor growth, even if treatment was initiated after the tumor had reached an appreciable size (102). Mouse interferon, administered exogenously in amounts which closely mimicked the interferon blood levels induced endogenously by poly(I).poly(C), failed to duplicate the antitumor effects of poly(I).poly(C). Hence, the role of interferon in the antitumor action of poly(I).poly(C) in nude mice with transplanted L-929 tumors can be eliminated. Since these results were obtained in thymusless mice, the role of thymus-dependent immunity can also be eliminated. What are the factors the antitumor activity of poly(I).poly(C) may be based upon ? Poly(I).poly(C) could exert a direct inhibitory effect on the metabolism and proliferation of tumor cells. Alternatively, it might act indirectly through activation of some host cells, e.g. macrophages.

Poly(I).poly(C) and other dsRNA indeed exert a variety of biological effects other than interferon induction (Table 9). Some of these effects may contribute to the antiviral efficacy of the dsRNA : e.g. the stimula-

Table 7. Experimental virus infections which are inhibited by poly(I).poly(C)*

Virus	Disease	Animal
Papovaviridae		
- Mouse polyoma	Tumor growth	Rat
- Simian virus 40	Tumor growth	Hamster
Adenoviridae		
- Human adenovirus 12	Tumor growth	Hamster
Herpetoviridae		
- Herpes simplex	Keratoconjunctivitis	Rabbit, man
	Encephalitis	Mouse
- Equine abortion	Encephalitis	Hamster
- Infectious bovine rhinotracheitis	Respiratory infection	Calve
Poxviridae		
- Vaccinia	Keratoconjunctivitis	Rabbit
	Skin lesions	Mouse
	Encephalitis	Mouse
- Fibroma	Tumor growth	Rabbit
Picornaviridae		
- Coxsackie B3	Myocarditis	Mouse
- Human rhinovirus 13	Respiratory infection	Man
- Foot-and-mouth disease	Encephalitis	Mouse
- Columbia SK	Encephalitis	Mouse
- Encephalomyocarditis	Encephalitis	Mouse
Togaviridae		
- Semliki forest	Encephalitis	Mouse
- Simian hemorrhagic fever	Encephalitis	Rhesus monkey
- Tick-borne encephalitis	Encephalitis	Mouse
- West Nile	Encephalitis	Mouse
Orthomyxoviridae		
- Influenza A, B	Respiratory infection	Mouse, baboon
Paramyxoviridae		
- Parainfluenza 1, 3	Respiratory infection	Mouse, hamster
- Pneumonia virus of mice	Respiratory infection	Mouse
Retroviridae		
- Moloney sarcoma	Tumor growth	Mouse
- Rous sarcoma	Tumor growth	Chicken
- Friend leukemia	Leukemia	Mouse
- Kirsten leukemia	Hemolytic anemia	Rat
Rhabdoviridae		
- Rabies	Encephalomyelitis	Rabbit
- Vesicular stomatitis	Encephalitis	Mouse

* Reviewed by De Clercq and Stewart (92). For more recent reports concerning the effects of poly(I).poly(C) on experimental viral infections, see, for example, refs. 93-96.

Table 8. Experimental tumors which are inhibited by poly(I).poly(C)*

	Host
Virus induced tumors	
Mouse polyoma	Rat
Simian virus 40	Hamster
Human adenovirus 12	Hamster
Fibroma	Rabbit
Moloney sarcoma	Mouse
Rous sarcoma	Chicken
Friend leukemia	Mouse
Kirsten leukemia	Rat
Chemically induced tumors	
9,10-dimethylbenzanthracene	Mouse
3-methylcholanthrene	Mouse
X-ray induced tumors	
Thymic lymphoma	Mouse
Transplanted tumors	
Reticulum cell sarcoma	Mouse
L-929 sarcoma	Mouse
Walker 256 carcinosarcoma	Mouse
Methylcholanthrene-induced fibrosarcoma	Mouse
Ehrlich ascites tumor	Mouse
Rauscher leukema virus-induced ascites tumor	Mouse
Polyoma	Mouse
B-16 malignant melanoma	Mouse
Mammary adenocarcinoma	Rat
L-1210 leukemia	Mouse
B-1237 lymphoma	Mouse
Spontaneous tumors	
Mammary adenocarcinoma	Mouse
AKR leukemia	Mouse

* Reviewed by De Clercq and Stewart (92). For more recent reports concerning the effects of poly(I).poly(C) on experimental tumors, see, for example, refs. 99-102. Inhibition of spontaneous leukemia in AKR mice has been obtained with poly(A).poly(U) (103, 104).

tory effects on macrophage and lymphocyte functions. Other effects are clearly undesirable : e.g. inhibition of DNA, RNA and protein synthesis (115-117). Rapidly growing cells such as the hematopoietic stem cells appear to be most vulnerable to the cytotoxic and cytostatic action of poly(I).poly(C) (118,119). Hence, the most characteristic features of the *in vivo* toxicity of dsRNAs are anemia, lymphopenia, thrombocytopenia, splenomegaly and thymic atrophy (120).

The spectrum of biological effects displayed by interferon is even more diverse than the effects exerted by dsRNAs (Table 9). To some extent, interferon duplicates the effects of dsRNAs, e.g. the inhibitory effect on virus multiplication and the stimulatory effect on macrophage activity. The antiviral activity of dsRNA is mediated by interferon. Whether other effects of dsRNA such as the activation of macrophages are also interferon-mediated, is not clear. To another extent, interferon antagonizes the action of dsRNA. For example, dsRNAs either increase or decrease the antibody response depending on the time of administration of the dsRNA in relation to the antigen (121). Interferons generally suppress the immune response (108). Poly(I).poly(C) is a potent mitogen for lymphocytes (120,122). Interferon, however, inhibits the mitogenic activity of materials such as phytohemagglutinin and concanavalin A (109,110).

Could the desirable (antiviral) effects of dsRNAs be uncoupled from their undesirable (toxic) effects, or do both covary ? Attempts to dissociate the toxic from the antiviral properties of poly(I).poly(C) by reduction of the molecular size of poly(I) or poly(C) (123), or by modification of the ribose-phosphate backbone (33) or bases (45,50) have not proven very successful. One may postulate, therefore, that the structural requirements underlying the interferon inducing activity of dsRNAs also govern many of the other biological responses of dsRNAs, including toxicity.

To design a polynucleotide interferon inducer having greater therapeutic efficacy than poly(I).poly(C) one may also speculate on possible differences in temperal requirements between the antiviral (interferon inducing) effects and other effects. If the time for triggering the antiviral or interferon response is indeed shorter than the time required for the induction of the other responses, one might eventually succeed in developing an interferon inducer which is superior to poly(I).poly(C). Mismatched analogues of poly(I).poly(C) such as poly(I).poly(C_{13},U) and poly(I).poly(C_{20},G) at least partially fulfil this premise (120,124).

Table 9. Diversity of effects exerted by polynucleotide inducers of interferon*

dsRNA	INTERFERON
induction of interferon → INTERFERON	induction of antiviral state (secondary antiviral protein ?)
pyrogenicity	inhibition of cell growth (antiproliferative activity)
inhibition of cell growth (antiproliferative activity)	priming or blocking of subsequent interferon production
inhibition of DNA, RNA and protein synthesis (antimetabolic activity)	stimulatory effect on macrophage activity
stimulatory effect on macrophage activity	immunosuppressive effect (inhibition of antibody production
immunoadjuvant effect (stimulation of antibody production)	anti-mitogenic effect (on lymphocytes)
mitogenic effect (on lymphocytes)	enhancement of cytotoxicity of dsRNA
"acute" and "chronic" toxicity	enhancement of immunolytic effect of lymphocytes on target cells
	enhancement of expression of surface (e.g. histocompatibility) antigens
	other effects on specific cell functions (stimulation of expression of some enzymes, suppression of others)

* Reviewed by De Clercq and Stewart (92,105) and Carter and De Clercq (106). For more recent reports concerning the breadth of interferon action, see, for example, refs. 107,108 (immunosuppressive effect of interferon, 109,110 (anti-mitogenic effect of interferon on lymphocytes), 111,112 (enhancement of expression of histocompatibility antigens), 113,114 (stimulation or inhibition of enzyme induction).

Poly(I).poly(C) has been administered topically to children with localized herpes zoster (125). It has also been given intranasally to volunteers prior to infection with rhinovirus and influenza virus (126). The results were disappointing. No reports have appeared so far on the systemic use of poly(I).poly(C) in human virus infections. In man, poly(I).poly(C) does not induce as high interferon levels as in rodents (127). This low interferon inducing activity could, in part, be ascribed to the presence in human plasma of a nuclease that degrades the dsRNA before it can reach its target cells (128). Recently, Levy *et al*. (129) prepared a mixture of poly(I).poly(C), poly-L-lysine and carboxymethyl-cellulose. The mixture, the exact nature of which was not defined, showed an increased resistance to hydrolysis by ribonucleases as compared to free poly(I). poly(C), and, in addition, stimulated the production of interferon in monkeys under conditions free poly(I).poly(C) failed to do so (129). Whether this mixture is also more effective in man or offers any other advantage over free poly(I).poly(C), e.g. a more favorable activity to toxicity ratio, remains to be seen. According to Ts'o *et al*. (124), polynucleotide duplexes with increased resistance to enzymatic hydrolysis should not, *a priori*, possess any advantage over poly(I).poly(C). Such duplexes are likely to be less selective since they persist longer in biological fluids than poly(I).poly(C), thus having a greater probability of inducing undesirable (toxic) side effects.

For prophylactic and/or therapeutic purposes interferon can be applied exogenously or induced endogenously. The latter requires the use of interferon inducers such as poly(I).poly(C). Because of the toxic side effects of interferon inducers, current interest has shifted toward the passive administration of interferon. Human interferon administered prophylactically to volunteers with rhinovirus or influenza virus infections did not yield dramatic results (130). The relative inactivity of interferon may have been related to the low dose of interferon used. With significantly higher doses, Jones *et al*. (131) demonstrated a therapeutic effect in patients with ulcerative herpetic keratitis. Interferon has also been reported to suppress the virus markers (DNA polymerase activity, core antigen expression) associated with hepatitis B virus infection in patients with chronic active hepatitis (132,133). These studies suggest that interferon may be useful in eradicating the virus carrier state of chronic hepatitis B patients.

While the latter findings point to the potential benefits of exogenous interferon in the therapy of human virus infections, they should, by no means, minimize the importance of the polynucleotide inducers of interferon. DsRNAs such as poly(I).poly(C) are by far the most potent inducers of interferon. Hence, they will remain indispensable to produce the (enormous) quantities of interferon that are needed for exogenous interferon administration. It should also be recognized that in several clinical situations including chronic active hepatitis and some neoplastic diseases the potentials of dsRNAs have not or not properly been explored. In view of the diversity of effects exhibited by dsRNAs and the differences in temporal requirements (120) that might determine these effects, extended studies with specifically "tailored" polymers, including mismatched poly(I).poly(C) analogues (60) and some poly(I) derivatives (66), would seem particularly warranted.

Acknowledgments.

I am indebted to Mrs. Janine Putzeys for editorial assistance.

REFERENCES

1 C.B. Anfinsen, S. Bose, L. Corley and D. Gurari-Rotman, Proc. Nat. Acad. Sci. USA, 71 (1974) 3139.
2 E. Knight Jr., J. Biol. Chem., 250 (1975) 4139.
3 W.J. Jankowski, M.W. Davey, J.A. O'Malley, E. Sulkowski and W.A. Carter, J. Virol., 16 (1975) 1124.
4 M.W. Davey, E. Sulkowski and W.A. Carter, Biochemistry, 15 (1976) 704.
5 M.W. Davey, E. Sulkowski and W.A. Carter, J. Virol., 17 (1976) 439.
6 E. Knight Jr., Proc. Nat. Acad. Sci. USA, 73 (1976) 520.
7 A. Billiau, V.G. Edy, H. Sobis and P. De Somer, Int. J. Cancer, 14 (1974) 335.
8 R.M. Friedman and J.M. Ramseur, Proc. Nat. Acad. Sci. USA, 71 (1974) 3542.
9 I. Gresser, M.-T. Bandu, D. Brouty-Boyé and M. Tovey, Nature, 251 (1974) 543.
10 J. Desmyter and W.E. Stewart II, Virology, 70 (1976) 451.
11 J. Vilcek, E.A. Havell, L.W. Mozes and B. Berman, Proceedings First Intersectional Congress of IAMS (Vol. 4), T. Hasegawa (ed.), Science Council of Japan (1975) 65.
12 M.-H.T. Lai and W.K. Joklik, Virology, 51 (1973) 191.
13 G.J. Atkins, M.D. Johnston, L.M. Westmacott and D.C. Burke, J. Gen. Virol., 25 (1974) 381.
14 A.K. Field, A.A. Tytell, G.P. Lampson and M.R. Hilleman, Proc. Nat. Acad. Sci. USA, 58 (1967) 1004.
15 A.K. Field, A.A. Tytell, G.P. Lampson and M.R. Hilleman, Proc. Nat. Acad. Sci. USA, 61 (1968) 340.
16 E. De Clercq, Topics in Current Chemistry, 52 (1974) 173.
17 E. De Clercq and T.C. Merigan, Nature, 222 (1969) 1148.

18 E. De Clercq, P.F. Torrence and B. Witkop, Proc. Nat. Acad. Sci. USA, 71 (1974) 182.
19 E. De Clercq, P.F. Torrence, P. De Somer and B. Witkop, J. Biol. Chem., 250 (1975) 2521.
20 P.F. Torrence and E. De Clercq, Submitted for publication (1976).
21 P.F. Torrence, E. De Clercq and B. Witkop, Biochemistry, 15 (1976) 724.
22 E. De Clercq, B. Janik and R.G. Sommer, Chem.-Biol. Interactions, 14 (1976) 113.
23 C. Colby and M.J. Chamberlin, Proc. Nat. Acad. Sci. USA, 63 (1969) 160.
24 J.F. Niblack and M.B. McCreary, Nature, 233 (1971) 52.
25 J.A. O'Malley, Y.K. Ho, P. Chakrabarti, L. DiBerardino, P. Chandra, D.A.O. Orinda, D.M. Byrd, T.J. Bardos and W.A. Carter, Mol. Pharmacol., 11 (1975) 61.
26 O.A. Aksenov, A.L. Timkowsky, O.N. Ageeva, E.M. Kogan, S.E. Bresler, A.A. Smorodinstev and N.S. Tikhomirova-Sidorova, Vop. Virus, 18 (1973) 345.
27 A.L. Timkovsky, O.A. Aksenov, S.E. Bresler, E.M. Kogan, A.A. Smorodintsev and N.S. Tikhomirova-Sidorova, Vop. Virus, 18 (1973) 350.
28 A.S. Novokhatsky, F.I. Ershov, A.L. Timkovsky, S.E. Bresler, E.M. Kogan and N.S. Tikhomirova-Sidorova, Acta Virol., 19 (1975) 121.
29 E. De Clercq, F. Eckstein and T.C. Merigan, Ann. N.Y. Acad. Sci., 173 (1970) 444.
30 P.F. Torrence and E. De Clercq, unpublished data (1976).
31 E. De Clercq, P.F. Torrence and B. Witkop, Biochemistry, 15 (1976) 717.
32 D.R. Black, F. Eckstein, J.B. Hobbs, H. Sternbach and T.C. Merigan, Virology, 48 (1972) 537.
33 D.R. Black, F. Eckstein, E. De Clercq and T.C. Merigan, Antimicrob. Ag. Chemother., 3 (1973) 198.
34 E. De Clercq, F. Eckstein and T.C. Merigan, Science, 165 (1969) 1137.
35 E. De Clercq, F. Eckstein, H. Sternbach and T.C. Merigan, Virology, 42 (1970) 421.
36 D.W. Hutchinson, M.D. Johnston and M.A.W. Eaton, J. Gen. Virol., 23 (1974) 331.
37 E. De Clercq and B. Janik, Biochim. Biophys. Acta, 324 (1973) 50.
38 P.F. Torrence, J.A. Waters, C.E. Buckler and B. Witkop, Biochem. Biophys. Res. Commun., 52 (1973) 890.
39 E. De Clercq, B. Zmudzka and D. Shugar, FEBS Letters, 24 (1972) 140.
40 T.C. Merigan and F. Rottman, Virology, 60 (1974) 297.
41 E. De Clercq, F.M. Rottman and D. Shugar, FEBS Letters, 42 (1974) 331.
42 D.L. Steward, W.C. Herndon Jr. and K.R. Schell, Biochim. Biophys. Acta, 262 (1972) 227.
43 P.F. Torrence, E. De Clercq, J.A. Waters and B. Witkop, Biochem. Biophys. Res. Commun., 62 (1975) 658.
44 P.F. Torrence, E. De Clercq, J.A. Waters and B. Witkop, Biochemistry, 13 (1974) 4400.
45 E. De Clercq, V.G. Edy, P.F. Torrence, J.A. Waters and B. Witkop, Mol. Pharmacol., in press (1976).
46 A.M. Bobst, P.F. Torrence, S. Kouidou and B. Witkop, Proc. Nat. Acad. Sci. USA, in press (1976).
47 S. Arnott, D.W.L. Hukins, S.D. Dover, W. Fuller and A.R. Hodgson, J. Mol. Biol., 81 (1973) 107.
48 P.F. Torrence, E. De Clercq and B. Witkop, Biochim. Biophys. Acta, in press (1976).

49 P. Faerber, K.-H. Scheit and H. Sommer, Eur. J. Biochem., 27 (1972) 109.
50 K. Reuss, K.-H. Scheit and O. Saiko, Communication at the Symposium "Synthetic nucleosides, nucleotides and polynucleotides" (Max-Planck-Institut für Biophysikalische Chemie, Göttingen, 3-5 May 1976).
51 E. De Clercq, M. Hattori and M. Ikehara, Nucleic Acids Research, 2 (1975) 121.
52 E. De Clercq, P.F. Torrence, T. Fukui and M. Ikehara, Nucleic Acids Research, 3 (1976) 1591.
53 E. De Clercq, P.F. Torrence and M. Ikehara, in preparation (1976).
54 E. De Clercq, P.F. Torrence, J. Hobbs, B. Janik, P. De Somer and B. Witkop, Biochem. Biophys. Res. Commun., 67 (1975) 255.
55 S.K. Arya, T.L. Helser, W.A. Carter and P.O.P. Ts'o, Mol. Pharmacol., in press (1976).
56 E. De Clercq, A. Billiau, M. Hattori and M. Ikehara, Nucleic Acids Research, 2 (1975) 2305.
57 R.J. Erickson and J.C. Grosch, Biochemistry, 13 (1974) 1987.
58 E. De Clercq, A. Billiau, J. Hobbs, P.F. Torrence and B. Witkop, Proc. Nat. Acad. Sci. USA, 72 (1975) 284.
59 A.A. Tytell, G.P. Lampson, A.K. Field, M.M. Nemes and M.R. Hilleman, Proc. Soc. Exp. Biol. Med., 135 (1970) 917.
60 W.A. Carter, P.M. Pitha, L.W. Marshall, I. Tazawa, S. Tazawa and P.O.P. Ts'o, J. Mol. Biol., 70 (1972) 567.
61 S.J. Mohr, D.G. Brown and D.S. Coffey, Nature, New Biol., 240 (1972) 250.
62 S. Matsuda, M. Kida, H. Shirafuji, M. Yoneda and H. Yaoi, Arch. Ges. Virusforsch., 34 (1971) 105.
63 E. De Clercq and P. De Somer, J. Virol., 9 (1972) 721.
64 E. De Clercq, W.E. Stewart II and P. De Somer, Virology, 54 (1973) 278.
65 J. Pitha and P.M. Pitha, Science, 172 (1971) 1146.
66 M.N. Thang, L. Bachner, E. De Clercq and B.D. Stollar, submitted for publication (1976).
67 M.I. Johnston, B.D. Stollar, P.F. Torrence and B. Witkop, Proc. Nat. Acad. Sci. USA, 72 (1975) 4564.
68 E. De Clercq and P. De Somer, J. Gen. Virol., 16 (1972) 435.
69 E. De Clercq and P. De Somer, J. Gen. Virol., 22 (1974) 271.
70 A.F. Wagner, R.L. Bugianesi and T.Y. Shen, Biochem. Biophys. Res. Commun., 45 (1971) 184.
71 J. Taylor-Papadimitriou and J. Kallos, Nature, New Biol., 245 (1973) 143.
72 P.M. Pitha and J. Pitha, J. Gen. Virol., 21 (1973) 31.
73 L. Bachner, E. De Clercq and M.N. Thang, Biochem. Biophys. Res. Commun., 63 (1975) 476.
74 D.W. Hutchinson and T.C. Merigan, J. Gen. Virol., 27 (1975) 403.
75 J. Vilcek, T.G. Rossman and F. Varacalli, Nature, 223 (1969) 682.
76 J. Vilcek, Ann. N.Y. Acad. Sci., 173 (1970) 390.
77 Y.H. Tan, J.A. Armstrong, Y.H. Ke and M. Ho, Proc. Nat. Acad. Sci. USA, 67 (1970) 464.
78 J. Vilcek and E.A. Havell, Proc. Nat. Acad. Sci. USA, 70 (1973) 3909.
79 J. De Maeyer-Guignard, E. De Maeyer and L. Montagnier, Proc. Nat. Acad. Sci. USA, 69 (1972) 1203.
80 S. Pestka, J. McInnes, E.A. Havell and J. Vilcek, Proc. Nat. Acad. Sci. USA, 72 (1975) 3898.
81 M.N. Thang, D.C. Thang, E. De Maeyer and L. Montagnier, Proc. Nat. Acad. Sci. USA, 72 (1975) 3975.

82 F.H. Reynolds Jr., E. Premkumar and P.M. Pitha, Proc. Nat. Acad. Sci. USA, 72 (1975) 4881.
83 M.H. Ng, B. Berman and J. Vilcek, Virology, 49 (1972) 322.
84 E.A. Havell, J. Vilcek, E. Falcoff and B. Berman, Virology, 63 (1975) 475.
85 Y.H. Tan, J.A. Armstrong and M. Ho, Virology, 45 (1971) 837.
86 Y.H. Tan, D.K. Jeng and M. Ho, Virology, 48 (1972) 41.
87 E.A. Havell and J. Vilcek, J. Cell Biol., 64 (1975) 716.
88 P.B. Sehgal, I. Tamm and J. Vilcek, Science, 190 (1975) 282.
89 P.B. Sehgal, I. Tamm and J. Vilcek, Virology, 70 (1976) 532.
90 P.B. Sehgal and I. Tamm, Proc. Nat. Acad. Sci. USA, 73 (1976) 1621.
91 J. Vilcek and M.H. Ng, J. Virol., 7 (1971) 588.
92 E. De Clercq and W.E. Stewart II, Medikon, 1 (1972) 331.
93 D.A. Stringfellow, J.C. Overall Jr. and L.A. Glasgow, J. Infect. Dis., 130 (1974) 470.
94 D.A. Stringfellow, J.C. Overall Jr. and L.A. Glasgow, J. Infect. Dis., 130 (1974) 481.
95 M.W. Harmon and B. Janis, J. Infect. Dis., 132 (1975) 241.
96 E. De Clercq and M. Luczak, J. Infect. Dis., 133 Suppl. (1976) A226.
97 E. De Clercq, M.R. Nuwer and T.C. Merigan, J. Clin. Invest., 49 (1970) 1565.
98 E. De Clercq and P. De Somer, J. Nat. Cancer Inst., 47 (1971) 1345.
99 E. De Clercq and W.E. Stewart II, J. Nat. Cancer Inst., 52 (1974) 591.
100 D.J. Slamon, J. Nat. Cancer Inst., 55 (1975) 329.
101 K. Elgjo and M. Degré, J. Nat. Cancer Inst., 54 (1975) 219.
102 E. De Clercq, submitted for publication (1976).
103 W.P. Drake, E.F. Cimino, M.R. Mardiney Jr. and J.C. Sutherland, J. Nat. Cancer Inst., 52 (1974) 941.
104 W.P. Drake, W.J. Pendergrast Jr., R.E. Kramer and M.R. Mardiney Jr., Cancer Res., 36 (1976) 1172.
105 E. De Clercq and W.E. Stewart II, in Selective Inhibitors of Viral Functions, W.A. Carter (ed.), CRC Press, Cleveland, Ohio (1973) 81.
106 W.A. Carter and E. De Clercq, Science, 186 (1974) 1172.
107 H.M. Johnson and S. Baron, Cellul. Immunol., 25 (1976) 106.
108 H.M. Johnson and S. Baron, IRCS J. Med. Sci., 4 (1976) 50.
109 P. Lindahl-Magnusson, P. Leary and I. Gresser, Nature, New Biol., 237 (1972) 120.
110 K.R. Rozee, S.H.S. Lee and J. Ngan, Nature, New Biol., 245 (1973) 16.
111 P. Lindahl, P. Leary and I. Gresser, Eur. J. Immunol., 4 (1974) 779.
112 D. Killander, P. Lindahl, L. Lundin, P. Leary and I. Gresser, Eur. J. Immunol., 6 (1976) 56.
113 D.W. Nebert and R.M. Friedman, J. Virol., 11 (1973) 193.
114 T. Matsuno, N. Shirasawa and S. Kohno, Biochem. Biophys. Res. Commun., 70 (1976) 310.
115 C.T. Teng, M.C. Chen and L.D. Hamilton, Proc. Nat. Acad. Sci. USA, 70 (1973) 3904.
116 A.M. Badger, S.R. Cooperband and J.A. Green, J. Nat. Cancer Inst., 49 (1972) 613.
117 J. Hunter, T. Hunt, R.J. Jackson and H.D. Robertson, J. Biol. Chem., 250 (1975) 409.
118 P. Jullien and J. De Maeyer-Guignard, Int. J. Cancer, 7 (1971) 468.
119 I. Martelly and P. Jullien, J. Nat. Cancer Inst., 53 (1974) 1021.
120 W.A. Carter, J. O'Malley, M. Beeson, P. Cunnington, A. Kelvin, A. Vere-Hodge, J.L. Alderfer and P.O.P. Ts'o, Mol. Pharmacol., 12 (1976) 440.

121 D. Collavo, B. Finco and L. Chieco-Bianchi, Nature, New Biol., 237 (1972) 154.
122 J.H. Dean, W.C. Wallen and D.O. Lucas, Nature, New Biol., 237 (1972) 218.
123 W.E. Stewart II and E. De Clercq, J. Gen. Virol., 23 (1974) 83.
124 P.O.P. Ts'o, J.L. Alderfer, J. Levy, L.W. Marshall, J. O'Malley, J.S. Horoszewicz and W.A. Carter, Mol. Pharmacol., 12 (1976) 299.
125 S. Feldman, W.T. Hughes, R.W. Darlington and H.K. Kim, Antimicrob. Ag. Chemother., 8 (1975) 289.
126 D.A. Hill, S. Baron, J.C. Perkins, M. Worthington, J. Van Kirk, J. Mills, A.Z. Kapikian and R.M. Chanock, J. Am. Med. Assoc., 219 (1972) 1179.
127 A.K. Field, C.W. Young, I.H. Krakoff, A.A. Tytell, G.P. Lampson, M.M. Nemes and M.R. Hilleman, Proc. Soc. Exp. Biol. Med., 136 (1971) 1180.
128 J.J. Nordlund, S.M. Wolff and H.B. Levy, Proc. Soc. Exp. Biol. Med., 133 (1970) 439.
129 H.B. Levy, G. Baer, S. Baron, C.E. Buckler, C.J. Gibbs, M.J. Iadarola, W.T. London and J. Rice, J. Infect. Dis., 132 (1975) 434.
130 T.C. Merigan, S.E. Reed, T.S. Hall and D.A.J. Tyrrell, Lancet, I (1973) 563.
131 B.R. Jones, D.J. Coster, M.G. Falcon and K. Cantell, Lancet, II (1976) 128.
132 H.B. Greenberg, R.B. Pollard, L.I. Lutwick, P.B. Gregory, W.S. Robinson and T.C. Merigan, New Engl. J. Med., 295 (1976) 517.
133 J. Desmyter, M.B. Ray, J. De Groote, A.F. Bradburne, V.J. Desmet, V.G. Edy, A. Billiau, P. De Somer and J. Mortelmans, Lancet, in press (1976).

IMMUNOLOGY OF SYNTHETIC POLYPEPTIDES

P. H. MAURER
Department of Biochemistry, Jefferson Medical College, Philadelphia, Pa.

SUMMARY

Parameters associated with the immune response have been elucidated in both random and inbred strains of animals employing protein models of random and sequential polymers of amino acids. The presentation will deal with studies emphasizing the elucidation of the genetic control of the immune response in guinea pigs and mice. Methods that have been employed to measure the responses will be outlined.

In guinea pigs immune response (Ir) genes have been identified which are dominant, expressed in thymus derived (T cell) lymphoid cells and are closely linked to the major histocompatibility complex (MHC). These genes control the immune responses (all or none) as measured both by cell mediated immunity (T cell) and antibody levels (B cell). The specificity of the responses of the two available inbred strains indicate the following: Only Strain 2 g.p. responds to the following random polymers poly(glu^{60}-lys^{40}), poly(glu^{60}ala^{40}), poly(glu^{58}ala^{38}tyr^{4}), poly(glu^{58}lys^{37}tyr^{5}). Only Strain 13 g.p. responds to poly(glu^{50}tyr^{50}), poly(glu^{80}phe^{20}), poly(glu^{80}val^{20}) and poly(glu^{75}leu^{25}). Increasing the mol% of tyr or phe in the polymers, i.e. poly(glu^{60}ala^{30}tyr^{10}) (GAT^{10}), poly(glu^{60}lys^{30}phe^{10}) allows the strain 13 g.p. to respond to these polymers. Unique recognition specificity was also shown with sequential highly helical polypeptides: Strain 13 g.p. responds to poly(Tyr-Glu-Ala-Gly) and poly(Phe-Glu-Ala-Gly) whereas strain 2 g.p. responds to poly(Tyr-Ala-Glu-Gly) and poly(Phe-Ala-Glu-Gly).

With mice where many more genetically characterized inbred, congenic and recombinant inbred strains are available, examples of the strain distribution patterns (SDP) (responders vs. nonresponders) will be presented with the random terpolymers GAT^{10}, GLPhe^{5}, poly(glu lys ala) and the sequential polymers poly(T-G-A-Gly) and poly(T-A-G-Gly). The T cell involvement and the dominant nature of the immune response gene(s) responsible for recognition of the polymers will be discussed. In contrast to the g.p. it has been possible to localize the gene(s) to a specific chromosomal locus in the mouse genome. The genes are linked to the MHC termed H-2 of the mouse. The "mapping" of the Ir genes to locus denoted as the "I" subregion of the H-2 complex will be demonstrated. Examples will be given of "all or none", unigenic, and multigenic "H-2 control" of the immune response as well as the importance of "complementation" of genes for eliciting immune response to a number of the polymers.

In this symposium, which is devoted to the structure and properties of biopolymers, it is exceptionally pertinent that a section has been included dealing with some medical applications. As an important aspect of the use of biopolymers might involve the host reactions to the polymers, an awareness should exist about the nature of immune responses that polymers of amino acids can elicit. Within this context I should like to present findings, predominantly of our laboratory, dealing with immunological studies performed with different kinds of synthetic polymers of α-L amino acids. The major contributions of other laboratories has therefore not been included in this presentation.

Since the 1950's when we were concerned about whether an immune response might be elicited against some simple polymers of α-L amino acids, there has been a phenomenal growth in the use of the polymers of amino acids, both random and sequential, to help elucidate a number of areas associated with the immune response.

It might be of interest to recall some of the history of the use of synthetic polymers of amino acids in immunology. When synthetic polymers of amino acids were being considered as possible plasma expanders, it was necessary to determine whether immune responses might be developed against the materials in experimental animals, including man. This could, of course, be detrimental to the host. Although a "problem free" synthetic plasma expander of amino acids has not yet been developed, one has learned a good bit about how to prepare polymers on the one hand, and that there are a number of parameters in polymers that might contribute to immunogenicity.

Without going through the long history of the various kinds of polymers of α-L amino acids that were studied, it would suffice to say that polymers prepared from the α N-carboxyanhydrides containing two or three different amino acids were immunogenic in outbred rabbits [1-3], guinea pigs [4,5], mice [6,7], etc., depending upon the amino acids involved and the percentage composition of the various amino acids in the random polymers. From the nature of the variations in response patterns obtained in outbred animals in repeated experiments it was soon realized that the results might be due to the "genetic makeup" of the animal which was governing the ability of the animal to recognize the polymer as "foreign". One of the major contributions of subsequent studies with polymers has been the finding that the immune response does have associated with it a very important "genetic parameter".

Because polymers of amino acids can be "tailored" to contain a more restricted range of "determinants" than exists in proteins, it is possible to present a host with fewer determinants against which an animal can respond. Although the initial immunological studies dealt with random bred guinea pigs, mice and rabbits, the greatest impact and contribution of synthetic polymers of amino acids to immunology has been the use of highly genetically characterized animals of a specific species. The immunological studies in inbred guinea pigs and mice with both random and sequential polymers will serve as illustrations.

For the non immunologists it is important to understand a few aspects of the immune response that might ensue following exposure of a host to an immunogen. In simplified terms - there are two interdependent paths in which the lymphoid system is involved in the immune response: one involves the behavior of thymic derived lymphoid cells, termed T cells, and the other the bone marrow derived lymphoid cells, termed B cells. Figure 1 shows the action of these lymphoid cells in conjunction with another class of cells known as macrophages. T cells can interact with antigen and the macrophage through the appropriate antigen receptors in presenting the antigen to the B cell antigen receptors (immunoglobulin). This phenomenon of the T cell-B cell collaboration leads to a series of proliferative cellular events and antibody formation. The antibody that is measured in the serum (humoral response) is a product of the B cells. The first aspect of the response shown, i.e. the responses of T cells following specific interaction with antigen, does not involve antibody formation, but rather a phenomenon of cell mediated immunity (CMI) and the production of sensitive or "educated" T cells. The manifestations of cell mediated immunity are several: In the immunized guinea pig the intradermal injection of the specific antigen results within 24 hrs (delayed reaction) in an erythematosis reaction associated with the invasion into the area of sensitized T cells. By a number of fractionation techniques it is possible to separate lymphoid cells into populations of the T cells and B cells. The 'in vitro'

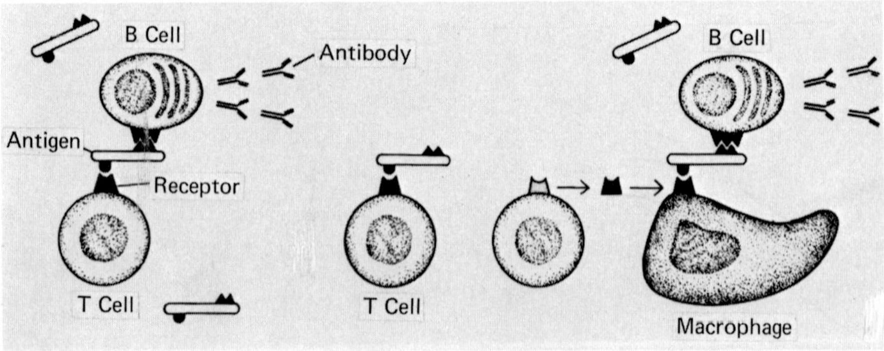

Fig. 1. _Left_: Antibody responses occur following simultaneous recognition of "carrier" determinants on the antigen by the specific T cell receptor and other antigenic determinants by the specific B cell receptor. _Right_: Pathway may involve reaction of antigen with specific T cell receptor, transfer of the receptor to a macrophage, which then allows interaction of antigen with the B cell receptor.

culture for several days of sensitive T cells with antigen leads to a proliferative response of the cells, and in the presence of tritium labeled thymidine one observes an increased incorporation of the tritiated thymidine into the DNA of the cells. This is the _in vitro_ correlate of the above cell mediated immunity [8].

The polymers that will be discussed below elicit immune responses which involve the interactions of both T cells and B cells. These polymers are therefore termed _T cell dependent antigens_. This is in contradistinction to some antigens such as polysaccharides which do not need T cells. These antigens have therefore been termed _T cell independent antigens_. Tables 1 and 2 list the random and sequential polymers referred to in this paper and the simplified nomenclature we have employed in the text. The polymers had molecular weights ranging from 20,000 to 70,000.

GUINEA PIG STUDIES

The advantage of the guinea pig for immunological studies is that this species can elicit both CMI and humoral responses. The responses were

Table 1

Random polymers of amino acids employed in this study

Copolymers	Nomenclature
Poly($Glu^{60}Lys^{40}$)	GL
Poly($Glu^{60}Ala^{40}$)	GA
Poly($Glu^{50}Tyr^{50}$)	GT
Poly($Glu^{80}Phe^{20}$)	G Phe
Poly($Glu^{75}Leu^{25}$)	G Leu
Poly($Glu^{80}Val^{20}$)	G Val

Terpolymers	Nomenclature
Poly($Glu^{57}Ala^{38}Tyr^{4}$)	GAT^{4}
Poly($Glu^{60}Ala^{30}Tyr^{10}$)	GAT^{10}
Poly($Glu^{60}Ala^{30}Lys^{10}$)	GAL^{10}
Poly($Glu^{57}Lys^{38}Tyr^{5}$)	GLT^{5}
Poly($Glu^{51}Lys^{34}Tyr^{15}$)	GLT^{15}
Poly($Glu^{57}Lys^{38}Phe^{5}$)	GL Phe^{5}
Poly($Glu^{54}Lys^{36}Phe^{10}$)	GL Phe^{10}
Poly($Glu^{54}Lys^{36}Ala^{10}$)	GLA^{10}
Poly($Glu^{36}Lys^{24}Ala^{40}$)	GLA^{40}
Poly($Glu^{24}Lys^{16}Ala^{60}$)	GLA^{60}
Poly($Glu^{57}Lys^{38}Leu^{5}$)	GL Leu
Poly($Glu^{56}Lys^{37}Ser^{7}$)	GL Ser
Poly($Glu^{55}Lys^{37}Pro^{8}$)	GL Pro

Superscripts refer to mol % of the α-L amino acid in the polymer.

Table 2

Sequential polymers of amino acids employed in this study

Polymer	Nomenclature
Poly($Tyr^{25}Glu^{25}Ala^{25}Gly^{25}$)	(T-G-A-Gly)
Poly($Phe^{25}Glu^{25}Ala^{25}Gly^{25}$)	(Phe-G-A-Gly)
Poly($Tyr^{25}Ala^{25}Glu^{25}Gly^{25}$)	(T-A-G-Gly)
Poly($Phe^{25}Ala^{25}Glu^{25}Gly^{25}$)	(Phe-A-G-Gly)

Superscripts refer to mol % of the α-L amino acid in the polymer.

therefore monitored by measuring the amount of antibody in the serum, the extent of the _in vivo_ delayed skin reactions, and the extent of antigen stimulation of sensitized T cells _in vitro_.

In guinea pigs, none of the many homopolymers of α-L amino acids that were tried were found to be immunogenic [4]. However, as the "complexity" and the number of possible structures (determinants) in the polymer was increased through the formation of random copolymers and terpolymers of amino acids, one elicited significant immune responses associated both with CMI and humoral antibody formation.

One of the techniques which has been used for measuring the level of antibody produced employed I^{125} labeled antigens. In the presence of specific antibody, radiolabeled antigen forms antigen-antibody complexes. If one then adds a reagent which can precipitate the antibody, this reagent will also precipitate the fraction of the radiolabeled antigen that is complexed with the antibody. The reagent we use is antibody directed against guinea pig gamma globulin (antibody). Following the precipitation of the antigen-antibody complexes, the _percent bound of added radiolabeled antigen_ is a quantitative index of the level of the humoral responses [9]. As is summarized in Table 3, when strain 2 and strain 13 guinea pigs, which are the only inbred guinea pig strains available, were immunized with a number of the random polymers, only the strain 2 guinea pig responded to GL [10,11], GA [10,11], GAT^4 [12], GLT^5 [10], $GLPhe^5$ [10] and the GLA [10] terpolymers. In contrast to this, the strain 13 guinea pigs that did not respond to these latter polymers, did respond to the GT, GPhe, GVal, GLeu [13] copolymers and to the terpolymers GAT^{10} [11], GLT^{15} [12] and $GLPhe^{10}$ [13]. The Fl guinea pigs (2 x 13) were able to respond to all of the polymers. That is the progeny of a (responder x nonresponder) guinea pig to each polymer were always responders. When Fl guinea pigs were backcrossed to responder animals, all of the progeny responded; and when the Fl guinea pigs were backcrossed to nonresponder animals, only 50% of the progeny responded. In classical genetic terms this indicated the dominant Mendelian unigenic control to respond against these simple polymers. The genes governing the responses were termed immune response (Ir) genes [14] and each polymer was shown to have its "own gene". In all of the guinea pig studies if no response ensued as measured by the serum antibody, there also was no delayed skin reaction and no _in vitro_ stimulation by antigen of the isolated T cells. Moreover, if one incorporated small inorganic molecules (haptenic groups), which by themselves cannot elicit

Table 3

Summary of responses of inbred strains of guinea pigs to random copolymers and terpolymers of amino acids[1,2]

Antigen	Strain of Guinea Pig		
	2	13	(2 x 13)F1
Copolymer			
GL	+	-	+
GA	+	-	+
GT	-	+	+
GPhe	-	+	
GVal	-	+	
GLeu	-	+	
Terpolymer			
GAT^{4}	+	-	+
GAT^{10}	+	+	+
GLT^{5}	+	-	+
$GLPhe^{5}$	+	-	+
GLT^{15}	+	+	
$GLPhe^{10}$	+	+	
GLA^{10}, GLA^{40}, GLA^{60}	+	-	+

[1] The original references present quantitative values.

[2] Guinea pigs were immunized with 100 μg - 500 μg of the polymers in complete Freund's adjuvant. Plus indicates responsiveness, minus indicates nonresponsiveness. "Nonresponders" responded to MBSA aggregates of the polymers.

immune responses, into the polymers, only those animals known to respond against the unconjugated polymer produced antibody not only against the polymer, but also against the haptenic grouping. The interpretation of the phenomenon is that the ability to respond against the whole polymer, which is the "carrier", governs the ability to respond against any haptenic grouping coupled into the polymer. The polymer reacts with the specific receptors on the T cells, which then present the "hapten" to the hapten specific B cell receptors in a form which leads to formation of anti-hapten antibody. This is an example of the concept of T cell help and T cell-B cell collaboration leading to antibody formation.

Another characteristic of the above immune responses is that the information and capacity to produce antibody against the various polymers resides in all of the guinea pigs, even the "nonresponders". However, in order to convert the "nonresponder" to a "responder" against the specific polymer, one has to employ another immunological procedure analogous to that just discussed. If GA, which is not immunogenic for strain 13 guinea pigs, is coupled or complexed with another immunogenic 'carrier', such as the foreign immunogenic protein methylated bovine serum albumin (MBSA), and then immunizes the strain 13 guinea pigs with the MBSA aggregate of GA or any other nonimmunogenic polymer, one can elicit definite antibody responses against the polymer [14], i.e. strain 13 guinea pigs can be made to produce antibody against GA, and strain 2 guinea pigs can be made to produce antibody against GT or GPhe, etc. However, in these situations, although antibody has been produced, no cell mediated immunity (T cell response) is present against the polymer in the "nonresponders". Here the MBSA has been used as an immunogenic carrier protein, and the polymer as a large "macromolecular hapten" for the "nonresponding" animals. These kinds of data have led to the dogma that the ability to respond against all of the polymers exists in both responder and nonresponder strains of guinea pigs, but that there is a defect in the nature of the recognition of the polymer by T cells for presentation to the immunocompetent B cells in the nonresponding animals. This defect has been bypassed by presenting the polymer on an immunogenic carrier wherein the "carrier" determinants are recognized by its specific set of T cells which then "present" the entire complex, including the polymer determinants to B cells that can recognize the polymer. Therefore, the genetic defect in the nonresponder appears to be at the T cell level, not the B cell [11].

Another technique for eliciting responses is shown by the ability of the strain 13 guinea pigs, that do not respond to GAT^{4} or GLT^{5}, to respond to GAT^{10}, GLT^{15} [12] and also $GLPhe^{10}$ [13]. This was accomplished by the creation of an increased amount of "GT" or GPhe" determinants to which the strain 13 guinea pig has been shown to respond. With the GLA terpolymers changing the mol composition of the amino acids did not alter the response patterns as the strain 13 guinea pig responds to neither GA nor GL determinants [13]. What is remarkable about the summary data is that it has shown the unique specificity of the recognition mechanisms of the immune system for slight differences between GAT^{4} and GAT^{10}, GLT^{5} and GLT^{15}, or GA, GLue and GVal.

This unique specificity has also been demonstrated with the sequential polymers of amino acids which we have recently been studying in depth. As shown in Table 4, only the strain 13 guinea pigs respond to the polymers (Tyr-Glu-Ala-Gly) [9] and (Phe-Glu-Ala-Gly) [15], and only the strain 2 guinea pigs respond to the (Tyr-Ala-Glu-Gly) [16] and (Phe-Ala-Glu-Gly)

[15]. However, both strains can produce antibody against the MBSA aggregates of the polymers. What is very intriguing is that these four polymers have exceedingly similar helical structures as measured by ORD and CD procedures, yet the changes in the positions of glutamic acid and alanine in the polypeptides have been detected by the T cell recognition system of the guinea pigs.

Table 4

Responses of inbred strains of guinea pigs to sequential polymers

Antigen	Strain of Guinea Pig 2	13
(T-G-A-Gly)	-	+
(Phe-G-A-Gly)	-	+
(T-A-G-Gly)	+	-
(Phe-A-G-Gly)	+	-

Of additional significance has been our findings that only the helical form [17] of the polypeptide (T-G-A-Gly) was immunogenic (Table 5). Not only was it necessary for the polymer to have helical structures to be immunogenic, but the immune response was indeed directed against the helical regions as indicated by the inability of the nonhelical oligopeptides to react with the antibody (inhibition studies) directed against the polymeric helical (T-G-A-Gly) (Table 6) [18].

In addition to the genetic differences in responses to the random and sequential polymers by the inbred strains of guinea pigs, the past few years have seen a most important development which indicated that not only are we dealing with immune response genes which are dominant, and are expressed at least in the thymus derived lymphoid cells, but that the recognition of specific macromolecules as immunogens by individual animals and inbred strains appears now to be governed by the products of individual dominant immune response genes located in the genome of a specific animal species in close relationship with genes of the major histocompatibility complex (MHC). These immune response genes have therefore been termed histocompatibility of <u>H-linked Ir genes</u> [19].

At this point some explanation of the MHC is in order. The major histocompatibility complex (MHC) which exists in higher vertebrates appears to have great biological significance as it has been maintained in

Table 5

Immune responses of strain 13 guinea pigs to (T-G-A-Gly) and its oligopeptide fractions

Immunogen	Molecular Weight	Molar Ellipticity θ_{222} (deg cm^2/dmol)	Response
(T-G-A-Gly)	53,300	-16,800	+
Fraction I	22,600	-15,800	+
Fraction II	9,700	-13,800	+
Fraction III	1,300	0	-
Fraction IV	438	0	-

Table 6

Inhibition of guinea pig anti$(T\text{-}G\text{-}A\text{-}Gly)_n$-$(T\text{-}G\text{-}A\text{-}Gly)_n$ reaction by homologous polymer and oligopeptide fractions

	Inhibitor Concentrations[a]				
	10	50	100	200	500
Inhibitors	% Inhibition				
$(T\text{-}G\text{-}A\text{-}Gly)_n$	69	82	100	93	94
Fraction I	47	75	84	89	90
Fraction II	22	49	60	80	87
Fraction III	0	0	0	0	0
Fraction IV	0	0	0	0	0

[a] Concentrations of inhibitors are expressed as the fold. 0.001 μg N $(T\text{-}G\text{-}A\text{-}Gly)_n$ equals one fold.

tact from species to species among the vertebrates. In man, this complex of genes has great clinical significance because of its role in disease resistance and in clinical transplantation. On the basis of information available, one may generalize a number of features which are probably common to all of the MCH's (Table 7). The presence of the specific MHC of a strain within a species can be identified employing specific antisera directed against the antigen associated with the specific gene complex.

These alloantisera are produced by immunizing one strain of guinea pig with the lymphoid cells of the other strain of guinea pig. One prepares what are termed _anti-2 sera_ and _anti-13 sera_ which can be used as immunological reagents to determine the nature of the H antigens present not only in inbred guinea pigs, but in Fl backcross and even random bred guinea

Table 7

Properties of major histocompatibility complexes

1.	Principal transplantation antigens of species
2.	Serologically detected antigens of lymphocytes on many tissues
3.	Immune response genes
4.	Major factors which stimulate graft versus host reactions
5.	Genetic polymorphism at many loci in complex

pigs which can have various proportions of 2 or 13 genes. These antisera react _in vitro_ with the approprite antigens on the lymphoid cells, and in the presence of complement (see presentation by B. Johnson) destroy the lymphoid cells.

It was then determined that the ability of either inbred or random bred guinea pigs to respond against a specific synthetic polymer of amino acid was related (linked) to the presence in the guinea pig genome of a specific profile of the antigens of the major histocompatibility complex. From the appropriate genetic breeding and immunization experiments with "responder" and nonresponder" inbred and random bred guinea pigs, which could be typed for their MHC antigens, it was soon learned that the ability to respond to the various random polymers was closely linked to the presence of these histocompatibility antigens in the genome, which were also expressed on T cells. In the presence of antisera directed against the specific H antigen, that is anti-2 sera or anti-13 sera, the specific proliferative responses of sensitized T cells, which follows their exposure _in vitro_ to antigens, was inhibited. This was only one bit of the evidence which indicated that the receptors on the lymphoid cells responsible for recognizing the specific antigen (polymers) are located in close proximity to the antigens which are coded for by the major histocompatibility genes [8].

The discovery of these specific H-linked Ir genes was one of the major contributions of synthetic polypeptides to immunology. It was associated with the important fact that one could present the host with antigens which had more restricted heterogeneity and specificity than did the usual kinds of protein antigens.

In summary, the immune responses of guinea pigs to the synthetic polymers of amino acids, the recognition of which is under the control of

H-linked Ir genes, has been characterized by delayed hypersensitivity (cell mediated immunity), by the *in vitro* correlate of cellular immunity, such as the specific blast transformation in the presence of specific antigenes, and by the synthesis of significant levels of specific antibody. Animals lacking the gene never develop delayed sensitivity and do not produce significant levels of antibody under these usual conditions of immunization. The activity of these immune response genes in guinea pigs has been responsible for the differences between responder and nonresponder animals.

By using various kinds of synthetic polymers of amino acids and other classes of antigens, it was possible in a short period of time to show that responses of other experimental animals in addition to guinea pigs to a wide variety of other antigens was also under the control of dominant H-linked Ir genes [11].

The next section will deal with some of our mouse studies.

MOUSE STUDIES

As with guinea pigs, our early studies with random bred mice indicated that random polymers could indeed behave as immunogens, one could obtain 'responders' and 'nonresponders' and by the appropriate genetic studies indicate that the ability to respond was a dominant autosomal Mendelian trait [6]. However, in contrast with the responses obtained in guinea pigs, the random bred mice (Swiss Webster mice) could not respond against the copolymers of two amino acids such as GL, GA or GT. However, when one employed terpolymers containing a limited amount of a third amino acid, i.e. GAT^4, GLA^5, GLT^5, etc., responses were elicited in a percentage of the random bred mice [6]. Rather than go through the many past and present studies with random bred mice, I should like to concentrate on studies with inbred mice.

Similar to what was found in guinea pigs, specific immune response genes controlling the responsiveness to synthetic polypeptides have been identified in mice. However, the mouse has been a more rewarding species for the study of H-linked Ir genes because of the availability of: 1) numerous inbred strains of mice which were initially developed by investigators concerned with tissue transplantation and cancer immunology, 2) congenic resistant strains which differ from similar mice only at the H-2 complex, and 3) strains with documented "recombination events" within the H-2 complex precisely in the region where we will show that the Ir genes are located. (Items 2 and 3 will be described in more detail below.) [20].

In order to understand much of the following, it will simplify matters to understand that the H-2 complex of the mouse is the most important major histocompatibility complex [21]. This H-2 locus is found on chromosome 17 of the mouse, and the genes in this area code for antigens which are most responsible for the rejection phenomenon which follows the transplantation of skin from one inbred mouse to another. The common features of the major histocompatibility complexes were shown in Table 7. The H-2 system as it is termed has extensive genetic polymorphism and it has been established that it is the major transplantation barrier in the mouse. The current view of the "genetic map" of a part of the H-2 gene complex on chromosome 17 is shown in Figure 2. Most of the traits which are associated with this complex are reflected in cell surface antigens and immune response or recognition events. It has been established that there is a linear order of discrete loci in this gene complex. The discrete loci, that is H-2K, Ir-lA, H-2D, etc., are defined on the basis of a number of criteria such as the determination of the functions or products which are demonstrably different from those of other genes in the complex. (For complete

Loci -	H-2K	I(Ir-IA)			Ss	H-2G	H-2D
Region or Subregion	K	IA	IB	IC	S	G	D
Genetic traits under control of specific complex							
	1. Serologically detectable H-2 alloantigens 2. Transplantation antigens	1. Immune response genes 2. T cell-B cell collaboration 3. Resistance to viral oncogenesis 4. I region (Ia) associated antigens 5. Mixed leukocyte and graft versus host reactions			1. Serum protein variants 2. Complement levels	1. Erythrocyte alloantigens	1. Serologically detectable H-2 alloantigens 2. Transplantation antigens

Fig. 2. Fine structure map of major histocompatibility complex (H-2) of the mouse.

discussion of this complex see ref. 21.) For this presentation it is important to know that many highly inbred strains exist that have the same major histocompatibility complex alleles (H-2) but differ in other minor histocompatibility alleles. The terminology of H-2 haplotype that has been assigned to the various inbred strains, i.e. a,b,d,k, etc., is based upon a classification employing alloantisera which can detect the nature of the unique gene products coded for mainly by the K and D regions of the complex which are present on lymphoid cells.

The strain distribution patterns of the inbred and congenic strains of mice against the random polymers GA [22], GAT^{4} [23], GAL^{10} [24] and GAT^{10} [23] are shown in Table 8. In all of our experiments mice have been immunized with 10 γ-100 γ of polymer. The last polymer has been studied in great detail by several laboratories. Rather than just list the 'responder' or 'nonresponder' status of the strains of mice as was done with the guinea pig, more quantitative values are given here, i.e. the % binding of a fixed amount of radiolabeled antigen by a 1:2 dilution of mouse serum.

Similar to the guinea pig studies, in all cases the F1 progeny of a (responder x nonresponder) were responders, which indicated the dominant nature of the gene control. From the data obtained with the appropriate backcross mice, such as the findings that (nonresponder x responder) F1 backcrossed to responders were all responders, and the F1 backcrossed to nonresponders led to 50% responders and 50% nonresponders, it was concluded that here too we were dealing with a classical Mendelian autosomal dominant single gene system (Table 9) [25].

However, the results obtained with congenic resistant strains (B10.S and B10.G) indicated that the immune responses were indeed linked with the major H-2 complex. (A congenic resistant strain is a strain which differs from another mouse strain only at the H-2 complex. Examples would be the B10.A mouse ($H\text{-}2^{a}$) which has the H-2 complex of an A mouse, although the rest of the genetic material and other H-antigens come from a C57BL/10 mouse. Similarly, a B10.S ($H\text{-}2^{s}$) mouse is congenic with the C57BL/10 ($H\text{-}2^{b}$) mouse.

As with the guinea pig, it was possible to convert 'nonresponders' to 'responders' by immunizing with the methylated bovine serum albumin aggregate of the specific polymer, i.e. immunizing mice of the $H\text{-}2^{q}$ or $H\text{-}2^{s}$ haplotypes, led to responses in these animals. This was inferential evidence that the defect in nonresponders was at the T cell level [25].

Table 8

Immune responses of inbred and congenic strains of mice to GA, GAT^{4}, GAT^{10} and GAL^{10}

Strain	H-2 Haplotype	Percent Antigen Bound ± S.E. GA	GAT^{4}	GAT^{10}	GAL^{10}
A/J	a	69 ± 8	71 ± 6	67 ± 3	60 ± 7
B10.A/SgSn*	a	57 ± 8	74 ± 6	80 ± 2	42 ± 9
C57BL/6J	b	68 ± 10	43 ± 7	86 ± 3	42 ± 3
C57L	b	43 ± 10	67 ± 9	83 ± 5	
BALB/c	d	57 ± 6	62 ± 7	81 ± 8	41 ± 11
DBA/2	d	63 ± 5	58 ± 9	86 ± 2	58 ± 10
A.CA*	f	85 ± 9	88 ± 2	86 ± 3	81 ± 5
C3H.JK*	j	0 ± 1		0 ± 0	
AKR/J	k	61 ± 8	73 ± 3	87 ± 1	61 ± 4
CBA/J	k	55 ± 5	51 ± 18		46 ± 5
C3H.NB/Sn*	p	0 ± 0	0 ± 0	1 ± 2	5 ± 6
P/J	p	2 ± 2	17 ± 3	2 ± 3	5 ± 6
DBA/1	q	7 ± 2	2 ± 3	1 ± 1	0 + 0
B10.Q*	q	4 ± 0		1 ± 2	
RIII	r	60 ± 18		76 ± 8	
SJL	s	42 ± 16	68 ± 12	2 ± 2	50 ± 3
A.SW*	s	61 ± 10	29 ± 12	0 ± 0	38 ± 11
B10.S*	s	66 ± 7	18 ± 5	0 ± 0	

*Congenic strain

Table 9

Antibody response to 10 γ GAT^{10}

Strain	No. of Mice	Responders	Nonresponders
C57BL/6	5	86 ± 3	
DBA/1	5		1 ± 1
(C57BL/6 x DBA/1)F1	5	91 ± 3	
(F1 x F1)F2	5	82 ± 8	
	5		8 ± 3
F1 x DBA/1	5	84 ± 3	
	5		5 ± 5
F1 x C57BL/6	7	69 ± 14	

Whereas with the two strains of guinea pigs studied the responses to the antigens which were H-linked and under Ir gene control were easily characterized both by delayed hypersensitivity and antibody synthesis, in the mouse one had to rely initially almost totally on antibody values. This is because the mouse generally exhibits poor cell mediated immunity. Recently, however, sensitive tissue culture techniques have been developed to measure the '*in vitro*' stimulation (blast transformation) by the antigen of sensitized mouse responder T cells. The T cell involvement and the dominant nature of the immune response genes responsible for recognition of these random polymers has now been shown conclusively [26].

Of considerable interest to this symposium is the different strain distribution patterns (SDP) noted with polymers that are structurally and chemically related. The 'nonresponders' to GA are mice of $H\text{-}2^p$ and $H\text{-}2^q$ haplotypes. These mice also do not respond to GAT^4, GAT^{10} or GAL^{10}. However, the mice of $H\text{-}2^s$ haplotype, which do respond to GA, GAT^4 and GAL^{10}, did not respond well to the GAT^{10}, which has 10 mol % tyrosine. Studies in several laboratories have indicated a possible explanation for these findings related to the following: Initial studies on T cells indicated that they are important for helping and "collaborating" with B cells in eliciting a humoral response against T independent antigens. Now we know that T cells can also be involved in suppressing immune responses [27]. By the identification of specific surface membrane antigens, it is known that there are indeed differences between the surface antigens of what has been termed "suppressor T" cells and "helper T" cells. In the GA-GAT^{10} system it appears that the T cells of nonresponding animals can recognize the GAT^{10} polymer, which leads to specific suppression of the immune response [28-30]. With mice of the $H\text{-}2^s$ haplotype, the presence in the polymers of runs of oligotyrosine residues can actually reduce the response. This is also shown by the fact that although GT is no immunogenic in any inbred strain of mouse [22], it can activate suppressor T cells in some strains of mice, such as those of $H\text{-}2^s$ haplotypes. This "suppressing" behavior of tyrosine incorporation into a macromolecule is in contrast to the usually ascribed enhancing role of tyrosine on the immune response.

An important difference between the guinea pig and mouse responses was that whereas with the guinea pig the responses could be characterized as "all or none", with the mouse, because of the large number of strains available, it was possible to note gradations in the level of antibody produced by many of the responding strains having the same H-2 haplotype.

This has indicated the influence of other non H-2 genes on the magnitude of the antibody response against several of the polymers, and was most marked with a number of the terpolymers of glutamic acid, lysine and a limited concentration of a "third" amino acid. Although quantitative values are not given for the various strains within a specific haplotype, the differences in response patterns of mice against the 'GL random terpolymers' containing 5-10 mol% of ala [31], tyr [32], phe [33], ser[34], leu[34], pro[35], are quite striking (Table 10). Although the antibody produced against each of these terpolymers reacts completely with the nonimmunogenic copolymer GL, the distinctive strain distribution pattern of responses observed reinforces the concept that one cannot always predict the immunogenicity of related synthetic polymers. Changing the mol composition of amino acids in a terpolymer, or substituting one amino acid for another in a polymerizing mixture, can indeed create new determinants and structures, both conformationally and linearly, that are recognized differently.

Table 10

Ir gene-controlled responses to random terpolymers containing glutamic acid lysine and "a third amino acid"

H-2 Haplotype	GLA^{5}	GLA^{10}	GLA^{20}	GLA^{40} GLA^{60}	GLT^{5}	GLT^{15}	$GLØ^{5}$	$GLØ^{10}$	GL Leu	GL Ser	GL Pro
a	+	+	+	+	-	-	-	-	+	+	-
b	-	+	+	+	-	-	-	-	-	-	-
d	+	+	+	+	+	+	+	+	+	-	-
f	+	+	+	+	-	-	-	-	-	+	-
j	+	+	+	+	+	+	+	+	+	-	-
k	+	+	+	+	-	-	-	-	+	+	-
p	-	-	-	+	-	-	-	+	-	-	-
q	+	-	-	+	+	+	+	+	-	-	-
r	+	-	-	+	+	+	+	+	+	-	-
s	+	+	+	+	-	-	-	-	-	-	+

The effect on the SDP of mice of changing the amino acid composition slightly is also shown with the polymers $GLPhe^{5}$ vs. $GLPhe^{10}$, GLA^{10} vs. GLA^{40} and GLA^{60}. Mice of the $H\text{-}2^{p}$ haplotype do not respond well to $GLØ^{5}$ but do respond to $GLØ^{10}$. Similarly, mice of $H\text{-}2^{p}$ and $H\text{-}2^{q}$ haplotypes that do not respond well to GLA^{10}, GLA^{20} and GLA^{30} do respond to GLA^{40} or GLA^{60}.

However, the determinants in GLA^{40} and GAL^{60} that are recognized by these mice have been shown to be "a unique set of determinants" which are not present in either GL or GA copolymers and which are quite helical. The immunochemical studies with antibody produced in these mice against GLA^{40} or GLA^{60} have shown the presence of antibody that cannot react with or be removed by GA and GL. The design of biopolymers for the eventual use or application in man must reckon with the exquisite and unpredictable nature of the immune system to recognize differently as "immunogenic" very closely related polymers.

A further concern to polymer chemists, and unfortunately to too few immunologists, is that variations in the method of preparation of copolymers and terpolymers can result in preparations which have similar amino acid compositions but different sequences and structures, and therefore possibly different immune response patterns. Influencing factors are related to the nature of the initiating polymerization agent, the solvent, and the extent to which the polymerization is carried out. Collaborative studies underway with Dr. James Anderson of Case Western Reserve involving the characterization, including immunogenetic studies of copolymers and terpolymers of similar amino acid compositions prepared under different conditions, might help further clarify this problem.

The exposure of the final polymer to "extreme" pH's may also alter the immune response. The ability of mice of $H\text{-}2^q$ haplotype to respond to "normal" GLT^5 (Table 10) is destroyed if the polymer is kept at pH 9.5 in Na_2CO_3 [36].

As with the guinea pig, the responses of mice to the sequential polymers of amino acids (T-G-A-Gly) and (T-A-G-Gly) were studied. The only responders to the (T-G-A-Gly) polymer were mice of $H\text{-}2^b$ haplotype, and to the (T-A-G-Gly) mice of $H\text{-}2^b$, $H\text{-}2^f$ and $H\text{-}2^r$ haplotypes. In all cases the nonresponders were converted to responders by immunizing with an MBSA aggregate of the polymer.

In general, it has been possible to inhibit the *in vitro* proliferative responses to all of the random and sequential polymers with alloantisera directed against the H-2 complex, and more specifically with the antisera directed against the I regions.

MAPPING OF GENE(S) CONTROLLING IMMUNE RESPONSES

It was indicated before that a major advantage of using mice for immunogenetic studies is that <u>recombinant inbred lines</u> of mice are available. These RI mice have initially been obtained by crossover events which have occurred within the chromosome of the H-2 locus during the breeding of F1 and backcross mice of two different strains. Rather than have the haplotypes of both parental strains at each locus in the H-2 region, as is usual for an F1 animal, the crossover and recombination events have led to the presence of the haplotype from one parent at one locus, and the haplotype from the other parent at another locus [21]. Table 11 contains some examples of these RI lines with the region of crossover indicated. By immunization of a number of these recombinant inbred strains, which contain responder and nonresponder alleles in different parts of the H-2 region and subregions, it has been possible to determine that the genes in the locus between K and S are responsible for controlling immune responses. This region has therefore been termed the I region.

Table 11

Genetic fine structure of selected intra H-2 recombinant strains

		H-2 Regions[a]					
	H-2		I				
Strain	Haplotype	K	I-A	I-B	I-C	S	D
A/J	a	k	k	k	[b] d	d	d
B10.A(4R)	h4	k	k	b	b	b	b
B10.A(3R)	i3	b	b	b	d	d	d
B10.A(5R)	i5	b	b	b	d	d	d
A.TL	tl	s	k	k	k	k	d
D2.GD	g4	d	d	b	b	b	b

[a] The letters under region indicate haplotypes of origin for that region.
[b] Indicates position of crossover.

This region has been shown to be exceptionally important not only because it contains the genes controlling the immune responses to a large number of antigens, but serological techniques it has also been shown to contain genes that code for a number of cell surface antigens present on both T cells and on B cells, i.e. I region associated genes (Ia).

Testing the responses of these RI lines of mice is termed "mapping" of the responder gene locus. Table 12 shows the early data that was obtained in mapping the Ir-GAT^{10} gene to the I region, and more specifically to a region within IA and IB [36]. Up until quite recently, all the known mouse Ir genes associated with T dependent polymers have been shown to be distinct from the genes coding for specificities controlled by the K and D regions, to 'map' in either the IA and IB regions, or as with the GLØ-GLT polymers to map in the IC region (Table 13), and to reflect unigenic control.

Table 12

Immune response patterns of recombinant strains of mice to GAT^{10}

Strain	H-2 Haplotype	Regions of the H-2 Complex K	IA	IB	IC	S	D	Percent Antigen Bound ± S.E.
A.TL	t1	s [a]	k	k	k	k	d	88 ± 10
A.TH	t2	s	s	s	s	s	d	8 ± 4
B10.T(6R)	y2	q	q	q	q	q	d	3 ± 3
B10.HTT	t3	s	s	s	k	k	d	3 ± 2
AQR	y1	q	k	k	d	d	d	71 ± 6

[a] Indicates position of crossover event

The major impetus for studies which have shown <u>multigenic I region control</u> was our finding that the responses to GLT-GLPhe polymers mapped to the IC subregion [32], but that a number of mouse strains having responder haplotypes (d) in the IC region and therefore predictably responders, were nonresponders; i.e. the AQR, B10.M(17R), E10.A and B10.A(2R) mice, which have the d responder allele in the IC region, did not respond to this polymer [36]. This led us and others to a series of studies which have now indicated that the responses to the GLPhe and GLT random polymers are under multigenic control [36,37,39].

This conclusion was obtained by immunization of a number of F1 mice that were the progeny of two nonresponding parents such as ($H\text{-}2^a \times H\text{-}2^b$). Indeed, the data shown in Table 13 indicated the need for two genes to elicit the response; one was in IA, the other in IC. Although this finding was at first surprising, it has been known for a number of years that the immune response is under the control of many genes and other examples exist for the need for two genes to "complement" each other to elicit a positive response [38,39]. This has led to the search for examples of "complementation" in responses against other immunogens which have since been found in a number of other systems, not only with random polymers of amino acids, but also with a class of polymers not discussed here, namely, the multichain polymers of amino acids such as (T, G)-A--L studied at the Weizmann Institute in Israel. With the multichain polymers, one gene is expressed at the T cell level and another at the B cell level [40]. However, so far with the random polymers of amino acids, it appears that both genes are expressed at the T cell level.

Table 13

Immune response patterns of recombinant strains of mice and F1 hybrids to $GLPhe^5$

Strain	H-2 Haplotype	Regions of the H-2 Complex K	I Region IA	IB	IC	S	D	Percent Antigen Bound ± S.E. $GLPhe^5$
HTG	g	d	d	d	d	d	[a]b	91 ± 7
C3H.OL	ol	d	d	d	d	k	k	73 ± 17
B10.T(6R)	y2	q	q	q	q	q	d	94 ± 13
B10.A(3R)	i3	b	b	b	d	d	d	60 ± 9
B10.A(5R)/SgSn	i5	b	b	b	d	d	d	42 ± 18
B10.S(9R)	t4	s	s	s	d	d	d	36 ± 7
AQR	yl	q	k	k	d	d	d	5 ± 3
B10.M(17R)	aql	k	k	k	d	d	f	11 ± 2
B10.A/SgSn	a	k	k	k	d	d	d	2 ± 3
B10.A(2R)	h2	k	k	k	d	d	b	2 ± 1
C57BL/6J	b	b	b	b	b	b	b	1 ± 2
(A/J x B10.A(2R)Fl	a/h	k/k	k/k	k/k	d/d	d/d	d/b	10 ± 3
(A/J x C57BL/6)Fl	a/b	k/b	k/b	k/b	d/b	d/b	d/b	47 ± 8
SJL/J	s	s	s	s	s	s	s	7 ± 4
(C57BL6J x SJL/J) Fl	b/s	b/s	b/s	b/s	b/s	b/s	b/s	41 ± 19
(A/J x SJL/J)Fl	a/s	k/s	k/s	d/s	d/s	d/s	d/s	45 ± 17

[a] Indicates position of crossover event

SUMMARY REMARKS

This presentation has dealt with only one area of the immunological studies that have employed synthetic polymers of amino acids, i.e. immunogenetics. By necessity it has been a somewhat superficial appraoch to the complex area of the genetic control of the immune response, and therefore some important recent developments have been omitted.

Other immunological and immunochemical studies with these polymers have been conducted by other colleagues in our laboratory to delineate the nature of antigenic determinants [41,42] in a number of the polymers against which antibody has been produced. In addition, because of the relative simplicity in structure of some of these polymers, they have also been used as probes for the nature of the antibody combining site [43,44].

Some of the more intriguing areas in present and future research are not just to prepare more polymers of amino acids, but rather to use polymers to obtain answers to important questions about: 1) the nature of the receptors on T cells that are responsible for the differences in 'responder' and 'nonresponder' status of animals; 2) the specificity of the T cell versus the B cell receptors; 3) the nature of the cellular collaboration between T and B cells; and 4) more about the factors that contribute to immunogenicity. In this regard and of importance to this symposium is the information that might be gained about the intrinsic properties of macromolecules which enhance immune responses so that one might be able to, hopefully, decrease the immunogenicity of polymers where necessary for human use.

ACKNOWLEDGMENTS

The author wishes to acknowledge the collaboration and contribution to these studies of a number of colleagues, especially Drs. Carmen Merryman and Allen Zeiger; and more recently, Drs. David Ganfield and Chang-Hai Lai.

The author's research referred to has been supported by NIH Research Grant AI 07825, National Cancer Institute Contract NOI-CB-43999, and American Cancer Society Grant IM-5E.

REFERENCES

1. P. H. Maurer, D. Subrahmanyam, E. Katchalski and E. R. Blout, J. Immunol. 83 (1959) 193.
2. P. H. Maurer, J. Immunol., 88 (1962) 330.
3. P. H. Maurer, B. F. Gerulat and P. Pinchuck, J. Immunol., 90 (1963) 381.
4. P. H. Maurer, J. Immunol., 90 (1963) 493.

5. P. H. Maurer and T. Cashman, J. Immunol., 90 (1963) 393.
6. P. Pinchuck and P. H. Maurer, J. Exp. Med., 122 (1965) 665.
7. P. Pinchuck and P. H. Maurer, J. Exp. Med., 122 (1965) 673.
8. E. Shevach, W. E. Paul and I. Green, J. Exp. Med., 137 (1972) 1207.
9. P. H. Maurer, G. Odstrchel and C. F. Merryman, J. Immunol., 111 (1973) 1018.
10. S. Ben Efraim and P. H. Maurer, J. Immunol., 97 (1966) 577.
11. B. Benacerraf (Editor), "Immunogenetics and Immunodeficiency", University Park Press, Baltimore, Md., 1975.
12. P. H. Maurer, Proc. Soc. Exptl. Biol. Med., 147 (1974) 640.
13. P. H. Maurer, unpublished observations.
14. B. Benacerraf, H. G. Bluestein, I. Green and L. Ellman, Prog. Immunol., 1 (1971) 485.
15. P. H. Maurer, A. R. Zeiger and C. H. Lai, unpublished observations.
16. A. R. Zeiger and P. H. Maurer, J. Immunol., 117 (1976) 708.
17. A. R. Zeiger, C. H. Lai and P. H. Maurer, Biopolymers, 14 (1975) 2281.
18. C. H. Lai and P. H. Maurer, unpublished observations.
19. D. H. Katz and B. Benacerraf (Editors), "Role of Products of Histocompatibility Gene Complexes in Immune Responses", Academic Press, 1976.
20. D. C. Shreffler and C. S. David, Advan. Immunol., 20 (1975) 125.
21. J. Klein (Editor), Biology of the Mouse Histocompatibility-2 Complex, Springer-Verlag, New York 1975.
22. C. F. Merryman and P. H. Maurer, J. Immunol., 116 (1976) 739.
23. C. F. Merryman and P. H. Maurer, J. Immunol., 108 (1972) 135.
24. P. H. Maurer and C. F. Merryman, Ann. Immunol. (Inst. Pasteur) 125C (1974) 189.
25. C. F. Merryman and P. H. Maurer, Medical Genetics Today, 10 (1974) 73.
26. R. H. Schwartz, L. Jackson and W. E. Paul, J. Immunol., 115 (1975) 1330.
27. R. K. Gershon, P. H. Maurer and C. F. Merryman, Proc. Soc. Nat. Acad. Sci. 70 (1973) 250.
28. J. A. Kapp, C. W. Pierce and B. Benacerraf, J. Exp. Med. 140 (1974) 172.
29. J. A. Kapp, C. W. Pierce, S. Schlossman and B. Benacerraf, J. Exp. Med. 140 (1974) 648.
30. J. A. Kapp, C. W. Pierce and B. Benacerraf, J. Exp. Med. 142 (1975) 50.

31. P. H. Maurer and C. F. Merryman, Immunogenetics 1 (1974) 174.
32. C. F. Merryman and P. H. Maurer, Immunogenetics 1 (1975) 549.
33. C. F. Merryman, P. H. Maurer and D. W. Bailey, J. Immunol. 108 (1972) 937.
34. P. H. Maurer and C. F. Merryman, unpublished results.
35. P. H. Maurer, C. F. Merryman and A. R. Zeiger, in The Role of Products of the Histocompatibility Gene Complex in Immune Responses, D. H. Katz and B. Benacerraf (Editors), Academic Press, p. 297.
36. C. F. Merryman, P. H. Maurer and J. H. Stimpfling, Immunogenetics 2 (1975) 441.
37. P. H. Maurer and C. F. Merryman, in Immune Reactivity of Lymphocytes, M. Feldman and A. Globerson (Editors), Plenum Press, N.Y. (1976), p. 531.
38. M. Zaleski and F. Milgrom, J. Immunol. 110 (1973) 1238.
39. M. Zaleski and J. Klein, J. Immunol. 113 (1974) 1170.
40. A. J. Munro and M. J. Taussig, Nature 256 (1975) 103.
41. H. C. McDonald, G. Odstrchel and P. H. Maurer, Immunochemistry 10 (1973) 119.
42. G. Odstrchel and P. H. Maurer, Immunochemistry 11 (1974) 15.
43. P. A. Liberti, W. A. Stylos and P. H. Maurer, Biochemistry 11 (1972) 3312.
44. H. J. Callahan, P. A. Liberti and P. H. Maurer, Immunology 25 (1973) 517.

POLYPEPTIDE COMPLEXES WITH COMPLEMENT+

BRIAN J. JOHNSON
Department of Microbiology
The Medical Center
University of Alabama
University Station, Birmingham, Alabama 35294

SUMMARY

Generally, the manifestation of various autoimmune diseases involve tissue destruction by complement. It seems possible therefore that the ability to control or lower complement levels *in vivo* would be therapeutically useful. To this end we have embarked on a program to find materials which could possibly be used as anticomplement drugs. One approach has been to investigate the complement fixing site of human IgG1 by chemical modifications of the immunoglobulin and also by the use of synthetic peptides. Another approach that has been used is based on the reports that polysaccharide polyanions can affect complement function [1-10]. Our interest in the immunochemical properties of polypeptides [11] prompted our curiosity to investigate the anticomplement activities of these materials and related polyanions.

INTRODUCTION

Preserved sheep blood and hemolysin were purchased from Cordis Laboratories, Miami, Florida. Peptides were synthesized by the solution technique using a method similar to that reported by Johnson and Trask [12]. All peptides possessed acceptable elemental and amino acid analysis. The synthesis of the following compounds used in this paper have been reported. Poly(L-tyrosyl-L-glutamyl-L-alanylglycyl)glycine IX [13]; (L-tyrosyl-L-glutamyl-L-alanylglycyl)$_4$glycine X [14]; (L-tyrosyl-L-glutamyl-L-alanylglycyl)$_3$glycine XI; poly(L-phenylalanyl-L-glutamyl-L-alanylglycyl) glycine XIII [15]; poly(L-tryptophanyl-L-glutamyl-L-alanylglycyl)glycine XIII [16]; poly(L-tyrosyl-γ-L-glutamyl-L-alanylglycyl)glycine XIV [17]; poly(L-tyrosyl-L-glutamylglycylglycyl)glycine XV [18]; poly(L-tyrosyl-L-glutamyl-L-alanyl-β-alanyl)glycine XVI [14]; poly(L-alanyl-L-glutamyl-L-alanylglycyl)glycine XVII [19]; poly(L-glutamic acid) XVIII was purchased

+Supported by Grant No. AI 11970 from the National Institutes of Health from Sigma Chemical Co. St. Louis, Mo. The random copolymers, poly(L-glutamic acid, L-alanine) XIX (50%; 50%); poly(L-glutamic acid, L-alanine, L-tyrosine) XX (48%:48%:4%) and poly(L-glutamic acid, L-alanine, L-tyrosine) XXI (45%:45%:10%) were gifts from Dr. P. Maurer, Jefferson Medical College, Philadelphia, Pa. Plasma containing IgG1, isolated from a patient with multiple myeloma, was a gift from Dr. W. T. Hammack, V. A. Hospital, Birmingham, Alabama.

Modification of Human IgG1 Using 2-Hydroxy-5-Nitrobenzyl Bromide

The reaction with 2-hydroxy-5-nitrobenzyl bromide was performed in a similar manner to that described by Allen and Isliker [20, 21]. To 4 ml of a 2.7% solution of the human IgG1 in 0.1 M sodium acetate buffer, pH 4.5, was added 0.2 ml of a 0.33 M dioxan solution of 2-hydroxy-5-nitrobenzyl bromide. The mixture was stirred for 1 hr at room temperature, centrifuged, and then dialyzed for 24 hr against 4 x 250 ml of 0.5 M sodium acetate. The number of moles of 2-hydroxy-5-nitrobenzyl bromide bound per mole of IgG1 was calculated by determining the absorbance at 410 nm (E_{410} = 18,000 $M^{-1}cm^{-1}$, pH > 12.0) [22] for the modified IgG1 and at 280 nm ($E^{1\%}_{1cm}$ = 13.6, pH 7.0) for IgG1. A mol. wt. of 150,000 was used for human IgG1.

Modification of Human IgG1 Using Tetranitromethane

The method of Atassi and Habeeb [23] was used for the nitration of human IgG1. To a solution of 1.8 μmoles of human IgG1 in 5 ml of 0.05 M tris(hydroxymethyl)aminomethane 1M NaCl, pH 8.0, was added 21.4 μmoles of tetranitromethane in 75 μl of ethanol. The mixture was stirred at room temperature for 2 hr, and then dialyzed for 24 hr against 4 x 250 ml of veronal-buffered saline, pH 7.4 containing 0.15 mM calcium and 1.0 mM magnesium ions [24]. The number of moles of nitrotyrosine associated with each mole of IgG1 was calculated by determining the absorbance at 280 nm for IgG1 ($E^{1\%}_{1cm}$ = 13.6 pH 7.0) and at 428 nm for nitro-IgG1 (E = 3800 M^{-1} cm^{-1}). Amino acid analysis of the modified IgG1 confirmed this result.

Modification of Human IgG1 Using 2,3-Butanedione Reagent

This modification of human IgG1 was done using the method of Grossberg and Pressman [25]. The reagent was prepared as described by these authors. To 2 ml of a 5.4% solution of human IgG1 in 0.001 M phosphate buffer, pH 7.5, was added 2 ml of the 2,3-butanedione reagent (O.D. = 0.5/ml). The pH of the reaction mixture was adjusted to 8.6. After 6 hr against 4 x 250 ml of veronal buffered saline, pH 7.4, containing 0.15 mM calcium and 1.0 mM magnesium ions [24]. Comparison of the amino acid analysis of acid hydrolysates of IgG1 and the modified IgG1 showed that half of the arginine residues of the immunoglobulin had been modified.

Poly(L-PhenylalanylMaleic Acid, Vinyl Methyl Ether)

To 1 g of the random copolymer maleic anhydride, vinyl methyl ether, dissolved in 50 ml of dimethylformamide was added 5 g (0.023 mole) of L-phenylalanine methyl ester hydrochloride and 3 g (0.03 mole) of triethylamine. The mixture was stirred overnight at room temperature and then dialysed against distilled water (4 x 1000 ml) at 4°. To the polymer solution was added 40 ml of 1N NaOH and the mixture was stirred at room temperature for 30 min. The pH of the solution was adjusted to 7.5 with hydrochloric acid and then it was dialyzed against distilled water (3 x 1000 ml) at 4°. Concentration by filtration through a PM-30 diaflo filtration membrane and then lyophilization yielded 0.65 g (65%) of a white fluffy powder. Amino acid analysis indicated 1.40×10^{-3} moles of phenylalanine bound per gram of polymer.

Identical conditions were used for the preparation of similar polymers listed in Table V. The amino acid moieties coupled were: O-tert-butyl-L-tyrosine methyl ester hydrochloride; L-histidine methyl ester hydrochloride; L-leucine methyl ester hydrochloride; L-proline methyl ester hydrochloride; glycine methyl ester hydrochloride; L-lysine methyl ester hydrochloride and L-tryptophanylglycine methyl ester. The amino acid analysis of these polymers indicated the amount of the respective amino acids bound per gram of polymer. These are reported in Table V.

Protein Adsorption of Latex Particles

To 0.25 ml aliquots of an aqueous 1.5% latex suspension, particle size 0.81 microns, was added 1 ml of the veronal buffered saline, pH 7.4, containing calcium and magnesium ions [24]. To each sample was added 20 μg of the pertinent human IgG1 or its modification, and the suspension was gently inverted through 180° every 5 sec, to insure complete mixing. Complete adsorption of the protein to the latex was accomplished by incubation for 16 hr at 4°.

Determination of the Complement Titre in Human Serum

To 1 ml of human serum was added 2 ml of veronal buffered saline containing Ca^{2+} and Mg^{2+} ions [24]. The mixture was incubated for 1 hr at 37°. A 1 ml aliquot of this was taken and appropriate dilutions were made in the ratios of 1:12, 1:15 and 1:20 using veronal buffered saline containing Ca^{2+} and Mg^{2+} ions. To 1 ml aliquots of each of these dilutions were added 5×10^8 sensitized sheep erythrocytes [24] and the volume of each tube was made up to 7.5 ml with veronal buffer. These suspensions were incubated for 1 hr at 37° and then centrifuged. The amount of lysis was obtained by measurement of the optical density of each of the supernatants at 541 nm. A von Krogh plot [24] of complement dilution against percentage of lysis of erythrocytes was made. The 50% hemolytic unit of complement, CH_{50}, defined as the quantity of complement required for 50% lysis was obtained from this van Krogh plot. Typically the determination was performed in triplicate and it was found that 1 ml of human serum contained 20 CH_{50} units.

Hemolytic Complement Fixation Assay

The method employed for the quantitative fixation test was based on that described by Mayer [24]. The protein-latex suspensions were centrifuged and to each of the pellets was added 1.2 ml of human complement, containing 20 CH_{50} units/ml. Each pellet was suspended in the complement and incubated for 2 hr at 37°. The suspensions were inverted every 15 min to insure reaction. After this incubation, the suspensions were centrifuged and 1 ml of the supernatant from each reaction mixture was removed and assayed for residual complement activity. The supernatants were diluted using veronal buffered saline containing Ca^{2+} and Mg^{2+} [24]. To 1 ml aliquots of each of these dilutions were added 5×10^8 sensitized sheep erythrocytes and the volumes made up to 7.5 ml with the veronal buffer. These suspensions were incubated for 1 hr at 37° and then centrifuged. The optical density of each of the supernatants were obtained at 541 nm. The following controls were run concurrently with the latex-protein samples: no latex-protein; latex only; sensitized sheep erhtyrocytes without complement; complement only; total lysis of the sensitized sheep erythrocytes.

Quantitation of Complement Consumption by Anticomplementary Materials

To each tube containing 1 ml of human serum was added 50-2000 μg aliquots of a particular material and the total volume of each tube was made up to 3 ml using veronal buffered saline containing Ca^{2+} and Mg^{2+} ions [24]. The mixtures were incubated for 1 hr at 37° and 1 ml of each solution was taken and diluted with buffer. To 1 ml aliquots of each dilution was added 5×10^8 sensitized sheep erythrocytes and the volume of each tube was made up to 7.5 ml with veronal buffer. The suspensions were incubated for 1 hr at 37° , centrifuged, and the amount of lysis determined by measuring the optical density of each supernatant at 541 nm. The number of CH_{50} units remaining in the human serum was obtained graphically from a von Krogh plot. The following controls were run concurrently with the polymer and peptide samples; no polymers or peptides, sensitized sheep erythrocytes and synthetic material; sensitized sheep erythrocytes without complement; complement only and total lysis of the sensitized sheep erythrocytes. The complement consuming capabilities of each of the synthetic materials was obtained by comparing the number of CH_{50} units remaining after incubation with the respective polymer of peptide to the number of CH_{50} units in the control. These results are shown in Tables I, III, IV, V and VI.

RESULTS

Alkylation of the tryptophan residues of human IgGl showed that 6.6 moles of the 2-hydroxy-5-nitrobenzyl moiety were bound per mole of IgGl. Nitration using tetranitromethane indicated 7.3 moles of nitrotyrosine per mole of human IgGl, and the 2,3-butanedione reagent modified 16 moles of arginine per mole of human IgGl.

In order to quantitate the complement fixing abilities of the human IgGl and its modifications, the method of Allan and Isliker [21, 22] was used. We chose as our standard 20 μg of human IgGl aggregated to 3.75 mg of latex. At this ratio of protein to latex all of the immunoglobulin was absorbed and this amount of IgGl routinely removed all of the measurable complement from human serum. It was found (see Table I) that compared to the native immunoglobulin the complement fixing ability of the alkylated material was reduced by 59% of the nitrated material by 72% and of the arginine-modified material by 77%. These observations imply that aromatic and arginine moieties play a role in the complement fixing properties of

Table 1. Anticomplementary Activity of Modified Human IgGl

Protein[a]	Amount of Modification[b]	Number of CH_{50} units removed/ μg of protein	Decrease in anticomplementary activity
Human IgGl	None	1	0
2-Hydroxy-5-nitrobenzyl-IgGl	6.6	0.41	59%
Nitro-IgGl	7.3	0.28	72%
2,3-Butanedione-IgGl	16.0	0.23	77%

[a]Amount of protein used: 20 μg adsorbed to 3.75 mg of latex.

[b]Moles of chemical modifier/mole of immunoglobulin.

the human IgG. This conclusion led us to investigate the complement fixing properties of a number of peptides incorporating the tryptophan residue and based on the sequences found in human IgG1 [26], human IgM [27], guinea pig IgG2 [28], rabbit IgG [29], and mouse IgG2a [30]. These are shown in Table II. The peptides shown in Table III were synthesized by the solution technique and each had acceptable elemental and/or amino acid analysis. The complement fixing properties of the peptides and also human IgG1 were obtained in solution because the peptides absorption to latex was insufficient. These results are shown in Figure 1.

Table II. Sequence Comparisons of Various Immunoglobulins Possessing the Tryptophan Residue Implicated in Complement Fixation

Immunoglobulin	Amino Acid Residues	Sequence
Human IgG1	271-281	Pro Gln Val Lys Phe Asn Trp Tyr Val Asp Gly
Human IgG1	307-317	Thr Val Leu His Gln Asn Trp Leu Asp Gly Lys
Guinea Pig IgG2	271-281	Pro Glu Val Gln Phe Thr Trp Phe Val Asp Asn
Rabbit IgG	271-281	Pro Glu Val Glx Phe Thr Trp Tyr Ile Asx Asx
Murine IgG2a	271-281	Pro Asp Val Gln Ile Ser Trp Phe Val Asp Ans
Human IgM	482-492	Ala Asp Val Phe Val Glu Trp Met Gln Arg Gly
Human IgM	522-532	Thr Val Ser Glu Glu Glu Trp Asn Thr Gly Gln

The peptide Tyr-Glu-Ala-Gly (VII) was not found to possess anti-complement activity when up to 2000 μg of the material per ml of human serum was used. However, it was found that polypeptides containing this

Table III. Complement Fixation Using Synthetic Peptides. Percentage of CH_{50} Units Removed from 1 ml of Human Serum by 2000 μg of Peptide

Assigned Number	Peptide	Percentage of CH_{50} Units Removed
I	Trp Leu	0
II	Z Val Glu Trp Gly	11
III	Tyr Val Glu Trp Gly Glu Arg Gly	34
IV	Glu Trp Tyr Glu Arg Gly	100
V	Z Asn Trp Tyr Val	47
VI	Tyr Glu Ala Gly	0
VII	Tyr Glu Arg Gly	24
VIII	Z Glu Trp Tyr Glu Arg(Tos) Gly	8

Table IV. Role of Molecular Size and Complement Consuming Abilities

Assigned Number	Material	M. Wt.	Percentage of CH_{50} Removed[a]
IX	$(Tyr\text{-}Glu\text{-}Ala\text{-}Gly)_nGly$	>50,000	100%
IX	$(Tyr\text{-}Glu\text{-}Ala\text{-}Gly)_nGly$	20-50,000	100%
IX	$(Tyr\text{-}Glu\text{-}Ala\text{-}Gly)_nGly$	10-20,000	100%
X	$(Tyr\text{-}Glu\text{-}Ala\text{-}Gly)_4Gly$	1,737	20%
XI	$(Tyr\text{-}Glu\text{-}Ala\text{-}Gly)_3Gly$	1,317	15%

[a]By, 2,000 μg of each polymer

repeating sequence of amino acids possessed anticomplement activity. This prompted an investigation of the molecular size of the polypeptide required for complement consumption. Thus the complement consuming abilities of various molecular weight fractions of the polymer IX and the oligopeptides X and XI were compared. The percentage of CH_{50} units removed by the various size materials are listed in Table IV. These results were obtained by comparison of the number of CH_{50} units/ml of human serum remaining after incubation with the pertinent material and of a control in the absence of such a material. It was found that there is a definite molecu-

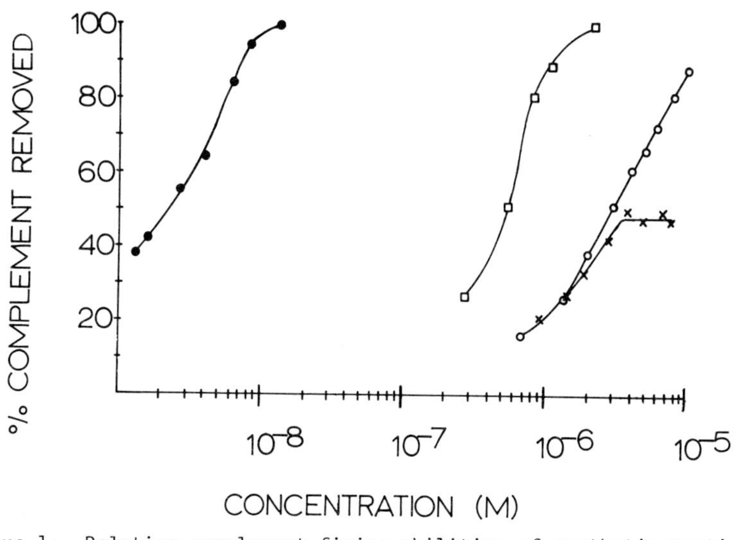

Figure 1: Relative complement fixing abilities of synthetic peptides. Where ● = human IgG1; x = Tyr-Val-Glu-Trp-Gly-Glu-Arg-Gly (III); □ = Glu-Trp-Tyr-Glu-Arg-Gly (IV); o = Z-Asn-Trp-Tyr-Val (V).

lar size requirement for polymer IX such that a minimum molecular weight of 10,000 daltons was required for 2000 μg of polymer IX to consume all the available CH_{50} units in a ml of human serum. For subsequent comparisons between widely different materials a molecular weight of over 30,000 daltons was used. By adopting this arbitrary standard it was felt that meaningful comparisons of the complement consuming abilities of the various materials could be made. This comparison is tabulated in Table V.

Table V. Complement Consuming Abilities of Polyanions

Assigned Number	Polymer[a]	Moles of Amino Acid Bound Per Gram of Polymer	Percentage of CH_{50} Removed[b]
IX	$(Tyr\text{-}Glu\text{-}Ala\text{-}Gly)_n Gly$	Equimolar	100
XII	$(Phe\text{-}Glu\text{-}Ala\text{-}Gly)_n Gly$	Equimolar	0
XIII	$(Trp\text{-}Glu\text{-}Ala\text{-}Gly)_n Gly$	Equimolar	25
XIV	$(Tyr\text{-}\gamma\text{-}Glu\text{-}Ala\text{-}Gly)_n Gly$	Equimolar	32
XV	$(Tyr\text{-}Glu\text{-}Gly\text{-}Gly)_n Gly$	Equimolar	30
XVI	$(Tyr\text{-}Glu\text{-}Ala\text{-}\beta\text{-}Ala)_n Gly$	Equimolar	100
XVII	$(Ala\text{-}Glu\text{-}Ala\text{-}Gly)_n Gly$	Equimolar	0
XVIII	$(Glu)_n$	Equimolar	0
XIX	$(Glu, Ala)_n$	Equimolar	0
XX	$(Glu, Ala, Tyr_4)_n$	$(3.6:6.7:0.25) \times 10^{-3}$	0
XXI	$(Glu, Ala, Tyr_{10})_n$	$(3.4:6.3:0.61) \times 10^{-3}$	0
XXII	$(Maleic\ Acid, Vinyl\ Me\ Ether)_n$	0	0
XXIII	$(Maleic\ Acid, Tyr, Vinyl\ Me\ Ether)_n$	1.44×10^{-3}	100
XXIV	$(Maleic\ Acid, Phe, Vinyl\ Me\ Ether)_n$	1.40×10^{-3}	100
XXV	$(Maleic\ Acid, His, Vinyl\ Me\ Ether)_n$	1.10×10^{-3}	0
XXVI	$(Maleic\ Acid, Leu, Vinyl\ Me\ Ether)_n$	1.20×10^{-3}	0
XXVII	$(Maleic\ Acid, Pro, Vinyl\ Me\ Ether)_n$	1.77×10^{-3}	0
XXVIII	$(Maleic\ Acid, Gly, Vinyl\ Me\ Ether)_n$	1.70×10^{-3}	0
XXIX	$(Maleic\ Acid, Lys, Vinyl\ Me\ Ether)_n$	1.67×10^{-3}	0
XXX	$(Maleic\ Acid, Trp\text{-}Gly, Vinyl\ Me\ Ether)_n$	1.35×10^{-3}	100
XXXI	$(Maleic\ Acid, Styrene)_n$	0	100
XXXII	$(Maleic\ Acid, Styrene, Tyr)_n$	0.75×10^{-3}	100

[a]Mol. wt. greater than 30,000 daltons

[b]By 2000 μg

DISCUSSION

In our investigation of compounds that could possible be used to lower *in vivo* complement levels it seemed reasonable to study the complement fixing site of human IgG1. Allan and Isliker [21, 22] have shown that the modification of a few tryptophan residues in rabbit IgG and its Fc fragment with 2-hydroxy-5-nitrobenzyl bromide lowered the anticomplementary activity of these materials in the presence of guinea pig complement. However, this modification did not significantly alter their Clq-binding capacities. It was found that a particular tryptophan residue of a cleavage peptide of rabbit Fc, was most probably responsible for this change in anticomplementary activity. It was suggested, from sequences of rabbit heavy chains [20], that the tryptophan residue at position 277 appeared to be involved in complement fixation. The reports of other investigators also indicate that the anticomplementary activity of IgG was associated with the Cγ2 domain of human [31], rabbit [29, 31], mouse [30, 31] and probably guinea pig [28] immunoglobulins.

We have found that alkylation of human IgG1 produced a reduction in its complement fixing abilities similarly nitration also lowered its ability to fix complement. This later observation suggests that either a

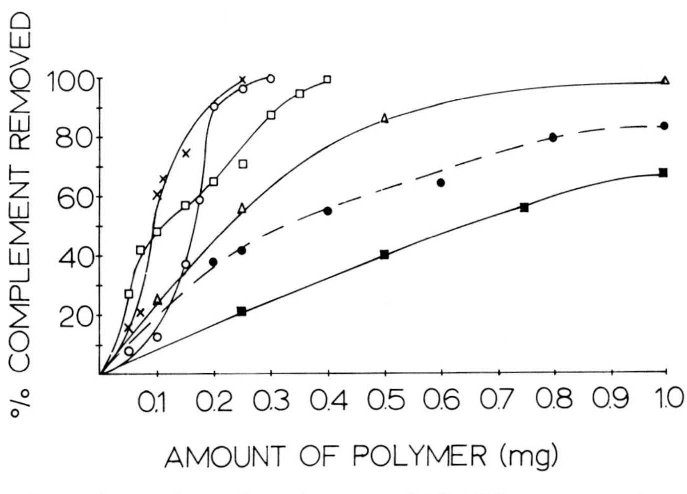

Figure 2. Percentage of complement consumed/ml of human serum by incremental amounts of polyanion. Where x = polyanion No. XXXII; o = polyanion No. XXIII; ◘ = polyanion XXXI; ▲ = polyanion No. XXIV; ● = human IgG1; ■ = polyanion No. IX.

tyrosine residue could possibly play a role in complement fixation, or that the tryptophan residue has been modified by the nitrating reagent. It is also possible that these modifications of the amino acid residues, could have caused secondary changes altering the complement fixing site.

Comparison of the sequences around the tryptophan residues of the Cγ2 domains of human IgG1 are shown in Table II. It is seen that there are two possible sites, based on tryptophan at position 277 and that at position 313. In order to differentiate between these sequences it was noted that the tryptophan at position 277 was adjacent to a tyrosyl residue whereas that at position 313 was not. One way of differentiating between these two sequences was to study the role of this tyrosyl residue. Interestingly, comparison of the corresponding sequences of the complement fixing human IgG1 with those of guinea pig IgG2, rabbit IgG and mouse IgG2a, also show that they all possess this aromatic sequence of tryptophan and tyrosine or phenylalanine.

The 2,3-butanedoine modification of human IgG1 implicates an arginine residue in the complement fixing area. However, comparison of the sequences of the human, rabbit, guinea pig and mouse complement fixing immunoglobulins does not reveal any arginine residues close to the active tryptophan moiety. The arginine deficiency in this region of the molecule and the evidence for its role in complement fixation by human IgG1 are hard to reconcile. However, inspection of the sequences of human IgM and the Cγ4 complement fixing domain of this immunoglobulin [27] show an arginine residue at position 491. Two tryptophan residues occur in this domain at positions 488 and 528, however neither of them are adjacent to either a tyrosine or a phenylalanine residue. Although the sequences surrounding these tryptophans show little homology to each other, there is some correlation between the arginine containing sequence 482-492 of human IgM and the sequences 271-281 of human IgG1, guinea pig IgG2, rabbit IgG and mouse IgG2a. These observations lead us to investigate the role of the tyrosine and arginine moieties in peptides which mimic those sequences found in human IgG1 and IgM.

The apparent necessity of the juxtaposition of the tyrosine and tryptophan residues was investigated by studying the complement fixing properties of peptides possessing the tryptophan moiety but were deficient in an adjacent tyrosyl residue. The peptides, I, II and III possessed little or no complement fixing properties. However, if the tyrosyl residue was adjacent to the tryptophan then enhanced complement fixing abilities were immediately apparent. This is shown by compounds IV and V as reported in Table III and Figure 1. The necessity of the juxtaposition of tryptophan and tyrosine residues is illustrated by comparison of compounds III and V. Compound III, which possesses non-adjacent tyrosine and tryptophan residues has little complement fixing ability. However, when these residues are adjacent, as in compound IV, then the

Table VI. Relative Abilities of Polyanions to Consume Complement

Assigned Number	Compound[a]	Moles of Amino Acid Bound per Gram of Polymer	Number of CH_{50} Units Removed per mg of material
-	Human IgG1	-	30.7
IX	$(\text{Tyr-Glu-Ala-Gly})_n\text{Gly}$	Equimolar	15.4
XXIII	$(\text{Maleic Acid, Vinyl Me Ether, Tyrosine})_n$	1.44×10^{-3}	58.9
XXIV	(Maleic Acid, Vinyl Me Ether, Phenylalanine)	1.40×10^{-3}	43.5
XXXI	$(\text{Maleic Acid, Styrene})_n$	0	87
XXXII	$(\text{Maleic Acid, Styrene, Tyrosine})_n$	0.75×10^{-3}	105.3

[a]Mol. Wt. greater than 30,000 daltons

peptide is highly active. This change in activity is further evident upon inspection of the dose-response curves in Figure 1. Compound III is only capable of removing half of the available complement from human serum. However, compound IV consumes all of the available CH_{50} units. This observation is confirmed by the ability of compound V to completely fix the complement in human serum. The sequence tryptophanyltyrosyl only occurs in the Cγ2 domain of human IgG1 at positions 277-278. This result tends to exclude the possibility that the tryptophan at position 313 is part of the complement fixing site of this immunoglobulin.

The role of the arginine residue was investigated by comparison of the peptides VI and VII, in which an alanine residue has been replaced by an arginine. Only the arginine containing peptide fixes complement under the conditions of our assay. Further evidence for this enhancing effect of the arginine residue is apparent when compounds IV and VIII are compared. Both of these peptides are identical, except that the amino groups of peptide VIII are blocked. The complement fixing properties of the unblocked peptide IV, is far greater than that if its analog VIII. Peptide IV mimics the sequence surrounding arginine 491 and tryptophan 488, an area which could possibly be the locus of the complement fixing site of human IgM. This peptide, IV, is the best complement fixing peptide shown in Figure 1. Although it is 200 times less effective than human IgG1, such peptides as IV should be advantageous for therapeutic use because of their low antigenicity.

An investigation of the complement fixing properties of the tetra-peptide (VII) indicated that it was inactive under the conditions of our assay. However, it was found that the oligopeptides X and XI, which possess the same repeating sequence of amino acids, did reveal complement fixing tendancies. It is well known that polyanions such as heparin [2-6], dextran sulfate [7, 8], carrogeenan [9] and cellulose sulfate [10] affect complement. It is suggested that a similar effect is occurring within the peptide series, VII, X and XI. Indeed, the polymer IX, which possesses the same repeating sequence of amino acids, is an even more efficient complement fixing material than any of the peptides VII, X and XI. See Table III and IV.

The observations that tryptophan and tyrosine containing polypeptide poly(Tyr-Glu-Ala-Gly) IX also consumed complement prompted the investigation of the role of aromatic moieties on the complement fixing properties of polyanions. For this purpose twenty-two polyanions were screened for anticomplement activity and these are shown in Table V.

It was found that those materials which possess aromatic moieties as well as the anionic groups consume complement. Such a criteria covers such diverse compounds as IX and XXXII. The former material has a polypeptide backbone and the aromatic, tyrosyl group is adjacent to the anionic glutamyl residue, whereas the latter material has a carbon backbone and the aromatic and anionic moieties are randomly arrayed. This requirement of an aromatic moiety is illustrated by comparison of compounds IX and XVII. Replacement of the tyrosyl residue in IX by an alanyl moiety as in polymer XVII produces a complement loss of complement consuming ability. Further evidence for the necessity of an aromatic residue is evident by comparison of compounds XXII, XXIII, XXIV and XXX. These materials are identical except that the polymers XXIII, XXIV and XXX have small amounts of aromatic residues attached to them. These residues are necessarily next to the anionic carboxyl residue since these polymers were synthesized from the precursor polymer, poly(maleic anhydride, vinyl methyl ether). Thus polymer XXIII could be thought of having repeating tyrosyl and carboxyl moieties along a carbon chain, somewhat analogous to compound IX. Although compounds XII, XX and XXI all have aromatic and anionic moieties none consume complement. Compounds XX and XXI are random copolymers. It is possible that either there is insufficient tyrosine incorporated into them in order to satisfy the aromatic requirement or that

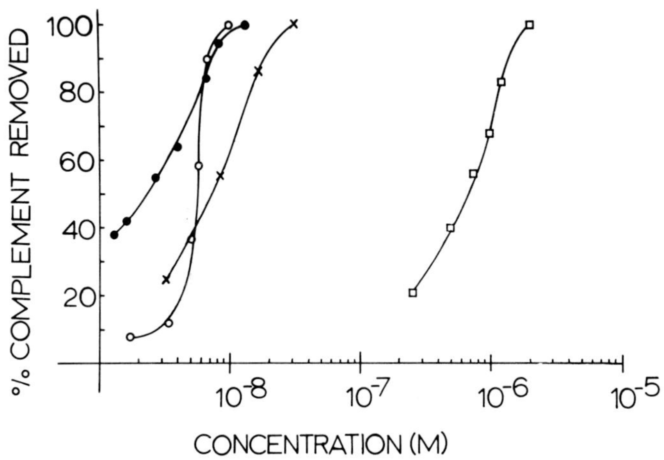

Figure 3. Percentage of complement removed/ml of human serum by incremental amounts of polyanion. Where ● = human IgG1; o = polyanion XXIII; x = polyanion XXIV and ◻ = polyanion IX.

the tyrosyl residues are not adjacent to the anionic carboxyl group of glutamic acid. We favor this latter explanation.

The next parameter that was studied was the relative potency of the various complement consuming materials. The dose response curves for compounds IX, XXIII, XXIV, XXXII are shown in Figure 2. The response curve for the human immunoglobulin IgG1 is included for comparative purposes. It is seen that compounds XXIII, XXIV, XXXI and XXXII all remove the available CH_{50} units when up to 1000 μg of the polymers were used. In contrast, IgG1 and polymer IX were found to be incapable of removing all of the available CH_{50} units at this concentration. Comparison of the tyrosyl and phenylalanyl derivatives XXIII and XXIV of the polymer, poly maleic acid vinyl methyl ether, indicate that the tyrosyl residue does have a greater complement consuming effect than the phenylalanyl moiety. Thus 300 μg of the tyrosyl analog XXIII removed all the available CH_{50} units whereas a 1000 μg of the phenylalanyl analog XXIV was required. This reduced activity of the phenylalanyl residue to endow complement consuming properties to a polymer correlates with our observation of the differences in biological activities of polymers IX and XII.

A weight basis comparison of the amounts of each polymer and the human immunoglobulin IgG1 to remove half of the available CH_{50} units/ml of human serum are given in Table VI. It is seen that polymer IX has half the capability of the human immunoglobulin IgG1 to consume complement whereas the other polymers are somewhat better. Comparison of the tyrosyl substituted polymer XXXII with the unsubstituted polymer XXXI further indicates the role of the tyrosyl residue. The complement consuming ability of the polymer is increased by 20% when this residue is incorporated.

A comparison of these anticomplementary polymers, on a molar basis, with that of human IgG1 is shown in Figures 3 and 4. It is seen that polymer IX is far less effective as a complement removing agent that the other polymers.

CONCLUSIONS

From this work it has been found that the complement fixing site of human IgG1 probably is centered around the tryptophanyltyrosyl sequence

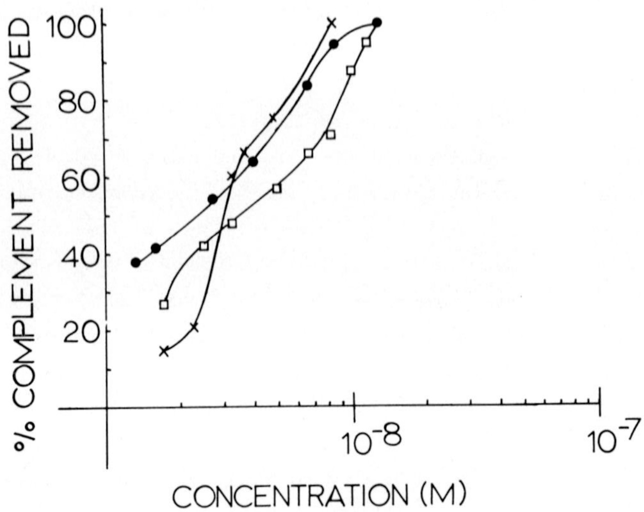

Figure 4. Percentage of complement removed/ml of human serum by incremental amounts of polyanion. Where ● = human IgG1; ◻ = polyanion XXXI and x = polyanion XXXII.

at position 277-278 of the Cγ2 domain of this immunoglobulin. Similarly the area, positions 487-491, of the Cμ4 domain of human IgM could be part of the complement fixing locus of this material. The small peptide Glu-Trp-Tyr-Glu-Arg-Gly has been found to be an effective anticomplement

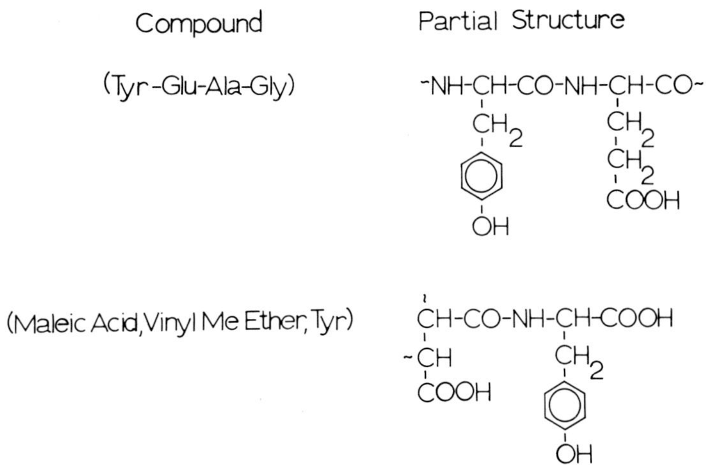

Figure 5. Comparison of the partial molecular structure of complement consuming polyanions.

material. Although it is 200 times less effective than human IgG1, such materials should be advantageous for <u>in vivo</u> use because of their expected low antigenicities.

It has also been found that polyanions of a sufficiently high molecular weight are good complement consuming materials provided that aromatic moieties are also included in their formulation. Comparison of the phenylalanyl, styrene and tyrosyl residues for providing this aromatic requirement indicates that the tyrosyl residue is probably the most effective.

The position of the aromatic moiety with respect to the anionic group would appear to be another important parameter that has to be satisfied. This is illustrated by comparison of the partial molecular structures of poly(L-Tyrosyl-L-glutamyl-L-alanylglycyl)glycine IX and poly(maleic acid, vinyl methyl ether, tyrosine) XXIII as shown in Figure 5. It is seen that both polymers contain a similar partial structure where the aromatic and anionic groups are adjacent to one another. A similar situation occurs with the polymers poly(maleic acid, styrene) XXXI and poly(maleic acid, styrene, tyrosine) XXXII.

Some of these high molecular weight materials have been shown to be antigenic [11]. However, if they could be used in vitro, they also may have therapeutic use.

It has been suggested by several investigators [33-35] that complement plays an immunopathological role in the inflammatory periodontal disease process. The gingival crevicular fluid from normal patients possesses the complement components C3, C4 and C5, and that from those with inflamed gingiva contains properidin factor B and electrophorectically converted C3 [36]. In addition, the functional nature of complement components in the crevicular fluid from periodontal patients has also been demonstrated [37].

It occurred to us that if the levels of complement could be lowered in the ginigival crevicular fluid then tissue damage and inflammation due to complement activation could be controlled. The composition of the gingival crevicular fluid possesses some similarity to serum [38, 39], and since the polymers IX, XXIV, XXXI and XXII are good serum complement removing materials it is suggested that these polymers may find therapeutic use for patients with periodontal disease.

REFERENCES

1. P. J. Baker, T. F. Lint, B. C. McLeod, C. L. Behrends and H. Gewurz, J. Immunol., 114(1975)554.
2. E. E. Ecker and P. Gross, J. Infect. Dis., 44(1929)250.
3. K. Lauenstein, H. G. Stedentopf and H. Fischer, Z. Naturforsch., 206(1965)575.
4. P. Gross and E. E. Ecker, Proc. Soc. Expt. Biol. Med., 26(1929)508.
5. E. E. Ecker and L. Pillemer, J. Immunol., 40(1941)73.
6. A. G. Osler, H. G. Randall, B. M. Hill and Z. Ovary, J. Exp. Med., 110(1959)311.
7. U. Hadding, M. Konig, M. Limbert, H. U. Schorlemmer and D. J. Bitter-Suermann, J. Immunol., 111 (1973)286.
8. U. Hadding, M. Konig, M. Limbert, H. U. Schorlemmer and D. J. Bitter-Suermann, Eur. J. Immunol., 3(1973)527.
9. T. Borsos, H. J. Rapp and C. Crisler, J. Immunol., 94(1965)662.
10. V. Eisen and C. Loveday, Br. J. Pharmacol., 39(1970)831.
11. B. J. Johnson, J. Pharm. Sci., 63(1974)313.

12. B. J. Johnson and E. G. Trask, J. Org. Chem., 33(1968)4521.
13. B. J. Johnson and E. G. Trask, J. Chem. Soc.,C(1969)2644.
14. B. J. Johnson and C. Cheng, J. Med. Chem., 17(1974)320.
15. B. J. Johnson and C. Cheng, J. Med. Chem., 14(1971)1238.
16. B. J. Johnson and N. Tsang, J. Med. Chem., 15(1972)488.
17. B. J. Johnson, J. Med. Chem., 15(1972)423.
18. B. J. Johnson and E. G. Trask, J. Med. Chem., 14(1974)251.
19. B. J. Johnson, C. Cheng and N. Tsang, J. Med. Chem., 15(1972)95.
20. R. Allen and H. Isliker, Immunochem., 11(1974)175.
21. R. Allan and H. Isliker, Immunochem., 11(1974)243.
22. T. E. Barman and D. E. Koshland, J. Biol. Chem., 242(1967)5771.
23. M. Z. Atassi and A. F. S. A. Habeeb, Biochemistry, 8(1969)1385.
24. E. A. Kabat and M. M. Mayer, in Experimental Immunochemistry, C. C. Thomas, Springfield, Ill, 1961, p. 149.
25. A. L. Grossberg and D. Pressman, Biochemistry, 7(1968)272.
26. G. M. Edelman, Biochemistry, 9(1970)3197.
27. A. Shimizu, C. Paul, H. Kohler, T. Chinoda and F. W. Putnam, Science, 182(1971)287.
28. D. E. Tracey and J. J. Cebra, Biochemistry, 13(1974)4796.
29. R. L. Hill, H. E. Lebovitz, R. E. Fellows and R. Delaney, in J. Killander (Editor), Proc. 3rd Nobel Symposium, Interscience, N. Y., 1967, p. 109-127.
30. J. M. Kehoe, A. Bourgois, J. D. Capra and M. Fougereau, Biochemistry, 13(1974)2499.
31. J. M. Kehoe and M. Fougereau, Nature, 224(1969)1212.
32. Z. Ovary, P. H. Saluk, L. Quijada and M. E. Lamin, J. Immunol., 116(1976)1265.
33. S. E. Mergenhagen, T. R. Tempel and R. Snyderman, J. Dent. Res., 49(1970)256.
34. R. Snyderman, J. Amer. Dent. Assoc., 87(1973)1020.
35. R. J. Genco, P. A. Mashimo, G. Krygier and S. A. Ellison, J. Periodont., 45(1974)330.
36. R. Attstrom, A. Laurel, U. Lahsson and A. Sjoholm, J. Periodon. Res., 10(1975)19.
37. F. J. Courts, R. J. Boakle, M. S. Silverman and H. H. Fudenberg, Clin. Immunol. Immunophath, "in press".
38. P. Brandtzaeg, Arch. Oral Biol., 10(1965)795.
39. E. J. Shillitoe and T. Lehner, Arch. Oral Biol., 17(1972)241.

INTERACTION OF SYNTHETIC POLYPEPTIDES WITH BLOOD

A. G. WALTON
Department of Macromolecular Science
Case Western Reserve University, Cleveland, Ohio 44106

SUMMARY

A series of synthetic biopolymers (polypeptides) have been synthesized in order to determine the role of individual amino acids in catalysing or inhibiting steps in the blood coagulation process.

Specifically the behavior of blood platelets and blood factors in contact with soluble and insoluble polypeptides is explored. A series of soluble cationic polyamino acids, random copolymers and sequential polypeptides have been studied in terms of their capability of eliciting platelet aggregation and release reactions. Insoluble random copolypeptides have also been examined for platelet compatibility and its mediation by adsorbed protein.

The interaction of blood factors with polypeptides shows that two of them (Hageman factor and prothrombin) can be activated with concommitant change in structure. Most of the others interact with, but are not activated by, polypeptides.

INTRODUCTION

Since the earliest period of polypeptide synthesis, attention has focussed on a) the ability of synthetic analogs to reproduce *structural features* of proteins and b) the ability of analogs to reproduce or induce *biological function* of proteins. In this article the latter objective is explored in terms of the interaction of both soluble and insoluble synthetic polypeptides with blood components. Information is reviewed concerning the factors which induce changes in one or more blood components, with its resultant effect on blood clotting.

Under normal biological circumstances, blood comes into contact with surfaces consisting of several components including adsorbed proteins, endothelial cells and perhaps substructure of elastoidal nature. When damaged, other proteinaceous components of vascular material are exposed to blood including collagen, proteoglycan, other minor proteins and various additional cellular materials. Under these circumstances, blood clotting occurs, much attention having been focussed on collagen as an inducer of the clotting process.

Thus under some circumstances blood is in contact with nonenzymic proteins which seem to be completely passive to blood coagulation and in others different proteins cause active clotting. It would seem that since similar peptide residues are present in both cases that secondary or higher order structure in the proteins holds the key to activity or passivity in blood interaction. (A word of caution is in order in that many of the proteins involved are glycoproteins, lipoproteins or proteoglycans and activity/passivity may be influences by the non protein moiety). Nevertheless synthetic polypeptides allow us the opportunity to study the role of peptide residues and their configuration, conformation and other aspects of structure, both in insoluble and soluble form.

The earliest reported studies of synthetic polypeptide interaction with blood involved the effect of the basic polyamino acids, poly-L-lysine, poly-D, L-lysine and poly-D, L-ornithine on blood clotting. These materials were found to retard blood clotting at a concentration of 100 μg/ml both *in vitro* [1] and *in vivo* in rabbits [2] and rats [3]. Very little work seems to have been performed using more complex polypeptides until quite recently with the exception of lysine/glutamic acid copolymers [4] which showed varying degrees of clot retardancy which was a minimum at a mole ration of 2:3. Interest in the basic polyamino acids has been related to modelling the behavior of basic protamines which act as strong clotting retardants [5] and are used in thrombolytic therapy.

There were no reports of the behavior of blood in contact with insoluble polypeptides until 1975 when a published report [6] showed that glutamate ester/leucine copolymers were apparently fairly compatible with whole blood. The relative compatibility assessed by a modified kinetic method increased in the order poly-γ-methyl-D-glutamate, copoly(γ-benzyl-L-glutamate:γ-methyl-D-glutamate) 1:4, copoly(γ-methyl-L-glutamate:L-leucine) 1:1, copoly(γ-benzyl-L-glutamate:L-leucine) 1:4, copoly(γ-benzyl-L-glutamate:L-leucine) 1:1, poly-γ-benzyl-L-glutamate.

In the following report we shall concentrate upon the interaction of synthetic polypeptides with single blood components rather than with the extremely complex combinations found in whole blood. The work is divided into essentially two sections, that involving the cellular components, particularly blood platelets and that involving blood factors or other blood proteins.

A. Synthetic Polypeptides and Blood Platelets

Blood platelets are present in blood to the extent of approximately 5 x 10^8/ml and are small (∿2μ diameter) non nucleated cells which are very sensitive to stimulus. Under the influence of various chemical agents they undergo morphologic changes, aggregation and specific release of calcium, A.D.P, serotonin (2 hydroxytryptamine) and various platelet factors. A schematic diagram of a platelet is shown in fig 1. The release of various chemicals causes a "snowball" effect on further aggregation and release, and activates enzymes involved in the blood clotting sequence. Consequently the behavior of platelets and specifically the release process are intimately involved in blood coagulation. Collagen is known to cause the release process in platelets, evidence pointing to the involvement of lysyl groups and ordered aggregates of molecules [7].

The release process as initiated by collagen involves two apparent stages of aggregation, primary aggregation induced by the collagen and secondary aggregation elicited by the released material, particularly calcium.

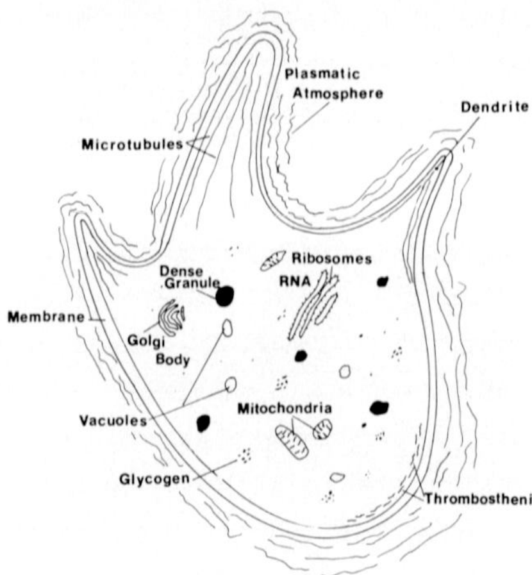

Fig 1. Schematic diagram of blood platelet showing some morphologic deformation (pseudopod or dendrite extension).

i) Soluble Polypeptides

It has been known for some time that poly-L-lysine causes platelet aggregation [8-14], the suggested mechanism involving the neutralization of the negative surface charge on the platelet membrane and bridging between adjacent platelets. While investigators agree that polylysine causes platelet aggregation, there has been some disagreement on whether it is capable of eliciting the platelet release reaction and a biphasic aggregation curve, as does collagen.

An interesting concept introduced recently is that the platelet has surface "recognition" sites which respond specifically to an active or passive substrate. For example Silberberg and coworkers* have incubated platelets with soluble collagen, gelatin, sequential copolypeptide and random copolypeptide collagen models and found a strong decrease in ability of the platelets to adhere to insoluble collagen and undergo the release reaction. Their conclusion is that the primary structure or even amino acid content is sufficient to block recognition sites. However since platelets interact with a large range of soluble and insoluble materials it is difficult, at this time to put the concept of recognition sites into a uniform theory of platelet adhesion and release.

In the following report, an analysis of platelet aggregation and release as induced by a series of poly (amino acids) and polypeptides containing cationic residues, is presented.

Space does not permit a detailed presentation of preparative methods but washed platelets were prepared by standard methods and incubated with tritium labelled serotonin. Aggregation and release were studied in buffered solutions containing 2.5 x 10^5 platelets/μl, NaCl(140 mM), EDTA(0.3 mM), Tris buffer(30 mM) and glucose (5 mM) at pH 7.4.

Aggregation was studied using a Payton dual channel aggregometer and solutions were assayed for released serotonin by removing platelets (2450 g for 15 mins at 2 °C) and counting radioactive H^3 in the supernatant.

The cationic polypeptides used were poly-L-lysine HBr (M. Wt. 85,000), poly-L-arginine HCl (M. Wt. 43,000) and poly-L-ornithine HBr (M. Wt. 120,000) each from Sigma Chemical Co.; poly-(N^{ε}-Me-L-lysine) HCl (M. Wt. 11,300), poly-(N^{δ}-Me-L-ornithine) HCl (M. Wt. 41,000) and poly-(N^{δ}-Et-L-ornithine) HCl (M. Wt. 28,000) kindly provided by Prof. Yang at the Cardiovascular Research Inst., San Francisco; poly(LysAla$_2$) (M. Wt. ∿10,000) and poly(Lys-Pro-Gly) (M. Wt. ∿20,000) from Biopolymer Corp.; poly(Lys-Ala-Glu) (M. Wt. ∿6000) provided by Prof. Goren, Calgary University; poly(Lys-Gly-Pro-Pro-Gly-Pro)$_n$ (M. Wt. ∿10,000) synthesized by G. F. Sigler [14]; and copoly(Lys:Ala) (ratio 1:1) provided by Prof. J. Anderson, Case Western Reserve University.

---

*A. Silberberg, I. R. Miller, F. A. Meyer, Z. Weisman and W. Weiss, Annual Contract Report to U.S.N.H.L.I. Aug. 1 (1976) Contract No. N01-HB-5-2907.

RESULTS AND DISCUSSION

Four of the polypeptides studied, polylysine, polyarginine, polyornithine and poly(Nδ-Et-L-ornithine) behaved in a similar manner causing biphasic aggregation that was a function of concentration. Fig 2 shows aggregation curves for polyarginine that indicate maximal aggregation in the region of 10 mg/ml. Above this concentration the aggregation process is suppressed and becomes monophasic. It has been suggested that the initial aggregation process is caused by the charge neutralization and binding of platelets [11, 12] and that the second is caused by endogenous materials such as calcium, A.D.P and fibrinogen released by the platelets. Addition of 3 mM EDTA does in fact strongly suppress the second phase of aggregation, supporting the concept that calcium is involved. The suppression of aggregation at high dosage of polycation suggests that excess of the polyamino acid may be complexing one or more of the agents normally released from platelets. Indeed there is evidence that fibrinogen is bound by polylysine [15].

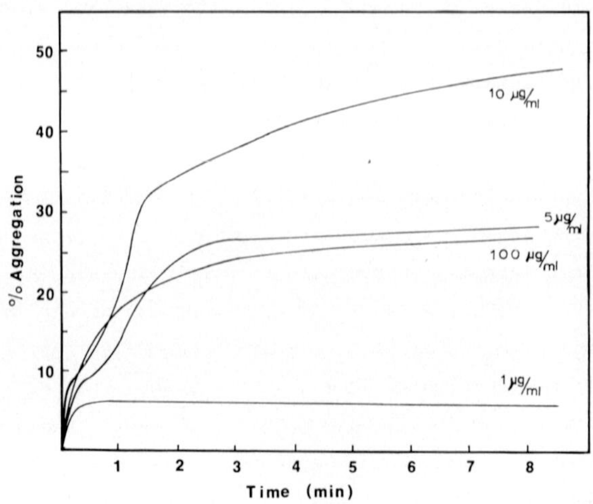

Fig 2. Aggregation of washed platelets as a function of dosage of poly-L-arginine. Maximum aggregation and biphasic response is shown at ∿10µg/ml.

In terms of their relative ability to cause release of serotonin from platelets, the three unmodified polyamino acids at 10 µg/ml gave approximately the same amount of release when tested with three different (donor) platelet preparations (four samples); polylysine 31-42%, polyarginine 32-45% and polyornithine 38-58%, however, the modified poly (N^{δ}-Et-ornithine) caused only 4-15% release. Presumably in this case the presence of the N ethyl group causes some charge screening.

The other N modified polyamino acids also displayed somewhat different behavior poly(N^{δ}-Me-L-ornithine) and poly(N^{ε}-Me-L-lysine) caused only one wave of aggregation in the 5-10 μg/ml concentration range. At 100 μg/ml all of the lysyl and ornithyl polypeptides gave less than 4% serotonin release, polyarginine caused 11-14% release at the same concentration. The possibility of non specific lysis of platelets by the polyamino acids was examined by ^{51}Cr labelling of the membranes which revealed that all of the cationic polyamino acids except polyarginine caused less than 7% at 100 μg/ml, polyarginine, caused 10-20% lysis. These data indicate that most of the serotonin released at high polyamino acid dosage probably originates from lysed platelets. At low (10 μg/ml) concentrations where 30% or more of platelets released serotonin, the process is probably specific release.

Contrary to the performance of the polyamino acids, the copolypeptides containing lysine produced less (copoly Lys:Ala) or no $(LysAla_2)_n$, $(Lys\text{-}Pro\text{-}Gly)_n$, $(Lys\text{-}Ala\text{-}Glu)_n$ $(Lys\text{-}Gly\text{-}Pro\text{-}Pro\text{-}Gly\text{-}Pro)_n$ aggregation or release. Whereas the sequential copolypeptides are of fairly low molecular weight, at least one of them has been shown [14] to have collagen-like structure [$(Lys\text{-}Gly\text{-}Pro\text{-}Pro\text{-}Gly\text{-}Pro)_n$]. We conclude therefore that although low molecular weight materials may lack the ability to cause aggregation [12], charge density is an equally important and perhaps more important factor.

ii) Insoluble Polypeptides

In view of the ability of cationic polyamino acids to induce aggregation and release from blood platelets, it is of some interest to know whether cationic peptide residues in insoluble films also cause similar platelet perambulations. Indeed the question arises whether there are combinations of amino acids which are particularly active or passive to platelet interaction. The possibility of developing tissue compatible polypeptides has been suggested [16, 17]. A series of random copolypeptide films containing L-leucine, L-lysine (and derivative), L-glutamic acid (and derivatives) and L-phenylalanine has now been studied in terms of washed platelet interactions and preliminary data reported [18].

Two types of experimental approach have been used. The first has involved the polypeptide substrate coated onto 0.5 mm glass beads. Serotonin labelled platelets were passed through the column in buffer containing NaCl (140 mM), EDTA (3.3 mM), Tris buffer (30 mM) and glucose (5 mM) at pH 7.4 and 37°C. On passing through the column, some platelets adhered to the substrate without undergoing release, others adhered with release and yet others flowed through without adhering. Counts of the supernatant and platelets passing through the column revealed the relation between adhesion and release as a function of platelet density, substrate composition and flow rate.

Substrates used in these experiments were random copolypeptides of two components designed to give a range of surface properties including positive and negative charge and highly hydrophic character. Specifically the copolypeptides were of the type $(A_xB_y)_n$ where A is γ-benzyl-L-glutamate, L-glutamic acid or L-lysine and B is L-leucine.

Examination of the surface coated beads by scanning electron microscopy visualized both the film and adherent platelets (fig 3). The platelets do not appear to aggregate significantly on the surface and though a large proportion of adsorbed platelets that do adhere undergo release reactions, the released moieties are presumably inhibited from causing aggregation and release of non adsorbed platelets by the EDTA in solution.

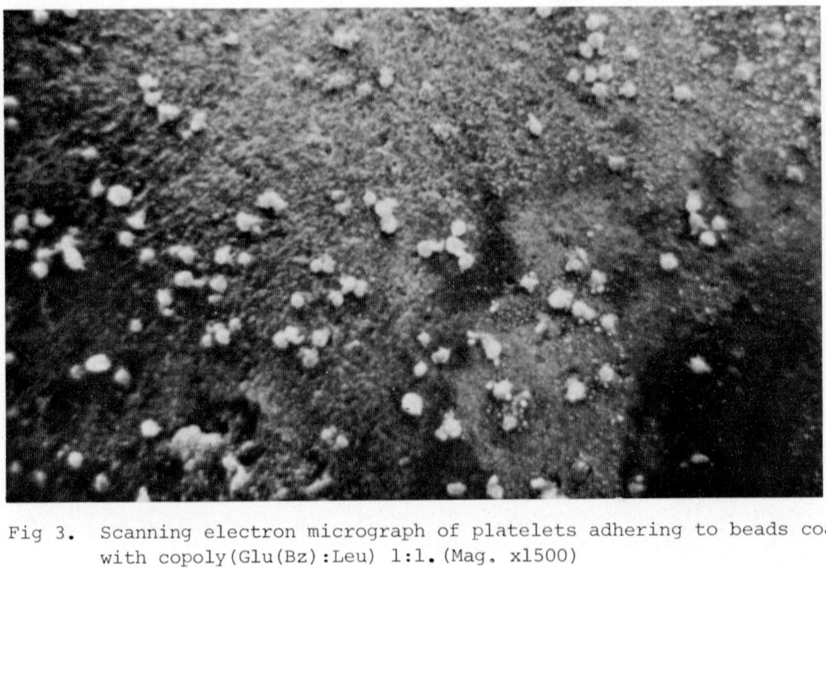

Fig 3. Scanning electron micrograph of platelets adhering to beads coated with copoly(Glu(Bz):Leu) 1:1. (Mag. x1500)

Initial experiments designed to measure the radioactive content of successive aliquots taken from the column revealed that most of the unreacted platelets and released serotonin passed through the column in the first four minutes after application of the platelets to the column. By fifteen minutes 99% of the eventual radioactivity had emerged from the column and all subsequent data were based on fifteen minutes delay from injection of platelets. To make sure that the adhesion was not nucleated by holes in the film, a series of runs was carried out in which the beads had been treated with various polypeptide concentrations (Benzylglutamate: Leucine 1:1). Table 1 shows the effect of such treatment.

Table 1

Effect of Polypeptide Film Formation Upon Adhesion and Release of Blood Platelets

Substrate	Solution Conc.≠ of G.L(1:1)*	%Adhesion[a]	%Release
Glass beads	0	30.4	12.6
" " + polypep.	0.5%	34.5	16.7
" " "	2%	23.9	16.9
" " "	5%	18.7	18.8

≠Solvent was chloroform

[a]Based on dosage of 5×10^8 platelets

*Copoly γ-benzyl-L-glutamate:L-leucine platelets

It is interesting to note that with heavier coating the number of platelets adhering decreased but the amount of release remained constant. The explanation of this strange effect is unknown, however it may be speculated that there are two different types of surface sites one of which leads to adhesion with release, the other without release. Scanning electron micrographs of films show that they are highly fibrous (fig 4). Precipitation from higher concentrations may be expected to produce finer, more closely woven fibers and perhaps a more uniform substrate. It is possible that some sites in the glass beads remain at low density of coated polypeptide, however scanning electron microscopy does not reveal specific holes or sites. It would seem then that surface morphology is an important parameter in defining platelet compatibility.

To examine further the relation between surface properties and uptake of platelets a series of columns with the same substrate treatment and surface area were exposed to increasing numbers of platelets. Table 2 shows that the uptake of platelets by the surface increases with dosage but the fraction of platelets undergoing release drops drastically at in-

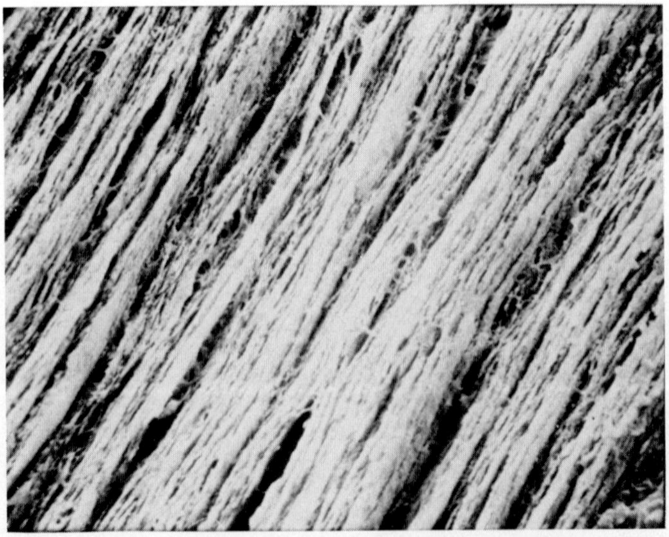

Fig 4. Scanning electron micrograph of fracture film of copoly(Glu(Bzl):Leu) 1:4 Mag X1500(by P. H. Geil and S. Barenberg).

creased coverage. This type of phenomenon has been observed and reported under slightly different conditions previously [18] and has been attributed by us to either a) the release of an entity or entities by platelets which adsorbs on the substrate passifying it to further platelet release (we shall show later that fibrinogen could be such an entity) or b) physical crowding of platelets prevents or modifies the morphologic changes, including pseudopod extension which in turn inhibits release.

Table 2. Effect of Platelet Dosage on Adhesion and Release of Blood Platelets-Substrate Glu(Bzl):Leu 1:1

Number of Platelets Applied	%Adhesion	%Release
5.1×10^9	5.5	0.66
1.7×10^9	22.6	2.72
7.8×10^7	57.8	39.6
3.5×10^7	59.9	36.0
1.6×10^7	67.4	51.9

The data in Table 2 show that the total number of adhering and releasing platelets becomes approximately constant at a surface coverage corresponding to 100 μ^2/platelet.

The effect of substrate composition on platelet adhesion and release is shown in Table 3. Perhaps surprisingly there seems to be very little difference in percent adhesion of platelets over a wide range of polypeptide composition varying from negatively charged glutamic acid containing materials to positively charged lysine containing peptides.

Table 3. Effect of Polypeptide Composition on Platelet Adhesion and Release

Substrate	Conformation	%Adhesion	%Release
Glass	α-helix	30.4	12.6
Glu(Bzl)	α-helix	19.1	18.4
Glu(Bzl):Leu 4:1	α-helix	27.2	18.1
Glu(Bzl):Leu 1:1	α-helix	42.8	20.8
Glu(Bzl):Leu 1:4	α-helix	33.0	18.1
Glu(OH):Leu 1:4	α-helix	41.4	25.8
Lys:Leu 1:99	α-helix	41.6	18.3
Lys:Leu 1:9	α-helix	32.1	16.6
Lys:Phe 1:4	α-helix	38.4	13.2

Conformation established by film C.D spectroscopy

(Standard deviation is ∿3-5%)

Evidently then, if synthetic polypeptides can mimic the behavior of vascular wall proteins, the overall composition is not the prime factor in controlling platelet behavior.

Various articles in the literature suggest that the ability of a substrate to adsorb albumin, γ-globulin or fibrinogen, controls the passivation of surfaces, with albumin layers providing a benign and the others, very active and blood incompatible surfaces. Accordingly we have undertaken a series of studies in which the above proteins and coacervated α-elastin have been precoated onto the polypeptide films. The results of these studies are presented in Table 4.

Table 4. Modification of Polypeptide/Platelet Interaction Preadsorbtion of Proteins

Substrate	Protein	%Adhesion	%Release
Glu(Bzl):Leu 1:1	Albumin	14.4	3.9
Glu(Bzl):Leu 1:1	Fibrinogen	9.8	4.6
Polycarbonate	Elastin	11.5	8.6
Glu(Bzl):Leu 1:1	γ-Globulin	16.1	5.6

It is notable that the adhesion and release of blood platelets are drastically reduced by each of the proteins which seem to be similar in behavior. (A small percentage of platelets are damaged physically on passing through the column as determined by ^{51}Cr labelling so the low protein release percentage may in fact arise from other sources). The fact that platelet adhesion and release in this work does not directly correlate with whole blood clotting studies seems to indicate that the ultimate behavior of platelets in whole blood is mediated by entities not present in the washed platelet system. (Whole blood tests using elastin as a substrate show it to behave very well as a substrate, whole blood tests with fibrinogen show 65% or more platelet release).

The main question arising from a comparison of polypeptides and adsorbed protein interaction of platelets is why they behave so differently. Our main proposition in this area is that adsorbed proteins form a mobile mediating layer between a solid substrate and highly variable platelet membrane. In order to avoid compression or expansion of the platelet membrane the interfacial tension created in the adsorption process must be dissipated. Otherwise platelet factors within or adjacent to the membrane may be released during the interfacial interaction, thus activating the contractile protein and initiating morphologic changes which lead to the release process. Such a situation is shown schemmatically in fig 5.

If indeed surface mobility is a criterion for platelet compatibility it should be possible to devise crosslinked gel-like polypeptides which provide a "gentle" interface for the platelet. As an initial attempt in this direction copoly Lys;Leu materials containing ratios of Lys:Leu which render them water soluble have been treated with thio glycollic acid to form -SH groups which are then oxidised in the solid state to form -S-S- bridges. Crosslinked materials are transparent films which are strongly water absorbing. Films of these materials have been exposed to washed, labelled platelet solutions as before and platelet release examined. Table 5 shows the relation between behavior and crosslink density. Again surprisingly at very low crosslink density, release is high (% adhesion is difficult to obtain with such a low surface area) but at 5% crosslink density the surface becomes quite passive. The reason for this change in properties with crosslink density may be that at low density, material leaches from the substrate cuasing high release (because of high ionic density). The effect is though noticeably higher for the copolymer than for copoly(Lys:Ala) 1:1 or even polylysine, though concen-

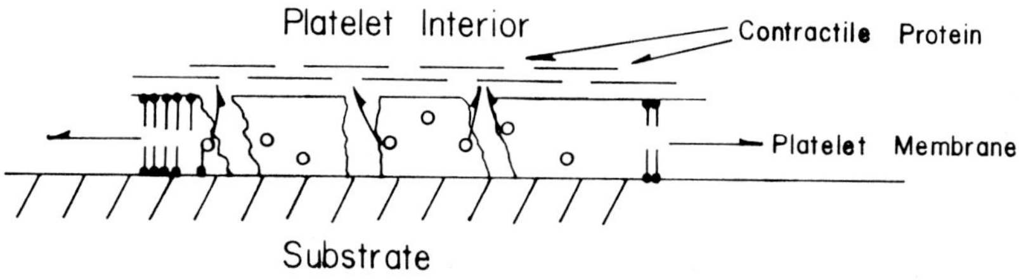

(a)

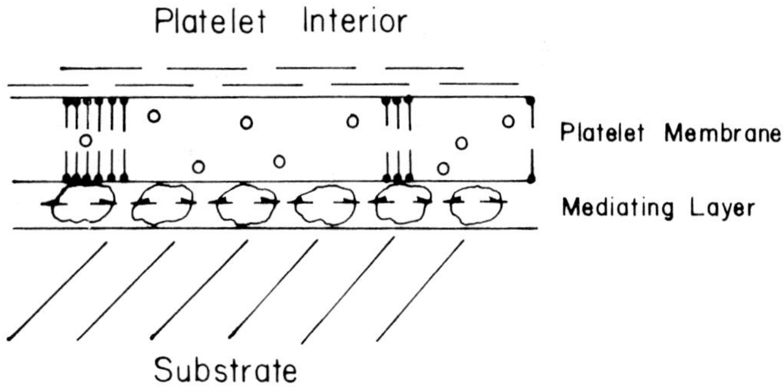

(b)

Fig 5. a) Distortion of a platelet membrane due to interfacial forces causing activation of the contractile mechanism.
b) Mediation of interfacial forces by a mobile passivating layer.

tration dependence may be involved. The other possibility seems to be that with light crosslinking the surface contains long freely moving charged chains.

Table 5. Interaction of Platelets with Crosslinked Polypeptide Hydrogels

Material	Substrate Comp.	Crosslink Density	%Release Run 1	Run 2
1	Copoly(Lys:Leu) 1:1	1%	98%	∼100%
2	" (Lys:Leu) 1:1	5%	∼0	3%
3	" (Lys:Leu) 4:1	5%	∼0	3%
4	Polylysine 10μg/ml (standard)		46%	58%
5	Copoly(Lys:Ala) 1:1 10μg/ml		23%	25%

There seems to be no defineable relation between the ability of soluble polypeptides to initiate aggregation and release and the properties of similar peptide groups in the solid state. In fact there is little or no relation between peptide content of a substrate and its ability to interact with platelets (at least with the limited combinations studied here). Surface mobility of adsorbed species and morphology of the surface appear to be the major controlling factors.

B. Interaction of Polypeptides with Blood Proteins

In this area two types of proteins may be distinguished, the active or inactive forms of blood factors taking part in the intrinsic and extrinsic clotting schemes, and non factors such as albumin and the globulins. The clotting schemes involve an enzymmatic cascade process which may be represented as follows:

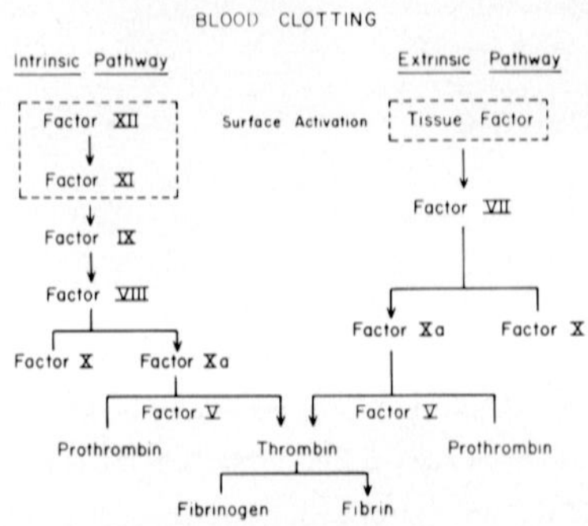

Properties for the well defined coagulation factors are presented in Table 6 (following Davie et al. (28)). The scheme is undoubtedly incomplete but represents an appropriate framework within which to examine the behavior of synthetic polypeptides.

The questions that may be asked are a) do polypeptides bind to blood factors, b) do polypeptides activate or deactivate enzymes or zymogens c) does the interaction between polypeptides and the blood factors throw any light on the mechanism of interaction of the blood factors?

These questions are not easily answered and until recently there was no information available concerning the structure of the blood factors themselves not to mention structural changes induced by polypeptides.

Table 6. List of Blood Factors

Roman Numeral Designation	Common Name	Molecular Weight
Factor I	Fibrinogen	340,000 (human,bovine)
Factor II	Prothrombin	68,700 (human) 72,000 (bovine)
Factor III	Tissue factor	220,000; 330,000(bovine)
Factor IV	Calcium ions	
Factor V	Proaccelerin	290,000-400,000(bovine)
Factor VII	Proconvertin	63,000 (human)
Factor VIII	Antihemophilic factor	1.1 million(human,bovine)
Factor IX	Christmas factor	55,400(bovine)
Factor X	Stuart factor	55,000(bovine)
Factor XI	Plasma Thromboplastin antecedent	160,000(human,bovine)
Factor XII	Hageman factor	90,000(human) 82,000(bovine)
Factor XIII	Fibrin stabilizing factor	300,000 (bovine plasma) 320,000 (human plasma) 146,000 (human platelets)

Factor XII and Its Mechanism of Activation

Factor XII, Hageman factor is converted to its active form XIIa by contact with a number of inorganic surfaces and also has been reported activated by collagen [19] wherein there appears to be some evidence that glutamate groups may be involved [7]. There seems little doubt that the activation process involves only a change in conformation of the protein. Such a suggestion has drawn support from circular dichroism studies which show the active form to resemble a high pH form in which the anionic groups are neutralized and the configuration of aromatic groups is changed [20, 21]. Attempts to induce activation of XII by the collagen analog polytripeptide poly(Pro-Gly-Glu) were found to be unsuccessful [22] perhaps because this particular sample was in a random conformation and, as with platelet activation it is conceivable that an aggregated form with high charge density is required.

A fairly definitive but preliminary indication that glutamate groups in the solid state can initiate Hageman factor activation is the correlation between changes in intrinsic fluorescence of Hageman factor in solution and adsorbed on glutamate copolypeptides. Fig 6 shows the correlation between tryptophan fluorescence of inactive HF, HF activated by ellagic acid, and HF partially activated by copoly(Glutamic acid:Leucine) 1:4.

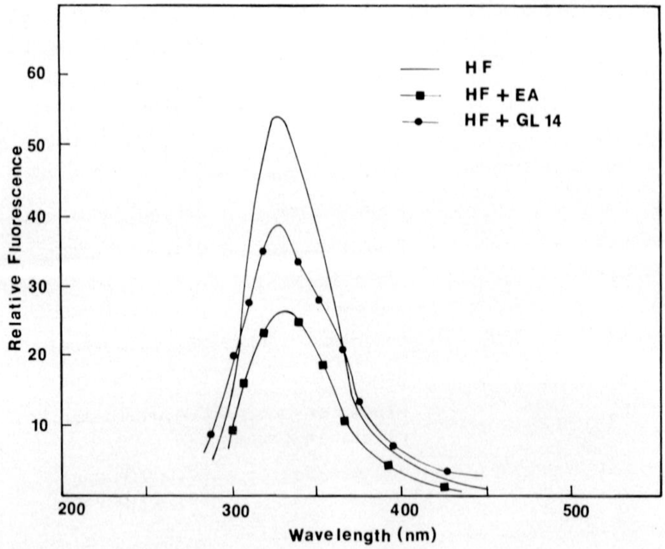

Fig 6. Intrinsic tryptophan fluorescence of Factor XII (HF) in non activated (solution) form, activated by ellagic acid (EA) and partially activated on copoly(Glu(Bzl):Leu) 1:4 particles (by B. D. Fair and W. B. Rippon).

The available evidence is then that changes in surface charge which can be induced by interaction with glutamyl polypeptides causes conformational changes revealing an active site. Since the aromatic part of the CD spectrum changes concurrently with activation, tyrosyl groups may be involved at or near the active site.

Factors XI and XIa

Although there has been a report on the structure of factor XI studied by CD spectroscopy [23], the available evidence indicates that the purity of preparations is not such that detailed structural implications should be drawn at this time. However the conversion of factor XI to the active enzyme XIa, by trypsin, suggests the important role of chain cleavage [24]. Factor XIIa does not seem to be able to activate XI suggesting that an additional factor may be required.

Factor IX

Factor IX is converted to the enzyme IXa by peptide cleavage induced by factor XIa. For new work, reported here, freshly prepared aqueous solutions of factor IX, Christmas factor, supplied by Dr. Earl Davie of Washington University was dialysed and examined by CD spectroscopy at 25°C pH 6.8-7.2, 0.05 mg/ml (fig 7). On storage in frozen solution at -20°C structural changes occur as indicated in the figure. These changes may be characterized as a loss of conformational integrity, probably a decrease in β-sheet content induced by some freeze denaturing process.

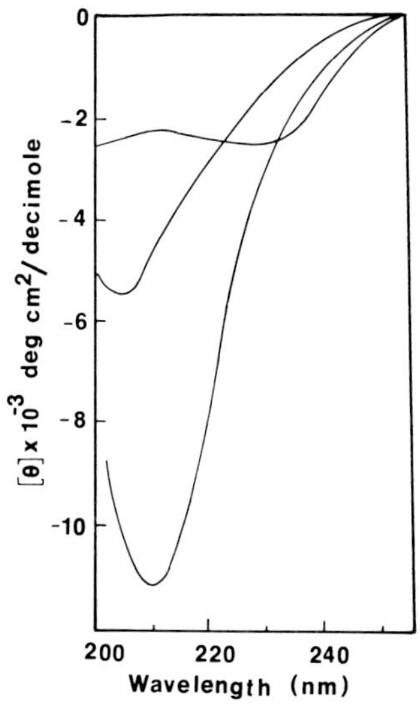

Fig 7. Circular dichroism spectra of Christmas factor (IX) as a function of time A, initial solution, B frozen solution after two weeks, C after four weeks in deep frozen solution.

Difference CD spectroscopy carried out using two week old solutions of factor IX mixed with poly(glutamic acid) or polylysine showed that the former interacted to produce a modified spectrum. (The difference CD method involves placing separate samples of A and B in the CD beam, running a spectrum of the two in series, then mixing and repeating. The two resultant spectra may be subtracted manually or by computer. The method has been used previously by Gelman et al. to evaluate glycosaminoglycan/polypeptide interactions [25, 26]. In the GAG case, the positive charge of polyanions e.g polylysine was neutralized by the GAG inducing an α-helical conformational change in the polylysine). When polyamino acids interact with proteins a change of conformation in one or both may occur. Although definitive interpretation of the change is not always possible, changes in the wavelength range > 230 nm originate from protein and increases in the magnitude (negative) ellipticity in the 210-220 nm range probably originate in the polyelectrolyte. A comparison of the curves in the current case (fig 8) indicates that the protein seems to partially renature in the presence of poly(glutamic acid).

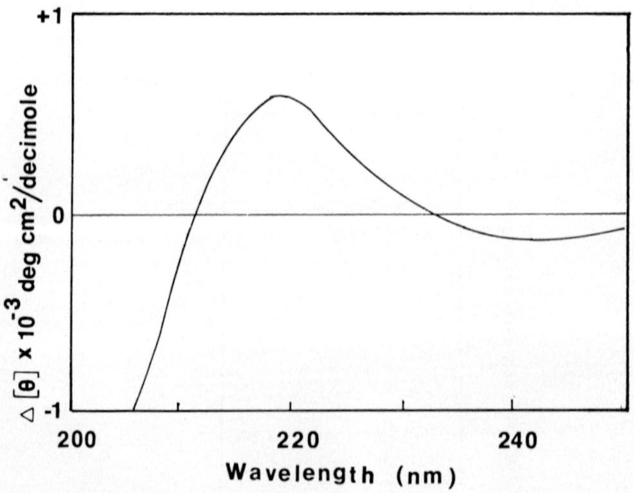

Fig 8. CD difference spectrum of poly(glutamic acid) 0.02 mg/ml and factor IX (0.05 mg/ml) showing change of conformation of one or both components.

The interaction of factor IX with poly(glutamic acid) is somewhat surprising when it is realized that the native factor contains virtually twice as many acidic residues (Asp + Glu) = 22%) as basic residues (Lys + Arg + 12%) [27]. It seems possible that the positively charged groups are clustered locally on the surface. Charge neutralization on the protein surface would normally be expected to induce additional structure, as it does here.

Stuart Factor X

No work concerning the interaction of polypeptides, or secondary structure of, factor VIII seem to have been reported, however factor IXa will activate factor X in the presence of calcium ions and phospholipid [28], a reaction which is strongly accelerated by factor VIII. This suggests that complementary protein aggregates form during the activation process.

Factor X was provided for the present work by Dr. D. Walz of Wayne State University.

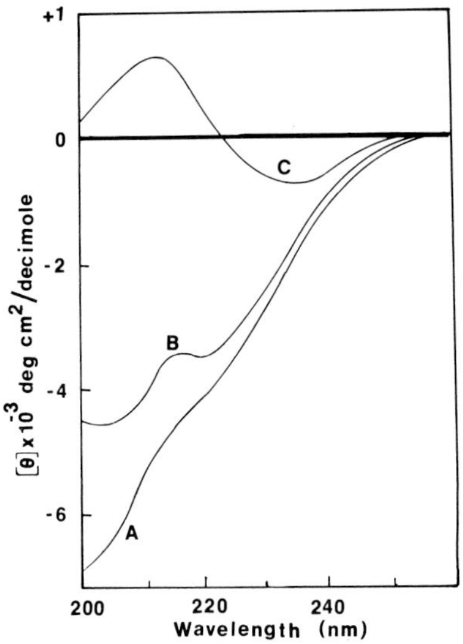

Fig 9. Circular dichroism spectra of bovine Factor X. Samples stored at ∿2°C. A original sample, B sample one week old, C two weeks old. Spectra at pH 6.8, 25°C.

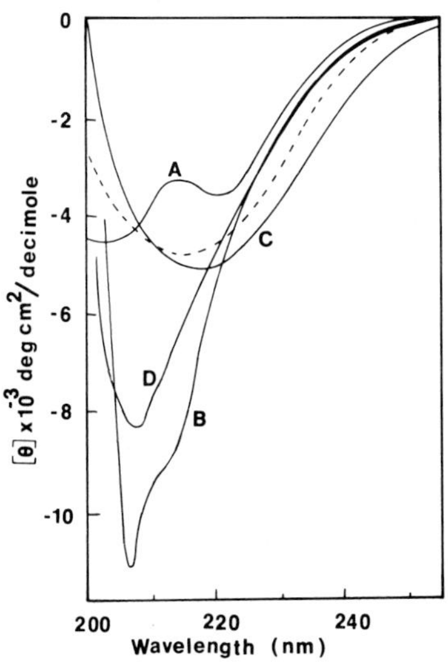

Fig 10. Illustration of the pH sensitivity of Factor X. All CD spectra are for material stored one week at 2°C and brought to 25°C. A, pH 6.8; B, pH 11.3; C solution B acidified to pH 2.6; D solution C, neutralized to pH 7.0. Dotted line change of D after one additional week at room temperature.

It also appears to lose structure when stored in cold aqueous solution (∿2°C) at a concentration of 0.02 mg/ml (fig 9), however factor X appears to be much more conformationally sensitive to pH than was factor IX (fig 10). Maximum ellipticity is achieved at high pH. The amino acid analysis of Stuart factor shows [27] a very similar composition to factor IX. Indeed not only are the amino acid analyses and structural characteristics (or lack thereof) very similar, but factor X also shows a charge interaction with polyglutamic acid by difference CD spectroscopy.

Factor X may be converted to active form(s) by trypsin treatment. We were unable to detect any conformational change during this process.

Factor V, Proaccelerin

Bovine factor V was obtained from Sigma Chemical Corp. in deep frozen phosphate buffer. It was dialysed and assayed by biuret determination. CD spectroscopy (fig 11) indicates remarkably little detailed structure. The material undergoes aggregation on standing above 50°C which gives rise to an increasing negative ellipticity at 205 nm.

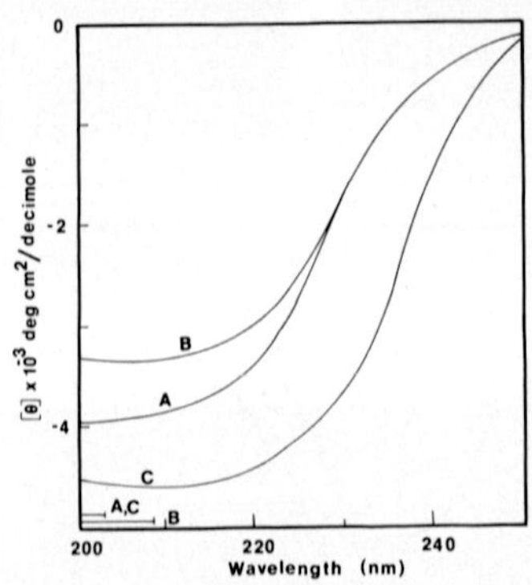

Fig 11. CD spectra of Factor V, Proaccelerin. A, pH 6.9; B, pH 11.3; C, pH 2.8. Return to pH 7 regenerates curve A. Bar lines indicate region of increased dynode voltage (≥80%) and particulate aggregation.

Difference CD of mixtures of V with poly(glutamic acid) and polylysine also revealed very little structural information but the former did give rise to a 25% decrease in ellipticity at about 205 nm.

Normally, factor Xa interacts with factor V to form a complex that acts as a catalyst in the activation of prothrombin [29]. The reaction requires the presence of calcium ion and phospholipid.

Factor II-Prothrombin

The interaction of prothrombin with polylysine not only leads to considerable changes in the CD spectrum, but is also known to cause prothrombin activation [30].

The bovine prothrombin used in this work was supplied by Sigma Chemical Corp., 14.3 units/ml in 0.14 M NaCl and 20 mM citrate. The protein content was determined by the Lowry method and solutions made up from dialysed material.

At 0.02 mg/ml of each component, there was no appreciable interaction with poly(glutamic acid) as determined by difference CD spectroscopy. However similar experiments with polylysine revealed considerable changes on mixing of the two components. Figure 12 shows the CD difference curves for three different concentrations of polylysine.

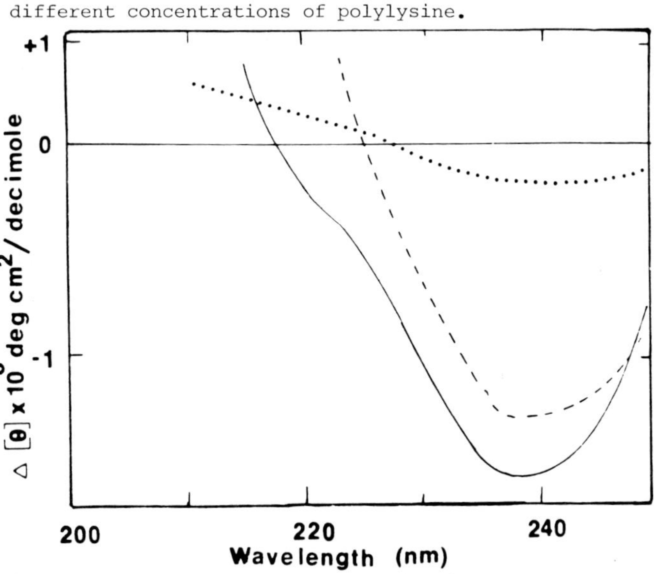

Fig 12. CD difference spectra of prothrombin with three concentrations of polylysine.... 1:1, —— 2:1, --- 4:1 protein in excess.

The maximum effect falls at an approximate ratio of 2:1 with protein in excess. The difference curves indicate an increase in ellipticity (negative) in the 240 nm region which must originate entirely from a change in conformation of the protein (or aggregation). The fact that there is some aggregation is indicated by an increased dynode voltage, however the effect appears to become dominant only below 210 nm.

Changes in the 240 nm region are almost certainly associated with changes in the relative arrangements of aromatic residues. Similar changes have been noted between active and heat denatured prothrombin [31]. The conformational changes induced by polylysine do not seem to be simply related to overall ionic charge (although there is an excess glutamic and aspartic acid in the protein since these residues account for only 20% of the total). There are, in addition to the anionic amino acids a few neuraminic acid groups in the sugar part of the protein which could cause additional binding.

Fibrinogen-Factor I

Aggregation of fibrinogen by polylysine to form insoluble complexes has been known for some time [15]. At neutral pH, the addition of polylysine to fibrinogen causes an increase in dynode voltage in the spectropolarimeter and anomalous scattering curves. However in the presence of 50 mM phosphate buffer the interaction seems much reduced and there are similar changes with both polylysine and polyglutamic acid.

It has been suggested [32] that basic polyamino acids accelerate the thrombin/fibrinogen reaction by removing acidic cleavage products by complexation. Polylysine does though also undergo interaction with thrombin causing minor conformational changes in the mixture.

Nonfactors Albumin and γ-Globulin

Albumin is known to interact with polylysine. No evidence for aggregation or change of conformation could be detected by CD spectroscopy for γ-globulin/polylysine mixtures.

DISCUSSION AND CONCLUSION

It appears that virtually all blood factors and proteins interact with, or aggregate with, either or both of polyanionic or polycationic peptides. In some cases (factors XII and II) activation of the factors occurs. A summary of reactions is given in table 7. A much more detailed study is necessary before conclusions may be reached concerning the mode of interaction but it seems likely that the main binding force between protein and polypeptide is ionic. This seems to suggest that there may be "patches" of ionic charge on the protein surface since often the net surface charge is unfavorable for polyion binding. If such is the case it may be that enzyme complexes are also held together, in major part, by ionic interaction.

Table 7. Summary of Blood Protein Interaction

Protein	Glutamate	Lysyl	Comment
Factor XII Hageman Factor	X		Polypeptide causes enzyme activation
Factor X Stuart Factor	X		Conformational change and/or binding
Factor IX Christmas Factor	X		
Factor V Proaccelerin	X	X	Mild binding with both
Factor II Prothrombin		X	Strong conformational effect in protein and activation
Thrombin		X	
Factor I Fibrinogen	X	X	
Serum Albumin		X	
γ-Globulin		X	

The failure of polylysine to accelerate whole blood clotting is initially quite surprising since it causes platelet and red cell aggregation, release of serotonin, ADP, Ca^{2+}, fibrinogen etc. from platelets, and activates prothrombin. It may however deactivate thrombin and fibrinogen and thus block the blood clotting mechanism at the penultimate stages.

In general it can be seen that both in solution and in the solid state,polypeptides can be synthesized which mimic many of the properties of proteins in the cardiovascular system. This includes factor activation and binding, cell agglutination and surface passivation. Although residue charge plays an important role, particularly in solution, the composition of polypeptide films seems to play a relatively minor part in their interaction with blood components,and features such as flexibility, hydration, surface mobility and substrate morphology become all important.

ACKNOWLEDGEMENT

Much of the new work reported in this paper was supported by grants from NIHL and the Heart Association of Northeast Ohio. The data were obtained by D. D. Solomon, W. Holaday, B. D. Fair and particularly Dr. J. Suchy, the materials provided from many individuals including Dr. J. M. Anderson, Dr. L. R. Treiber, M. VanDress and T. Hayashi. Helpful comments and assistance from Dr. D. Cowan are gratefully acknowledged.

REFERENCES

1. A. deVries, A. Schwager and E. Katchalski, Biochem. J., 49(1951)10.
2. J. R. Rubin, R. R. Becker and M. A. Stahmann, Proc. Soc. Expt. Biol. Med., 82(1953)231.
3. A. deVries, J. D. Feldman, O. Stein, V. Stein and E. Katchalski, Proc. Soc. Exptl. Biol. Med., 82(1953)237.
4. E. R. Blout, S. Farber, G. D. Fasman, E. Klein and M. Narrod in "Polyamino Acids, Polypeptides and Proteins", M. A. Stahmann Ed., Univ. of Wisconsin Press, Madison, Wisc. (1962) p. 379.
5. A. F. Portmann and W. D. Holden, J. Clin. Invest., 28(1949)1451.
6. T. Komai and Y. Nose, Jinko-Zohki, 4(1975)114.
7. H. L. Nossel, G. D. Wilner and E. C. LeRoy, Nature, 221(1969)75.
8. C. Jenkins, M. Packham, R. Kinlough-Rathbone and J. Mustard, Blood, 37(1971)395.
9. P. Massini, L. C. Metcalf, U. Nat and E. Luscher, Haemostasis, 3(1971)395.
10. L. C. Metcalf and D. Lyman, Thromb. Res., 5(1974)709.
11. Th. Pfleiderer and R. Brossmer, Thromb. Diath. haemorrh., 18(1967)674.
12. R. Keller, C. Muller-Eckhardt and F. Kayer, Int. Arch. Allergy, 33(1967)239.
13. W. Schneider, W. Kubler and R. Gross, Thromb. Diath. haemorrh., 19(1968)307.
14. G. F. Sigler, C. R. McMillin and A. G. Walton, Biopolymers (in press).
15. R. V. Rice, M. A. Stahmann and R. A. Alberty, J. Biol. Chem., 209(1954)105.
16. P. Parks, O. Malhoton, D.F. Gibbons and J.M. Anderson, Abst. 2nd Ann. Meeting, Soc. Biomaterials, Philadelphia(1976) p 96.

17. J. M. Anderson, D. F. Gibbons, R. L. Martin, A. Hiltner and R. Woods, J. Biomed. Mater. Res., 5(1974)197.

18. D. D. Solomon, D. H. Cowan and A. G. Walton in "A. C. S. 50th Colloid and Surface Science Symposium", J. Colloid and Interface Sci. (in press).

19. C. G. Cochrane, S. D. Revak, B. S. Aikin and K. D. Wuepper in "Inflammation: Mechanism and Control" ed. I. H. Lepow, P. A. Ward, Academic Press (1972) p. 119.

20. C. R. McMillin and A. G. Walton, J. Colloid Interface Sci., 48(1974) 1312.

21. C. R. McMillin, H. Saito, O. D. Ratnoff and A. G. Walton, J. Clin. Invest., 54(1974)1312.

22. A. G. Walton in "Structure of Fibrous Proteins", Ed. E. D. T. Atkins and A. Keller, Butterworths, London (1974) p. 139.

23. C. R. McMillin, A.G. Walton, H. Saito and O. Ratnoff, Biophys J. (Abst) 15, 147a(1975)

24. H. Saito, O. D. Ratnoff, J. S. Marshall and J. Rensky, J. Clin. Invest., 52(1973)850.

25. R. A. Gelman, W. B. Rippon and J. Blackwell, Biochem. Biophys. Res. Commun., 48(1972)708.

26. R. A. Gelman and J. Blackwell, Conn. Tiss. Res., 2(1973)31.

27. K. Fujikawa, M. H. Coan, D. L. Enfield, K. Titeusi, L. H. Ericsson and E. W. Davie, Proc. Nat. Acad. Sci., U. S. A., 71(1974)427.

28. E. W. Davie and K. Fujikawa, Ann. Rev. Biochem., 44(1975)799.

29. J. Jesty and M. P. Esnouf, Biochem. J., 131(1973)791.

30. K. D. Miller, W. H. Copeland and J. F. McGaraahan, Proc. Soc. Expt. Biol. and Med., 108(1961)117.

31. O. P. Malhotra and W. B. Rippon (awaiting publication).

32. H. A. Scheraga, Ann. N. Y. Acad. Sci., 75(1958)189.